Charles Seale-Hayne Library
University of Plymouth
(01752) 588 588
LibraryandITenquiries@plymouth.ac.uk

1. DeVerle P. Harris: *Mineral Resources Appraisal: Mineral Endowment, Resources, and Potential Supply: Concepts, Methods, and Cases*

2. J. J. Veevers (ed.): *Phanerozoic Earth History of Australia*

3. Yang Zunyi, Cheng Yuqi, and Wang Hongzhen (eds.): *The Geology of China*

4. Lin-Gun Liu and William A. Bassett: *Elements, Oxides, and Silicates: High-Pressure Phases with Implications for the Earth's Interior*

5. Antoni Hoffman and Matthew H. Nitecki (eds.): *Problematic Fossil Taxa*

6. S. Mahmood Naqvi and John J. W. Rogers: *Precambrian Geology of India*

7. Chih-Pei Chang and T. N. Krishnamurti (eds.): *Monsoon Meteorology*

8. Zvi Ben-Avraham (ed.): *The Evolution of the Pacific Ocean Margins*

9. Ian McDougall and T. Mark Harrison: *Geochronology and Thermochronology by the $^{40}Ar/^{39}Ar$ Method*

10. Walter C. Sweet: The Conodonta: *Morphology, Taxonomy, Paleoecology, and Evolutionary History of a Long-Extinct Animal Phylum*

11. H. J. Melosh: *Impact Cratering: A Geologic Process*

12. J. W. Cowie and M. D. Brasier (eds.): *The Precambrian-Cambrian Boundary*

13. C. S. Hutchinson: *Geological Evolution of South-East Asia.*

14. Anthony J. Naldrett: *Magmatic Sulfide Deposits*

15. D. R. Prothero and R. M. Schoch (eds.): *The Evolution of Perissodactyls*

16. M. Menzies (ed.): *Continental Mantle*

17. R. J. Tingey (ed.): *Geology of the Antarctic*

18. Thomas J. Crowley and Gerald R. North: *Paleoclimatology*

19. Gregory J. Retallack: *Miocene Paleosols and Ape Habitats in Pakistan and Kenya*

20. Kuo-Nan Liou: *Radiation and Cloud Processes in the Atmosphere: Theory, Observation and Modeling*

21. Brian Bayly: *Chemical Change in Deforming Materials*

22. A. K. Gibbs and C. N. Barron: *The Geology of the Guiana Shield*

23. Peter J. Ortoleva: *Geochemical Self-Organization*

24. Robert G. Coleman: *Geologic Evolution of the Red Sea*

25. Richard W. Spinrad, Kendall L. Carder, and Mary Jane Perry: *Ocean Optics*

26. Clinton M. Case: *Physical Principles of Flow in Unsaturated Porous Media*

27. Eric B. Kraus and Joost A. Businger: *Atmosphere-Ocean Interaction, Second Edition*

28. M. Solomon and D. I. Groves: *The Geology and Origins of Australia's Mineral Deposits*

29. R. L. Stanton: *Ore Elements in Arc Lavas*

30. P. Wignall: *Black Shales*

31. Orson L. Anderson: *Equations of State for Solids in Geophysics and Ceramic Science*

32. J. Alan Holman: *Pleistocene Amphibians and Reptiles in North America*

33. P. Janvier: *Early Vertebrates*

34. David S. O'Hanley: *Serpentinites: Recorders of Tectonic and Petrological History*

35. Charles S. Hutchison: *South-East Asian Oil, Gas, Coal and Mineral Deposits*

36. Maarten J. de Wit and Lewis D. Ashwal (eds.): *Greenstone Belts*

37. Tina Niemi, Zvi Ben-Avraham, and Joel R. Gat: *The Dead Sea: The Lake and Its Setting*

38. J. Alan Holman: *Pleistocene Amphibians and Reptiles in Britain and Europe*

39. Thomas J. Crowley and Kevin C. Burke: *Tectonic Boundary Conditions for Climate Reconstructions*

Tectonic Boundary Conditions for Climate Reconstructions

Edited by

Thomas J. Crowley

Kevin C. Burke

New York   Oxford

Oxford University Press

1998

Oxford University Press

Oxford   New York
Athens   Auckland   Bangkok   Bogotá   Buenos Aires   Calcutta
Cape Town   Chennai   Dar es Salaam   Delhi   Florence   Hong Kong   Istanbul
Karachi   Kuala Lumpur   Madrid   Melbourne   Mexico City   Mumbai
Nairobi   Paris   São Paulo   Singapore   Taipei   Tokyo   Toronto   Warsaw

and associated companies in
Berlin   Ibadan

Published by Oxford University Press, Inc.
198 Madison Avenue, New York, New York 10016

Oxford is a registered trademark of Oxford University Press

Library of Congress Cataloging-in-Publication Data
Tectonic boundary conditions for climate reconstructions /
edited by Thomas J. Crowley, Kevin C. Burke.
p.   cm.
ISBN 0-19-511245-8
1. Paleoclimatology—Mathematical models.
2. Geodynamics—Mathematical models.
3. Boundary value problems.
I. Crowley, Thomas J., 1948–   .   II. Burke, Kevin C., 1929–   .
QC84.T43   1998
551.6'09'01—dc21   97-23987

9 8 7 6 5 4 3 2 1
Printed in the United States of America
on acid-free paper

PREFACE

In the last ten to fifteen years there has been a movement to break down old disciplinary boundaries in the geosciences in order to develop a more unified view of the earth as an integrated system. Much of this effort has been stimulated by developments in the atmosphere and ocean sciences that are studying the effect of man's impact on the environment. However, solid earth sciences also have a role to play in the Earth System Science/Global Change Programs. Efforts to integrate solid and fluid elements of the geosciences have not progressed as rapidly as other elements of these programs.

In this volume we present examples of how integrating solid earth and climate studies can lead to better understanding of both disciplines. We focus on the role of tectonic boundary conditions for paleoclimate reconstructions. The climate modeling community has become interested in this subject because it provides a means to test their climate and ocean models under radically altered but still realistic boundary conditions, and because the dramatic nature of past climate change is a subject of intrinsic scientific interest.

A significant problem has developed with respect to modeling climate change on tectonic time scales. Comparison of modeling results with paleo-data sometimes indicate significant discrepancies. Three possible explanations for these discrepancies are: (1) there are some fundamental inadequacies in the model physics; (2) the paleo-data are open to reinterpretation; or (3) the tectonic boundary conditions for the paleo-reconstructions were not correctly specified.

In this volume we address the third possibility by exploring the importance of tectonic boundary conditions for climate model simulations and also the limitations of our knowledge of some of these boundary conditions. Examples include uncertainties with respect to continental configurations, elevations of continents, ocean "gateways," shape of the seafloor, extent of shallow epicontinental seas, and effect of changing rock type through time on atmospheric CO_2 levels. Although this list is not exhaustive, it does cover many problems of interest in both disciplines and will hopefully stimulate more interest in exploring the remaining uncertainties in both the fields of geodynamics and paleoclimatology.

We were stimulated to develop this book by discussions with the Geodynamics Committee of the Board on Earth Sciences and Resources of the United States National Academy of Sciences. Our thanks especially go the Robin Brett and William Dickinson, successive chairmen of the Geodynamics Committee. We also thank Joyce Berry of Oxford University Press for her support of our effort, Ann Galloway for invaluable assistance on tracking and preparation of manuscripts, and the many reviewers who consistently provided timely and thoughtful comments on the authors' manuscripts.

College Station, Texas   T.J.C.
Houston, Texas   K.C.B.
June 1997

CONTENTS

Contributors

DR. ERIC J. BARRON
Earth System Science Center
Pennsylvania State University
248 Deike Building
University Park, PA 16802

DR. ROBERT A. BERNER
Department of Geology and Geophysics
Yale University
New Haven, CT 06520-8109

DR. KAREN L. BICE
Earth System Science Center
Pennsylvania State University
248 Deike Building
University Park, PA 16802

DR. KEVIN C. BURKE
Department of Geosciences
University of Houston
Houston, TX 77004

DR. CLEMENT G. CHASE
Department of Geosciences
University of Arizona
Tucson, AZ 85721-0077

DR. THOMAS J. CROWLEY
Department of Oceanography
Texas A&M University
College Station, TX 77843-3146

DR. ANDREW D. CUNNINGHAM
Department of Geology and Geophysics
Rice University
Houston, TX 77005

DR. PETER G. DeCELLES
Department of Geosciences
University of Arizona
Tucson, AZ 85721-0077

DR. ANDRÉ W. DROXLER
Department of Geology and Geophysics
Rice University
Houston, TX 77005

DR. DAVID S. DUNCAN
Department of Marine Sciences
University of South Florida
St. Petersburg, FL 33701

DR. PETER J. FAWCETT
Department of Earth and Planetary Science
University of New Mexico
Albuquerque, NM 87131

DR. LISA M. GAHAGAN
Institute for Geophysics
The University of Texas at Austin
4412 Spicewood Springs Road, Bldg. 600
Austin, TX 78759-8500

DR. KATHRYN M. GREGORY-WODZICKi
Lamont-Doherty Earth Observatory
  of Columbia University
Palisades, NY 10964-8000

DR. PAMELA HALLOCK
Department of Marine Sciences
University of South Florida
St. Petersburg, FL 33701

DR. T. MARK HARRISON
Department of Earth and Space Sciences
Institute of Geophysics and Planetary Physics
University of California, Los Angeles
Los Angeles, CA 90095

DR. WILLIAM W. HAY
Department of Geological Sciences
Campus Box 250
University of Colorado
Boulder, CO 80309

DR. ALBERT C. HINE
Department of Marine Sciences
University of South Florida
St. Petersburg, FL 33701

DR. LAWRENCE A. LAWVER
Institute for Geophysics
The University of Texas at Austin
4412 Spicewood Springs Road, Bldg. 600
Austin, TX 78759-8500

DR. KIRK A. MAASCH
Department of Geological Sciences
University of Maine
Orono, ME 04469-5711

DR. BETTE L. OTTO-BLIESNER
Climate and Global Dynamics
National Center for Atmospheric Research
Boulder, CO 80303

DR. JUDITH T. PARRISH
Department of Geosciences
University of Arizona
Tucson, AZ 85721-0077

DR. WILLIAM H. PETERSON
Earth System Science Center
Pennsylvania State University
248 Deike Building
University Park, PA 16802

DR. DOUGLAS N. REUSCH
Department of Geological Sciences
University of Maine
Orono, ME 04469-5711

DR. EDWARD ROBINSON
Department of Geology
University of the West Indies
Kingston, Jamaica

DR. ERIC ROSENCRANTZ
Institute for Geophysics
The University of Texas at Austin
4412 Spicewood Springs Road, Bldg. 600
Austin, TX 78759-8500

DR. DAVID B. ROWLEY
Department of Geophysical Sciences
University of Chicago
5734 S. Ellis Ave.
Chicago, IL 60637

DR. FREDERICK J. RYERSON
Institute of Geophysics and Planetary Physics
Lawrence Livermore National Laboratory
Livermore, CA 94550

DR. CHRISTOPHER N. WOLD
Platte River Associates, Inc.
2790 Valmont Road
Boulder, CO 80304

DR. JAMES D. WRIGHT
University of Maine
Department of Geological Sciences
Institute for Quaternary Studies
Orono, ME 04469

DR. AN YIN
Department of Earth and Space Sciences
Institute of Geophysics and Planetary Physics
University of California, Los Angeles
Los Angeles, CA 90095

DR. A. M. ZIEGLER
Department of Geophysical Sciences
University of Chicago
5734 S. Ellis Ave.
Chicago, IL 60637

PART I

INTRODUCTION

CHAPTER 1

Significance of Tectonic Boundary Conditions for Paleoclimate Simulations

Thomas J. Crowley

Since the advent of the plate tectonic revolution geologists and paleoclimatologists have been interested in the relationship between plate tectonic changes and climate. In particular, they have sought to determine whether changes in positions of continents were associated with glacial periods in the Phanerozoic (Fig. 1.1). Somewhat independently, climate modelers have become interested in testing their climate models to determine how well they respond to altered boundary conditions. Donn and Shaw (1977) were among the first to try to model this relationship and provided some support for the concept that drift of continents into high latitudes could trigger glaciation.

Although the Donn and Shaw study was enlightening, the beginning of a more continuous series of modeling efforts occurred in 1981, when Eric Barron and colleagues (Barron et al., 1981) utilized an energy balance model (see next section) to try to determine the origin of ice free conditions during the Cretaceous thermal maximum (100 Ma, million years ago). They followed this effort with a number of other important papers, laying out the basic principles of the problem and using more sophisticated general circulation models (GCMs) in the process (e.g., Barron, 1985; Barron and Washington, 1982, 1985). Since that time many different research groups have been involved in modeling climate change on tectonic time scales (e.g., Crowley et al., 1986; Covey and Barron, 1988; Kutzbach and Gallimore, 1989; Maier-Reimer et al., 1990; Manabe and Broccoli, 1990; Oglesby, 1991; Chandler et al., 1992; Moore et al., 1992a; Valdes and Sellwood, 1992; Otto-Bliesner, 1993; Covey et al., 1994; Pollard and Schulz, 1994; Bice, 1997; Wilson et al., 1994).

In the course of these investigations modelers

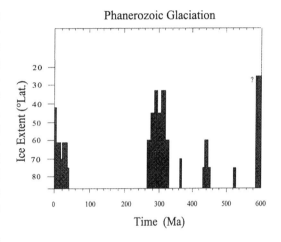

Phanerozoic Glaciation

Fig. 1.1. Latitudinal extent of glaciation versus time for the Late Precambrian and Phanerozoic. This summary is based on evidence from tillites and other direct evidence from glaciation; occasional evidence for alpine glaciation is excluded. The Phanerozoic data are modified from several primary sources (Crowell, 1982; Frakes and Francis, 1988; Frakes et al., 1992) and include additional information on the late Ordovician (Brenchley et al., 1994) and Cambrian (Briden et al., 1993; Bertrand-Sarfati et al., 1995) glaciations. Precambrian data are from Hambrey and Harland (1981) and Knoll and Walter (1992). The question mark refers to uncertainty in latitudinal extent of Precambrian glaciers. Note that there are considerable additional uncertainties with respect to some of these data and some second-order details are excluded. But the results nevertheless reflect, in the author's opinion, the prevailing view of paleoclimatologists as to the present state of our knowledge of the history of continental glaciation during the last 600 million years.

have tried to test their models with geologic data. Although agreement, usually of a semiquantitative nature, was often found, a number of notable model-data discrepancies occurred. These discrepancies have led researchers to inquire whether there are fundamental flaws in their models or whether some of the geologic data is open to reinterpretation.

A third possible reason for model-data discrepancies involves the input boundary conditions for the climate model simulations. If boundary conditions are not correct, then the model simulation will produce erroneous results. Such a possibility has led to the present volume, which includes modeling and observational chapters exploring the uncertainties of tectonic boundary conditions and some implications thereof with respect to their effect on modeled climate. The purpose of this chapter is to provide some introductory material concerning the general importance of tectonic boundary conditions for climate model simulations. Examples of past work will highlight how changes in such boundary conditions can result in significant changes in model response.

A Short Introduction to Climate Models

Although it is beyond the scope of this chapter to discuss in detail the structure of atmosphere and ocean models, a few orientation comments are useful. The reader is referred to chapters 1 and 2 of Crowley and North (1991) for more discussion on the matter; references therein can guide the interested reader to even more detailed descriptions of climate models.

A key concept to any discussion is the "hierarchy of climate models" (Schneider and Dickinson, 1974). This concept refers to the idea that the climate system can be modeled by an increasingly complex series of models which try to simulate the many processes in the climate system with ever more explicit detail.

A great deal of effort has gone into production of large scale GCMs of the atmosphere and ocean. Despite their complexity, GCMs of either the atmosphere or the ocean share a small number of common features. Global models predict fluid motion by solving Newton's third law for motion on a rotating sphere. An additional critical term involves heat transfer. Atmospheric models use the First Law of Thermodynamics as a starting point, while ocean models determine the effect of temperature and salinity on buoyancy. The buoyancy term strongly influences horizontal motion and plays a key role in determining the vertical stability of the water column. Where the

thermohaline system is unstable (usually in winter), overturn occurs and water sinks to various depths. Outflow of deep water in the two key sites of sinking (subpolar North Atlantic and Antarctica) has a significant effect on the horizontal advection of heat at the surface. For example, in the North Atlantic, deep water outflow must be compensated by shallow water inflow in order to conserve volume. As a consequence, the climates of northwest Europe are warmer than other high-latitude regions.

The basic equations are usually coupled with an equation of state and solved in either spectral or gridpoint space. Numerous parameterizations are used in models to account for the effects of clouds, ocean eddies, and other subgrid scale processes. The models include many adjustable parameters and do a reasonable job of simulating a number of large scale features of the atmosphere and ocean circulation. Attempts to test such models against the geologic record have also met with a fair level of success (cf. Crowley and North, 1991).

The effects of geography on climate can also be simulated with relatively simple models termed energy balance models (EBMs). Although there are many versions of EBMs (cf. Crowley and North, 1991, chapter 1), the form most often used for tectonic studies involves two dimensional (latitude–longitude) EBMs that allow for explicit inclusion of geography (North et al., 1983; Hyde et al., 1990). As the term implies, EBMs are thermodynamic models which calculate temperature as a function of the planet's energy receipt, as it varies seasonally and latitudinally and as it is affected by the local reflectivity (albedo) and land-sea distribution. This latter term is critical for paleoclimate simulations, because the different heat capacities of land and water result in significantly different seasonal cycles of temperature on land. The larger the landmass, the larger the seasonal cycle. Even though many climate feedbacks are not included in EBMs, extensive comparisons of EBMs with GCMs indicate comparable sensitivities to changes in different boundary conditions (Crowley et al., 1991). That is, the magnitude (sensitivity) of the EBM and GCM responses are similar (Hyde et al, 1989; Crowley et al., 1991). The general agreement indicates that, to a first approximation, seasonal cycle changes over land are almost linearly related to changes in the forcing terms (Crowley et al., 1991). The agreement also indicates that EBMs can be used to make first-order estimates of the seasonal temperature response to changes in some boundary conditions (e.g., landmass size, carbon dioxide forcing, response to orbital insolation changes).

In addition to a hierarchy of climate models, there is also a hierarchy of credibility of climate model results (Crowley and North, 1991). That is to say, some model results are more believable than others. For example, climate models do a fairly good job of simulating the seasonal cycle of temperatures and the location of zonal wind and precipitation belts. Likewise, ocean models have some success with simulation of the main subtropical gyres and general patterns of vertical overturn in the ocean. But most climate models do only a fair job of quantitatively simulating regional precipitation patterns correctly (e.g., Lau et al., 1996). Derivative fields, such as precipitation-minus-evaporation (P-E) and runoff, are even less well simulated. Ocean models may err on fine scales (e.g., upwelling patterns). Evaluation of paleo-model runs therefore requires not only knowledge of the model and boundary conditions, but also an appreciation of the strengths and limitations of the models. If proper appreciation of these limitations is maintained, meaningful results can be obtained from paleo-simulations and lead to real insight into the operation of important processes in the climate system.

Examples of the Way in Which Tectonic Changes Can Influence Climate

Uncertainties in tectonic boundary conditions potentially affect our ability to validate a number of key conclusions from climate model simulations. This section includes some examples from past modeling work on the effect of tectonic boundary conditions on climate. The survey is divided into sections illustrating both examples and needs for improved realism of model simulations.

Continental Configuration

As mentioned in the introduction, the role of changing land-sea distribution has long been suspected as a important factor responsible for long-term changes in climate. The basic idea is that movement of continents into high latitudes would result in

accumulation of snow cover and glaciation. Other ideas (such as effect of geography on the seasonal cycle) were less well developed until climate models were introduced into the debate.

Despite the dramatic changes in geography during the Phanerozoic, incorporation of geographic changes into climate models sometimes yielded surprising conclusions as to its effect. For example, Barron (1985) examined the effect of Tertiary plate tectonic position on global average temperatures and found a very small change in model-simulated global average temperature for the interval 60–20 Ma. Fawcett (1994; cf. Fawcett and Barron, Chapter 2) obtained a similar response for global temperatures spanning the last 250 million years (Fig. 1.2). The reason for this response is that, although geography plays a large role in determining the seasonal cycle of temperature (see below), mean annual changes due to geography are often much smaller (any warming in summer is usually offset by greater cooling in winter). The work of Barron and colleagues indicated that increased levels of atmospheric carbon dioxide are required to better reconcile models of past climates with observations (Barron and Washington, 1985; cf. Sloan and Rea, 1995; Barron et al., 1995).

Additional studies indicated that model response on a regional and seasonal scale can sometimes be dramatically different from the global temperature response. For example, EBM calculations by Crowley et al. (1986) and Hyde et al. (1990) suggested that plate motion induced changes in summer

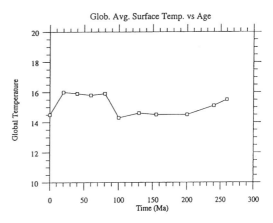

Fig. 1.2. Computed global mean annual surface temperatures for the last 250 million years due to changes in geography. After Fawcett (1994).

temperature over Greenland and Antarctica may have contributed to ice buildup on these landmasses (Fig. 1.3). Here it is assumed that subfreezing summer temperatures represent a critical boundary condition for glacial inception. For Greenland, the opening of the Norwegian Sea and the northward drift of Greenland contributed to the cooling trend in the model, for both processes reduced the amplitude of summer warming in Greenland. For Antarctica, cooling resulted from the separation of Antarctica from Australia. For the latter experiment, however, different results were obtained with different geographies (Fig. 1.4). This latter result in particular highlights the importance of accurate specification of geography for modeling the effect of Australia–Antarctica on summer temperatures.

Experiments for supercontinents (e.g., Fig. 1.5) indicate an even larger effect of geography on the seasonal cycle of landmasses (Crowley et al., 1989; Kutzbach and Gallimore, 1989). Such large temperature changes would also have had a significant effect on the monsoon system (Kutzbach and Gallimore, 1989; Parrish, 1993) and on the distribution of biota (Ziegler, 1990). But the simulated regional pattern of such changes may be affected by uncertainties in the locations of continents. For example, two different modeling groups have compared climate model results for the Late Jurassic (Kimmeridgian, about 150 Ma) with geologic data

EFFECT OF GEOGRAPHIC CHANGE ON SNOW/ICE AREA

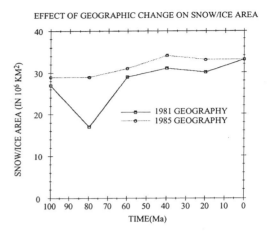

Fig. 1.4. Illustration of how uncertainties in geography can lead to different climate conclusions. Figure compares the differences in model-predicted Southern Hemisphere area covered by sea ice and summer snow over the past 100 million years, one example using the geography of Barron et al. (1981) (solid line), the other (dashed line) that of Barron (1985, 1987).

(Moore et al., 1992a; Valdes and Sellwood, 1992). Both of these reconstructions place northern North America at about 60°N paleolatitude. Yet some new paleomagnetism results suggest a more northerly (~15°C) location of North America at that time (van Fossen and Kent, 1992; van der Voo, 1992; Muttoni et al., 1996). Obviously, discrepancies between models and paleoclimatic data for that time could reflect uncertainties in the paleolatitudinal specification of North America and perhaps other land masses.

Greater uncertainties arise with respect to location and even area of the continents when Paleozoic reconstructions are modeled. For example, the apparent polar wander paths for Gondwana in the mid-Paleozoic is open to considerable uncertainty. Incorporation of different possibilities into an EBM (Fig. 1.6) yield significantly different solutions and have implications for the interpretation of the regional climate history of Gondwana (cf. Parrish, 1990).

The problem of land-sea distribution becomes even more uncertain when the Late Precambrian is concerned. Paleogeographic changes for this time interval are at present in a healthy stage of debate (e.g., Dalziel, 1991, 1997; Hoffman, 1991; Kirschvink et al., 1997). Climate model simulations (Crowley and Baum, 1993) suggest there is a

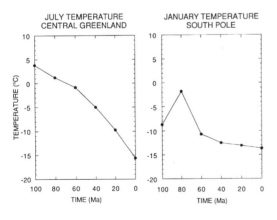

Fig. 1.3. Evidence that seasonal changes in geography may sometimes have a significant climate effect. Results show calculated summer temperature changes on Greenland and Antarctica due to changes in land-sea distribution during the last 100 million years. Based on results presented in Hyde et al. (1990). (From Crowley and North, 1991)

KAZANIAN 255 MA (ANNUAL CYCLE)

Fig. 1.5. Effect of a supercontinent on the seasonal cycle. Figure shows the EBM-calculated annual range of temperatures for a Late Permian Pangea (Kazanian, 255 Ma). This range is larger than the present range for Asia. Base map, with topography, is from Scotese (1986). (After Crowley et al., 1989). *Reprinted with permission of Geological Society of America. From T. Crowley, W. Hyde, and D. Short, "Seasonal cycle variations on the supercontinent of Pangaea," Geology 17, 457–460.*

possibility that Late Precambrian glaciations could be due in part to lower solar luminosity, but the results are critically dependent on choice of geographic boundary conditions for the model simulation. For example the reconstruction of Bond et al. (1984) provided a relatively good agreement between areas where the model predicted low summer temperatures and where evidence for glaciation has been found (Crowley and Baum, 1993).

Changes in geography can also influence the large-scale ocean circulation (e.g., Bice, 1997). Barron and Peterson (1991) utilized a simplified ocean model to calculate that changes in geography can influence the strength of the oceanic gyres (cf. Seidov, 1986) and loci of deepwater formation. Kutzbach et al. (1990) modeled the effect of the giant Paleozoic and early Mesozoic Panthalassa Ocean and simulated strong western boundary currents along the east coast of continents that could have significantly influenced the latitudinal range of subtropical biota. Archer (1996) has recently calculated that there may have been unusual

tides in the Panthalassa Ocean, including some examples of a tidal seiche in the "Tethys Embayment" that may be flanked by eastern islands and microcontinents (cf. Fig. 1.5). This is an unusual high tide with a period different than the standard lunar/solar periods and caused by resonance of the standard tides with basin geometry. But this conclusion obviously hinges on the credibility of the eastern barrier proposed by Scotese (1986).

Sea Level

Another critical boundary condition for climate models is sea level. Even if the location of the continents is well known, the accuracy of shoreline estimates will have a significant effect on the seasonal cycle of temperatures on land (the thermal inertia effect). Such a response has a direct bearing on a major paleoclimate problem for tectonic time scales—the question of "equable climates" (Crowley et al., 1989; Sloan and Barron, 1990). This concept is often not well defined but connotes low seasonal cycles with mild winter temperatures. The concept

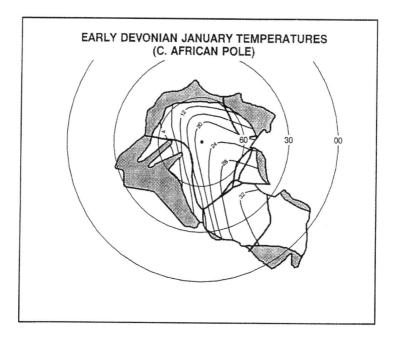

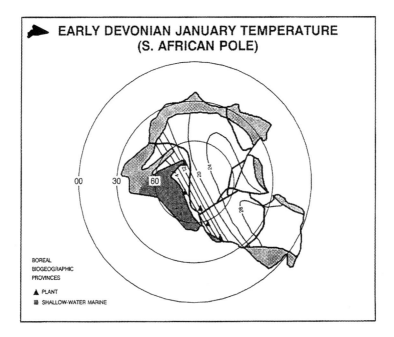

Fig. 1.6. Alternate geography simulations of January temperature for the Early Devonian (~390 Ma) of Gondwana. Because the pole position is not well known, two different reconstructions are figured. (a): simulation with central African (mid-Devonian) pole of Hurley and van der Voo (1987); (b) simulation with South African pole of Bambach et al. (1981). The lower panel also shows boreal biogeographic provinces as defined by the shallow-water marine (Malvinokaffric realm) and plant distributions from Ziegler et al. (1981) and Boucot and Gray, (1983). Top figure is from Crowley et al. (1987); bottom figure is from Crowley and North (1991). *Reprinted by permission from Nature vol. 329. pp. 803-807, copyright (c) 1987 Macmillan Magazines Ltd.*

is used to explain poleward displacements of thermophilic biota during warm time periods.

Changes in sea level can also affect precipitation because of the relation between the seasonal cycle, land-sea distribution, and precipitation. As discussed by Kutzbach and Gallimore (1989), assembly of a large landmass results in a large seasonal cycle of temperature, but precipitation decreases because much of the moisture is removed from of the atmosphere in coastal regions, leaving interiors dry. The reverse can occur during times of high sea level. The extent of continental flooding can also affect geochemical cycling, and geochemical models suggest a relationship between weathering and atmospheric CO_2 levels (Berner, 1994).

An extreme illustration of the effect of sea level involves the Late Ordovician (440 Ma), where one paleogeographic reconstruction (Scotese and Golonka, 1992) has less than half the land area as present. When a climate model was run with high CO_2 levels for this reconstruction (Crowley and Baum, 1995), it simulated large increases in average runoff for individual land grid points. Yet because the total land area was so small, total runoff to the ocean was less than present. This resulted in a response opposite that predicted for the weathering response in a high-CO_2 world (Berner, 1994).

If the extent of epeiric seaways is not correctly specified in climate models, then models will predict an inaccurate seasonal cycle. This response may result in significant discrepancies between models and data. For example, Yemane (1993; cf. Taylor et al., 1992; Rayner, 1995) suggested that one reason paleoclimate model predictions for the Late Permian disagree with observations is because an insufficient area of epeiric seaways was stipulated for that time period. Similarly, Valdes et al. (1996) suggest that a large eastward extension of the Late Cretaceous Western Interior Seaway on North America may have resulted in milder winter temperatures than simulated by earlier modeling efforts. This result is supported by a new Cretaceous reconstruction presented by Ziegler and Rowley in this volume (Chapter 7). The area of epeiric seas can also influence currents and tidal ranges in the seaway (e.g., Ericksen and Slingerland, 1990).

Orography

Changes in the elevation of landmasses may also have affected climate, both directly and through its effect on atmospheric CO_2 levels (e.g., Ruddiman and Kutzbach, 1989; Raymo et al., 1988; Raymo and Ruddiman, 1992). Climate model experiments have long indicated that orographic changes have a significant effect on the development of the Asian monsoon (e.g., Manabe and Terpstra, 1974; Kutzbach et al., 1989). Comparisons of simulated changes for the Cenozoic (Ruddiman et al., 1989) indicate that regional trends in climate in many places are consistent with the overall increase in continental elevations in the Cenozoic. A note of caution has been injected into this discussion by Molnar and England (1990), who point out that climate change can affect orography through changes in erosion and isostatic compensation. Thus, the cause-affect interpretation can be reversed (cf. Small and Anderson, 1995). Reviews in this volume by Chase et al. (Chapter 4) and Harrison et al. (Chapter 5) provide an updated summary of the long-term history of uplift in western North America and the Himalayas.

Elevation effects are also important for correct simulation of supercontinent climates. Moore et al. (1992b) found a significantly different model response for runs with different topographic specifications. Similarly, Otto-Bliesner (1993, this volume, Chapter 5) demonstrated that stipulation of the elevation of the Allegheny-Hercynian chain along the main Laurussia-Gondwana suture has a significant effect on the location of rain bands in the equatorial region (Fig. 1.7). In fact, Dewey and Burke (1973; cf. Menard and Molnar, 1988) have suggested that a Tibetan Plateau-scale feature may have existed in western Europe at that time (location was approximately 15°N and 10°E on reconstruction in Fig. 1.5). For a later time and on a finer spatial scale, elevation of the Triassic continental rift system should have influenced orographic uplift and subsequent adiabatic warming in the rift valleys (Hay et al., 1982). This response, coupled to Milankovitch forcing, may have affected the wet-dry sequences of sediments found in Triassic rift basin deposits (Olsen, 1986; Crowley et al., 1992). Here again, specification of the correct elevation will affect model results.

Even minor changes in elevation may result in different types of climate (cf. Hay and Wold, Chapter 6). For example, Crowley and Baum (1995) indicated that elevation differences of 300–500 m for the Late Ordovician (440 Ma) location of

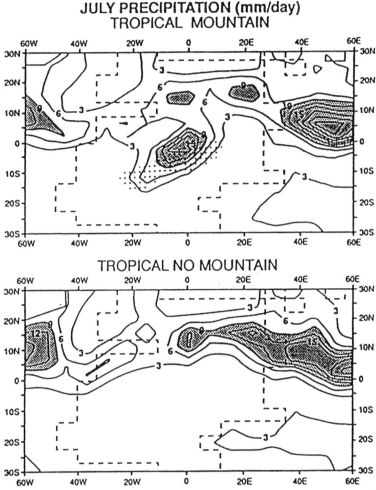

Fig. 1.7. Effect of elevation on location of rainbelts for the Allegheny-Hercynian chain in the mid-Carboniferous. Results show simulated July precipitation (mm/day) for a run with mountains and another run with no mountains. (From Otto-Bliesner, 1993)

Gondwana can lead to two different climate solutions—one with no summer snowcover and one with a permanent summer snowcover that could lead to development of an ice cap. The preceding illustration may represent an atypical case; the results nevertheless highlight the importance of orography. Although it is unrealistic to expect that geologists could refine their paleoelevations to within 300–500 m, Forest et al. (1995) have suggested a method that at times could constrain elevations to within 700 m.

Ocean Gateways

Marine geologists have long been fascinated with the potential effect of "ocean gateway" changes on

the oceanic circulation (Berggren and Hollister, 1974; Haq, 1981). This expression refers to the opening and closing of narrow straits between the ocean basins—for example, the Tethys, the central American isthmus, the Drake Passage, the Australia–Antarctic passage, the Indonesian straits, and the Greenland–Faeroes Ridge. The changes affect not only the migration of terrestrial organisms but also the exchange of heat and salinity between ocean basins. In addition, gateways may have significantly affected patterns of sediment distribution in the world ocean (e.g., Maier-Reimer et al., 1990; Heinze and Crowley, 1997).

Some sensitivity experiments with ocean models provided more insight into the importance of ocean gateways. An open Tethys Seaway has been the subject of two modeling studies (Barron and Peterson, 1990; Bush, 1997), with the coupled ocean-atmosphere model run agreeing with geological reconstructions (e.g., Haq, 1981). This response occurred because Tethys is far enough north to have come under the influence of westerly winds. Maier-Reimer et al. (1990; Mikolajewicz and Crowley, 1996) calculated that late Cenozoic closure of the central American isthmus (Droxler et al., Chapter 8) would have completely rearranged the circulation of the surface and deep waters of the Atlantic (Fig. 1.8). This response followed from the fact that at present, the North Atlantic is significantly saltier than the other oceans. Advection of this saline water into high latitudes and subsequent cooling in winter results in a vigorous thermohaline circulation which acts as a positive feedback to further feed warm waters into the subpolar North Atlantic. Export of this deepwater across the equator and into the world ocean not only affects the preservation of sediment in different ocean basins (Berger, 1970) but also affects heat exchange between the Northern and Southern Hemispheres. For example conservation of volume constraints result in an import of warmer surface-to-intermediate waters from the South Atlantic to the North Atlantic as a result of the export of deep North Atlantic Deep Water (NADW). Consequently, the Southern Hemisphere loses a significant amount of heat in this exchange (Crowley, 1992).

In their model experiment (Maier-Reimer et al., 1990) an open central American isthmus resulted in a free exchange of lower salinity North Pacific water with the North Atlantic. This response resulted in a reduction of North Atlantic salinities and a collapse of the thermohaline circulation. Similar experiments with closure of the Drake Passage resulted in very large changes in outflow of deep/bottom water from the Southern Ocean (Mikolajewicz et al., 1993). The modeled responses to Drake and Panamanian changes are consistent with the overall pattern of deepwater circulation inferred for the Cenozoic (Woodruff and Savin, 1989; Delaney, 1990; Katz and Miller, 1991).

Changes at other gateways should also have influenced the ocean circulation. For example, Wright (Chapter 9) and Wright and Miller (1996) note that subsidence of the Greenland-Faeroes Ridge may also have regulated NADW production. Opening of the Tasman Gateway enabled a deep connection between

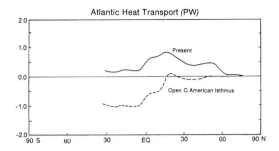

Fig. 1.8. Comparison of poleward ocean heat transport in the Atlantic for an ocean GCM simulation with present boundary conditions and one with an open Central American Isthmus. Based on results in Maier-Reimer et al. 1990. 1 petawatt (PW) = 1 x 10^{15} W. (From Crowley and North, 1991)

the Southern Indian and South Pacific Oceans (e.g., Kennett, 1977). At present, 40% of the total input of deepwater to the world's ocean occurs through this passage (Warren, 1981). Constriction of the Indonesian straits may have affected exchange of warm surface water between the Pacific and Indian Basins (Kennett et al., 1985; Hirst and Godfrey, 1993), thereby potentially influencing the heat content of waters in the North Pacific and, by extension, perhaps the climatic evolution of the high latitudes of the Northern Hemisphere.

There may also have been some important gateway changes during the Paleozoic. The opening between Laurussia and Gondwana that existed until about the Early Carboniferous (about 340 Ma) would have facilitated throughflow from the Tethys to the eastern paleo-Pacific (Panthalassa Ocean). Closure of these straits should have rearranged the tropical oceanic circulation and perhaps explain some significant changes in the global carbon cycle at that time (Grossman, 1994). Similarly, a possible shallow barrier along about 110°E paleolongitude for the Late Permian (Fig. 1.5) may have prevented exchange of deep waters between the eastern Tethys and the western Panthalassa Oceans. This feature could also have affected the surface circulation. For example, in the absence of such a barrier Kutzbach et al. (1990) simulate a strong east–west flow along the equator from the paleo-Pacific to the paleo-Tethys. The presence of shallow barriers may have significantly blocked this transport and affected the salinity of waters west of the barriers.

Although geologists know that gateway openings and closings have occurred, the key link between the geodynamic and paleoclimatic communities involves the details. For example, how well constrained is the opening of Drake Passage to 28–32 Ma (Barker and Burrell, 1977; Lawver and Gahagan, Chapter 10)? Modeling studies indicate that even opening of a shallow connection may have significantly influenced Antarctic outflow (Mikolajewicz et al., 1993). Since this shallow connection could have developed close in time to a significant global ice volume event near the Eocene/Oligocene boundary at about 32–34 Ma (e.g., Zachos et al., 1992), the climate event could be related to initial opening of the passage. Similar refinement of the time of opening/closing of other barriers is also important.

Bathymetry

Specification of seafloor bathymetry will be an increasingly important problem for paleoceanographic simulations. Bottom topography exerts a strong effect on the flow of water masses. Besides ocean gateways, the locations and elevations of mid-ocean ridges and oceanic plateaus will need to be known. Information on sill depth although difficult to obtain, would also be helpful (cf. Collins et al., 1996). Although less has been done from the modeling viewpoint on the significance of bathymetry, two simulations show striking examples of the potential effect on sill depth on an ocean model. Mikolajewicz et al. (1993) examined the joint effect of the Drake and Panamanian gateways and found two different solutions for Pacific bottom water flow that depended on specification of the sill depth of the Panama isthmus. For a bathyal sill depth (3000 m), deepwater coursing up the South Atlantic basin was prevented from entering the Pacific. For an abyssal sill depth (4100 m), the deep Atlantic water flowed into the Pacific, across the entire North Pacific, and came down the coast of Asia and the western South Pacific as a boundary current that flowed in a direction opposite to that of the flow pattern with a Panamanian sill depth at the shallower level.

In addition to sill depth changes, the general shape of the ocean basin, including height and extent of mid-ocean ridges, can also influence ocean circulation. In this volume Bice et al. (Chapter 11)

discuss an example of the effect of bathymetry on the circulation of an Eocene ocean model. Although it is at present unrealistic to expect geodynamicists to reconstruct all aspects of past bathymetry to levels that may be needed for some model studies, attention to the topic and improvement of existing reconstructions is essential if more reliable simulations are to be performed.

Effect of Tectonic Boundary Conditions on Atmospheric CO_2 Levels

Changes in solid earth boundary conditions can also affect climate through their effect on atmospheric CO_2 levels (e.g., Berner et al., 1983). For example, heat flow (and CO_2 production) can vary as a function of the rate of seafloor spreading or the total length of the ridges. These may have varied with time (Kominz, 1984). Changes in mantle plume activity may also have varied with time. Several authors have suggested higher levels of plume activity for the mid-Pacific plate during the mid-Cretaceous (Arthur et al., 1985; Larson, 1991; Tarduno et al., 1991), possibly due to a standstill of Pacific plate motion with respect to the spin axis (Tarduno and Sager, 1995). Such activity could have significantly affected atmospheric CO_2 levels (cf. Caldeira and Rampino, 1991). However, Heller et al. (1996) have challenged some of these interpretations; clarification of this problem is important for improving estimates of past atmospheric CO_2 levels.

As discussed earlier in this chapter, and in Otto-Bliesner (1995), land-sea distribution, and uncertainties thereof, can also affect climate model estimates of atmospheric weathering rates and runoff, which also can affect CO_2 levels (Berner, 1994). In Chapter 12 in this volume, Berner assesses some of the potential impact of such changes. In Chapter 13, Reusch and Maasch suggest that variations in exposure of mafic rocks along arc terranes could also significantly influence CO_2 levels. Distribution of other rock types can also be important (Bluth and Kump, 1991). In addition to the potential effect of orographic induced weathering increases on drawdown of atmospheric CO_2 (Raymo and Ruddiman, 1992), degassing from metamorphism along suture zones may lead to significant levels of CO_2 release (Kerrick and Caldeira, 1993).

Summary

Tectonic variations influence climate in a variety of ways, some of which have been touched on in this review. For example, changes in land-sea distribution influence the seasonal cycle of temperature and precipitation, which in turn affect processes as diverse as ice sheet growth and carbon dioxide levels. Changes in orography can have a very large effect on precipitation and CO_2 levels. Opening and closing of ocean gateways can dramatically influence the surface and deepwater circulation patterns in the world ocean. Future work in these areas requires fruitful collaborations between geodynamicists, paleoclimatologists, geochemists, and modelers.

Acknowledgments

This work was supported by NSF grant ATM-9529109. I thank E. Barron, K. Burke, and an anonymous reviewer for their comments on an earlier draft.

References

Archer, A. W., 1996, Panthalassa: Paleotidal resonance and a global paleocean seiche, *Paleoceanography, 11,* 625–632.

Arthur, M. A., W. E. Dean, and G. E. Claypool, 1985, Anomalous ^{13}C enrichment in modern marine organic carbon, *Nature, 315,* 216–218.

Bambach, R. K., C. R. Scotese, and A. M. Ziegler, 1981, Before Pangea: The geographies of the Paleozoic world, in *Paleontology and Paleoenvironments, vol. 1,* B. J. Skinner (ed.), pp. 116–128, William Kaufmann, Los Altos, Calif.

Barker, P. F., and J. Burrell, 1977, The opening of the Drake Passage, *Mar. Geol., 25,* 15–34.

Barron, E. J., 1985, Explanations of the Tertiary global cooling trend, *Palaeogeogr., Palaeoclimatol., Palaeoecol., 50,* 45–61.

Barron, E. J., 1987, Eocene equator-to-pole surface ocean temperatures: A significant climate problem? *Paleoceanography, 2,* 729–739.

Barron, E. J., and W. M. Washington, 1982, Cretaceous climate: A comparison of atmospheric simulations with the geologic record, *Palaeogeogr., Palaeoclimatol., Palaeoecol., 40,* 103–133.

Barron, E. J., and W. M. Washington, 1985, Warm Cretaceous climates: High atmospheric CO_2 as a plausible mechanism, in *The Carbon Cycle and Atmospheric CO_2: Natural Variations Archean to Present,* E. T. Sundquist and W. S. Broecker (eds.), pp. 546–553, *Geophys. Mono. Ser., vol. 32,* AGU, Washington, D.C.

Barron, E. J., and W. H. Peterson, 1990, Mid-Cretaceous ocean circulation: Results from model sensitivity studies, *Paleoceanog., 5,* 319–337.

Barron, E. J., and W. H. Peterson, 1991, The Cenozoic ocean circulation based on ocean general circulation model results, *Palaeogeogr., Palaeoclimatol., Palaeoecol., 83,* 1–28.

Barron, E. J., S. L. Thompson, and S. H. Schneider, 1981, An ice-free Cretaceous? Results from climate model simulation, *Science, 212,* 501–508.

Barron, E. J., P. J. Fawcett, W. H. Peterson, D. Pollard, and S. L. Thompson, 1995, A "simulation" of mid-Cretaceous climate, *Paleoceanography, 10,* 953–962.

Berger, W. H., 1970, Biogenous deep-sea sediments: Fractionation by deep-sea circulation, *Deep Sea Res., 81,* 31–43.

Berggren, W. A., and C. D. Hollister, 1974, Paleogeography, paleobiogeography, and the history of circulation in the Atlantic Ocean, in Studies in Paleo-Oceanography, W. W. Hay (ed.), pp. 126–186, *Soc. Econ. Paleontol. Mineral. Spec. Pub. 20.*

Berner, R. A., 1994, GEOCARB II: A revised model of atmospheric CO_2 over Phanerozoic time, *Am. J. Science, 294,* 56–91.

Berner, R. A., A. C. Lasaga, and R. M. Garrels, 1983, The carbonate-silicate geochemical cycle and its effect on atmospheric carbon dioxide over the last 100 million years, *Am. J. Sci., 283,* 641–683.

Bertrand-Sarfati, J., A. Moussine-Pouchkine, B. Amard, and A. A. K. Ahmed, 1995, First Ediacaran fauna found in western Africa and evidence for an Early Cambrian glaciation, *Geology, 23,* 133–136.

Bice, K. L., 1997, *An Investigation of Early Eocene Ocean Warming Using a High Resolution General Circulation Model,* Ph. D. Dissertation, Pennsylvania State University.

Bluth, G. J. S., and L. R. Kump, 1991, Phanerozoic paleogeology, *Am. J. Science, 291,* 284–308.

Bond, G. C., P. A. Nickeson, and M. A. Kominz, 1984, Breakup of a supercontinent between 625 Ma and 555 Ma: New evidence and implications for continental histories, *Earth Planet. Sci. Lett., 70,* 325–345.

Boucot, A. J., and J. Gray, 1983, A Paleozoic Pangaea, *Science, 222,* 571–581.

Brenchley, P. J., J. D. Marshall, G. A. F. Carden, D. B. R. Robertson, D. G. F. Long, T. Meidla, L. Hints, and T. F. Anderson, 1994, Bathymetric and isotopic evidence for a short-lived Late Ordovician glaciation in a greenhouse period, *Geology, 22,* 295–298.

Briden, J. C., E. McClelland, and D. C. Rex, 1993, Proving the age of a paleomagnetic pole: The case of the Ntonya Ring Structure, Malawi, *J. Geophys. Res., 98,* 1473–1479.

Bush, A. B. G., 1997, Numerical simulation of the Cretaceous Tethys circumglobal current, *Science, 275,* 807–810.

Caldeira, K., and M. R. Rampino, 1991, The mid-Cretaceous super plume, carbon dioxide, and global warming, *Geophys. Res. Lett., 18,* 987–990.

Chandler, M., D. Rind, and R. Ruedy, 1992, Pangaean climate during the Early Jurassic: GCM simulations and the sedimentary record of paleoclimate, *Geol. Soc. Am. Bull., 104,* 543–559.

Collins, L. S., A. G. Coates, W. A. Berggren, M.-P. Aubry, and J. Zhang, 1996, The late Miocene Panama isthmian strait, *Geology, 24,* 687–690.

Covey, C., and E. Barron, 1988, The role of ocean heat transport in climatic change, *Earth-Science Rev., 24,* 429–445.

Crowell, J. C., 1982, Continental glaciation through geologic times, in *Climate in Earth History,* W. H. Berger and J. C. Crowell (eds.), pp. 77–82, Natl. Acad. Press, Washington, D.C.

Crowley, T. J., 1992, North Atlantic Deep Water cools the southern hemisphere, *Paleoceanography, 7,* 489–497.

Crowley, T. J., and G. R. North, 1991, *Paleoclimatology,* Oxford Univ. Press, N.Y.

Crowley, T. J., and S. K. Baum, 1993, Effect of decreased solar luminosity on Late Precambrian ice extent, *J. Geophys. Res., 98,* 16723–16732.

Crowley, T. J., and S. K. Baum, 1995, Reconciling Late Ordovician (440 Ma) glaciation with very high (14X) CO_2 levels, *J. Geophys. Res., 100,* 1093–1101.

Crowley, T. J., D. A. Short, J. G. Mengel, and G. R. North, 1986, Role of seasonality in the evolution of climate over the last 100 million years, *Science, 231,* 579–584.

Crowley, T. J., J. G. Mengel, and D. A. Short, 1987, Gondwanaland's seasonal cycle, *Nature, 329,* 803–807.

Crowley, T. J., W. T. Hyde, and D. A. Short, 1989, Seasonal cycle variations on the supercontinent of Pangaea, *Geology, 17,* 457–460.

Crowley, J. T., S. K. Baum, and W. T. Hyde, 1991, Climate model comparison of Gondwanan and Laurentide glaciations, *J. Geophys. Res., 96,* 9217–9226.

Crowley, T. J., K.-Y. Kim, J. G. Mengel, and D. A. Short, 1992, Modeling 100,000 year climate fluctuations in pre-Pleistocene time series, *Science, 255,* 705–707.

Dalziel, I.W.D., 1991, Pacific margins of Laurentia and East Antarctica–Australia as a conjugate rift pair: Evidence and implications for an Eocambrian supercontinent, *Geology, 19,* 598–601.

Dalziel, I.W.D., 1997, Neoproterozoic-Paleozoic geography and tectonics: Review, hypothesis, environmental speculation, *Geol. Soc. Am. Bull., 109,* 16–42.

Delaney, M. L., 1990, Miocene benthic foraminiferal Cd/Ca records: South Atlantic and western equatorial Pacific, *Paleoceanog., 5,* 743–760.

Dewey, J. F., and K. C. A. Burke, 1973, Tibetan, Variscan, and Precambrian basement reactivation: Products of continental collision, *J. Geol., 81,* 683–692.

Donn, W. L., and D. M. Shaw, 1977, Model of climate evolution based on continental drift and polar wandering, *Geol. Soc. Am. Bull., 88,* 390–396.

Ericksen, M. C., and R. Slingerland, 1990, Numerical simulations of tidal and wind-driven circulation in the Cretaceous interior seaway of North America, *Geol. Soc. Am. Bull., 102,* 1499–1516.

Fawcett, P., 1994, Simulation of climate-sedimentary evolution: A comparison of climate model results to the geologic record for India and Australia, Ph.D. thesis, Pennsylvania State University, Pa.

Forest, C. E., P. Molnar, and K. A. Emanuel, 1995, Palaeoaltimetry from energy conservation principles, *Nature, 374,* 347–350.

Frakes, L. A., and J. E. Francis, 1988, A guide to

Phanerozoic cold polar climates from high-latitude ice-rafting in the Cretaceous, *Nature, 333,* 547–549.

Frakes, L. A., J. E. Francis, and J. Syktus, 1992, *Climate Modes of the Phanerozoic,* Cambridge Univ. Press, New York.

Grossman, E. L., 1994, The carbon and oxygen isotope record during the evolution of Pangea: Carboniferous to Triassic, in *Pangea: Paleoclimate, Tectonics, and Sedimentation During Accretion, Zenith, and Breakup of a Supercontinent,* G. D. Klein (ed.), pp. 207–228, Special Paper 288, Geological Society of America, Boulder, Colo.

Hambrey, H. A., and W. B. Harland (eds.), 1981, *Earth's Pre-Pleistocene Glacial Record,* Cambridge Univ. Press, Cambridge.

Haq, B. U., 1981, Paleogene paleoceanography: Early Cenozoic oceans revisited, Proceedings 26th International Geological Congress, Geology of Oceans symposium, Paris, July 7–17, 1980, *Oceanol. Acta,* 71–82.

Hay, W. W., J. F. Behensky, Jr., E. J. Barron, and J. L. Sloan II, 1982, Late Triassic-Liassic paleoclimatology of the proto-central North Atlantic rift system, *Palaeogeogr., Palaeoclimatol., Palaeoecol., 40,* 13–30.

Heinze, C. and T. J. Crowley, 1997, Sedimentary response to ocean gateway circulation changes, *Paleoceanography, 12,* 742–754.

Heller, P. L., D. L. Anderson, and C. L. Angevine, 1996, Is the middle Cretaceous pulse of rapid sea-floor spreading real or necessary? *Geology, 24,* 491–494.

Hirst, A. C., and J. S. Godfrey, 1993, The role of Indonesian throughflow in a global ocean GCM, *J. Phys. Oceanogr., 23,* 1057–1086.

Hoffman, P. F., 1991, Did the breakout of Laurentia turn Gondwanaland inside-out?, *Science, 252,* 1409–1412.

Hurley, N. F., and R. van der Voo, 1987, Paleomagnetism of upper Devonian reefal limestones, Canning Basin, Western Australia, *Geol. Soc. Amer. Bull., 98,* 138–146.

Hyde, W. T., T. J. Crowley, K.-Y. Kim, and G. R. North, 1989, Comparison of GCM and energy balance model simulations of seasonal temperature changes, 18,000 B.P. to present, *J. Clim.* 2, 864–887.

Hyde, W. T., K.-Y. Kim, T. J. Crowley, and G. R.

North, 1990, On the relation between polar continentality and climate: Studies with a nonlinear energy balance model, *J. Geophys. Res., 95,* 18,653–18,668.

Katz, M. E., and K. G. Miller, 1991, Early Paleogene benthic foraminiferal assemblages and stable isotopes in the Southern Ocean, in *Proceedings of the Ocean Drilling Program, vol. 114,* P. F. Ciesielski, Y. Kristoffersen et al. (eds.), pp. 481–512, Ocean Drilling Program, Texas A&M University.

Kennett, J. P., 1977, Cenozoic evolution of Antarctic glaciation, the circum-Antarctic ocean, and their impact on global paleoceanography, *J. Geophys. Res., 82,* 3843–3860.

Kennett, J. P., G. Keller, and M. S. Srinivasan, 1985, Miocene planktonic foraminiferal biogeography and paleoceanographic development of the Indo-Pacific region, pp. 197–236, *Geol. Soc. Am., Mem. 163.*

Kerrick, D. M., and K. Caldeira, 1993, Paleoatmospheric consequences of CO_2 released during early Cenozoic regional metamorphism in the Tethyan orogen, *Chem. Geol., 108,* 201–230.

Kirschvink, J. L., R. L. Ripperdan, and D. A. Evans, 1997, Evidence for a large-scale reorganization of Early Cambrian continental masses by inertial interchange true polar wander, *Science, 277,* 541–545.

Knoll, A. H., and M. R. Walter, 1992, Latest Proterozoic stratigraphy and earth history, *Nature, 356,* 673–678.

Kominz, M. A., 1984, Oceanic ridge volumes and sea-level change—An error analysis, in *Interregional Unconformities and Hydrocarbon Accumulation,* J. S. Schlee (ed.), pp. 108–123, Am. Assoc. Petrol. Geol. Mem. 36, Tulsa, Okla.

Kutzbach, J. E., and R. G. Gallimore, 1989, Pangean climates: Megamonsoons of the megacontinent, *J. Geophys. Res., 94,* 3341–3357.

Kutzbach, J. E., P. J. Guetter, W.F. Ruddiman, and W. L. Prell, 1989, Sensitivity of climate of late Cenozoic uplift in southern Asia and the American west: Numerical experiments, *J. Geophys. Res., 94,* 18,393–18,397.

Kutzbach, J. E., P. J. Guetter, and W. M. Washington, 1990, Simulated circulation of an idealized ocean for Pangaean time, *Paleoceanography, 5,* 299–317.

Larson, R. L., 1991, Latest pulse of earth: Evidence

for a mid-Cretaceous superplume, *Geology, 19,* 547–550.

Lau, K.-M, J. H. Kim, and Y. Sud, 1996, Intercomparison of hydrologic processes in AMIP GCMs, *Bull. Am. Meteorol. Soc., 77,* 2209–2227.

Maier-Reimer, E., U. Mikolajewicz, and T. J. Crowley, 1990, Ocean general circulation model sensitivity experiment with an open Central American isthmus, *Paleoceanography, 5,* 349–366.

Manabe, S., and T. B. Terpstra, 1974, The effects of mountains on the general circulation of the atmosphere as identified by numerical experiments, *J. Atmos. Sci., 31,* 3–42.

Manabe, S., and A. J. Broccoli, 1990, Mountains and arid climates of middle latitudes, *Science 247,* 192–195.

Menard, G., and P. Molnar, 1988, Collapse of a Hercynian Tibetan Plateau into a late Palaeozoic European basin and range province, *Nature, 334,* 235–237.

Mikolajewicz, U., and T. J. Crowley, 1997, Response of a coupled ocean/energy balance model to restricted flow through the Central American isthmus, *Paleoceanography, 12,* 429–441.

Mikolajewicz, U., E. Maier-Reimer, T. J. Crowley, and K.-Y. Kim, 1993, Effect of Drake and Panamanian gateways on the circulation of an ocean model, *Paleoceanography, 8,* 409–426.

Molnar, P., and P. England, 1990, Late Cenozoic uplift of mountain ranges and global climate change: Chicken or egg?, *Nature, 346,* 29–34.

Moore, G. T., D. N. Haysahida, C. A Ross, and S. R. Jacobson, 1992a, Paleoclimate of the Kimmeridgian/Tithonian (Late Jurassic) world: I. Results using a general circulation model, *Palaeogeogr., Palaeoclimatol., Palaeoecol., 93,* 113–150.

Moore, G. T., L. C. Sloan, D. N. Haysahida, and N. P. Umrigar, 1992b, Paleoclimate of the Kim meridgian/Tithonian (Late Jurassic) world: II. Sensitivity tests comparing three different paleotopographic settings, *Palaeogeogr., Palaeoclimatol., Palaeoecol., 95,* 229–252.

Muttoni, G., D. V. Kent, and J. E. T. Channel, 1996, Evolution of Pangea: Paleomagnetic constraints from the Southern Alps, Italy, *Earth Planet. Sci. Lett., 140,* 97–112.

North, G. R., J. G. Mengel, and D. A. Short, 1983, Simple energy balance model resolving the seasons and the continents: Application to the as tronomical theory of the ice ages, *J. Geophys. Res., 88,* 6576–6586.

Oglesby, R. J., 1991, Joining Australia to Antarctica: GCM implications for the Cenozoic record of Antarctic glaciations, *Clim. Dyn., 6,* 13–22.

Olsen, P. E., 1986, A 40-million-year lake record of early Mesozoic orbital climatic forcing, *Science, 234,* 789–912.

Otto-Bliesner, B. L., 1993, Tropical mountains and coal formation: A climate model study of the Westphalian (306 Ma), *Geophys. Res. Lett., 18,* 1947–1950.

Otto-Bliesner, B. L., 1995, Continental drift, runoff, and weathering feedbacks: Implication from climate model experiments, *J. Geophys. Res., 100,* 11537–11548.

Parrish, J. T., 1990, Gondwanan paleogeography and paleoclimatology, in *Antarctic Paleobiology: Its Role in the Reconstruction of Gondwana,* T. N. Taylor and E. L. Taylor (eds.), pp. 16–26, Springer-Verlag, New York.

Parrish, J. T., 1993, Climate of the supercontinent Pangea, *J. Geology, 101,* 215–233.

Pollard, D., and M. Schulz, 1994, A model for the potential locations of Triassic evaporite basins driven by paleoclimatic GCM simulations, *Glob. Planet. Change, 9,* 233–249.

Raymo, M. E., and W. F. Ruddiman, 1992, Tectonic forcing of late Cenozoic climate, *Nature, 359,* 117–122.

Raymo, M. E., W. F. Ruddiman, and P. N. Froelich, 1988, Influence of late Cenozoic mountain building on ocean geochemical cycles, *Geology, 16,* 649–653.

Rayner, R. J., 1995, The palaeoclimate of the Karoo: Evidence from plant fossils, *Palaeogeogr., Palaeoclimatol., Palaeoecol., 119,* 385–394.

Ruddiman, W. F., and J. E. Kutzbach, 1989, Forcing of Late Cenozoic Northern Hemisphere climate by plateau uplift in southern Asia and the American west, *J. Geophys. Res., 94,* 18409–18427.

Ruddiman, W. F., W. L. Prell, and M. E. Raymo, 1989, Late Cenozoic uplift in southern Asia and the American west: Rationale for general circulation modeling experiments, *J. Geophys. Res., 94,* 18379–18391.

Schneider, S. H., and R. E. Dickinson, 1974, Climate modeling, *Rev. Geophys. Space Phys, 12,* 447–493.

Scotese, C. R., 1986, Phanerozoic reconstructions: A new look at the assembly of Asia, *University of Texas Institute for Geophysics Technical Report 66*, 54 pp.

Scotese, C. R., and J. Golonka, 1992, *Paleogeographic Atlas: Paleomap Project*, University of Texas-Arlington, Arlington, Tex.

Seidov, D. G., 1986, Numerical modelling of the ocean circulation and paleocirculation, in *Mesozoic and Cenozoic Oceans*, K. Hsü (ed.), pp. 11–26, *Geodynamic Series, 15*, Am. Geophys. Union, Washington, D.C.

Sloan, L. C., and E. J. Barron, 1990, "Equable" climates during earth history, *Geology, 18*, 489–492.

Sloan, L. C., and D. K. Rea, 1995, Atmospheric carbon dioxide and early Eocene climate: A general circulation modeling sensitivity study, *Palaeogeogr., Palaeoclimatol., Palaeoecol., 119*, 275–292.

Small, E.E., and R.S. Anderson, 1995, Geomorphically driven Late Cenozoic rock uplift in the Sierra Nevada, California, *Science, 270*, 277–280.

Tarduno, J. A., and W. W. Sager, 1995, Polar standstill of the Mid-Cretaceous Pacific plate and its geodynamic implications, *Science, 269*, 956–959.

Tarduno, J. A., W. V. Sliter, L. Kroenke, M. Leckie, H. Mayer, J. J. Mahoney, R. Musgrave, M. Storey, and E. L. Winterer, 1991, Rapid formation of Ontong Java Plateau by Aptian mantle plume volcanism, *Science, 254*, 399–403.

Taylor, E. L, T. N. Taylor, and R. Cúneo, 1992, The present is not the key to the past: A polar forest from the Permian of Antarctica, *Science 257*, 1675–1677.

Valdes, P. J., and B. W. Sellwood, 1992, A palaeoclimate model for the Kimmeridgian, *Palaeogeogr., Palaeoclimatol., Palaeoecol., 95*, 47–72.

Valdes, P. J., B. W. Sellwood, and G. D. Price, 1996, Evaluating concepts of Cretaceous equability, *Palaeoclimates: Data and Modelling, 2*, 139–158.

van der Voo, R., 1992, Jurassic paleopole contro

versy: Contributions from the Atlantic-bordering continents, *Geology, 20*, 975–978.

van Fossen, M. C., and D. V. Kent, 1992, Paleomagnetism of the Front Range (Colorado) Morrison Formation and an alternative model of Late Jurassic North American apparent polar wander, *Geology, 20*, 223–226.

Warren, B. A., 1981, Deep circulation of the world ocean, in *Evolution of Physical Oceanography*, B. A. Warren and C. Wunsch (eds.), pp. 6–41, MIT Press, Mass.

Wilson, K. M., D. Pollard, W. W. Hay, S. L. Thompson, and C. N. Wold, 1994, General circulation model simulations of Triassic climates and preliminary results, in *Pangea: Paleoclimate, Tectonics and Sedimentation During Accretion, Zenith and Breakup of a Supercontinent*, G. D. Klein (ed.), pp. 91–116, Special Paper 288, Geol. Soc.Amer., Boulder, Colo.

Woodruff, F., and S. M. Savin, 1989, Miocene deepwater oceanography, *Paleoceanography, 4*, 87–140.

Wright, J. D., and K. G. Miller, 1996, Control of North Atlantic Deep Water circulation by the Greenland-Scotland Ridge, *Paleoceanography, 11*, 157–170.

Yemane, K., 1993, Contribution of Late Permian palaeogeography in maintaining a temperate climate in Gondwana, *Nature, 361*, 51–54.

Zachos, J. C., J. R. Breza, S. W. Wise, 1992, Early Oligocene ice-sheet expansion on Antarctica: Stable isotope and sedimentological evidence from Kerguelen Plateau, southern Indian Ocean, *Geology, 20*, 569–573.

Ziegler, A. M., 1990, Phytogeographic patterns and continental configurations during the Permian period, in *Palaeozoic Palaeogeography and Biogeography*, W. S. McKerrow and C R. Scotese (eds.), pp. 363–379, Geol. Soc. Mem., 12.

Ziegler, A. M., R. K. Bambach, J. T. Parrish, S. F. Barrett, E. H. Gierlowski, W. C. Parker, A. Raymond, and J. J. Sepkoski, 1981, Paleozoic biogeography and climatology, in *Paleobotany, Paleoecology, and Evolution*, K. J. Niklas (ed.), pp. 231–266, Praeger Scientific, New York.

PART II

ROLE OF CONTINENTAL CONFIGURATION

CHAPTER 2

The Role of Geography and Atmospheric CO_2 in Long Term Climate Change: Results from Model Simulations for the Late Permian to the Present

Peter J. Fawcett and Eric J. Barron

Over the geologically long timescale (tens to hundreds of millions of years) Earth's climate has varied considerably from periods of extreme warmth to periods of major glaciations (the so-called "greenhouse" and "icehouse" climates of Fischer, (1982). These marked variations in climate have been attributed to a variety of factors including both terrestrial and extraterrestrial processes. Since the general acceptance of plate tectonic theory, changes in geography which encompass both continental positions and orography, or elevation, have been frequently cited as an important long term climate forcing factor (e.g., Crowell and Frakes, (1970); Donn and Shaw, (1977)). The atmospheric concentration of carbon dioxide is another frequently cited forcing factor, especially in view of today's fossil fuel emissions and the global warming debate.

While changes in geography and in atmospheric CO_2 are considered to be important over the long-term, the degree to which each forcing factor has contributed to climate change in the geologic record is difficult to sort out. A simplified representation of the climate system, in this case a general circulation model (GCM), allows us to individually vary these forcing factors as boundary conditions to the GCM so that their relative importance can be determined in a systematic way. This type of approach is known as sensitivity testing. We focus here on geography and CO_2, although other forcing factors have undoubtedly played a role in long-term climate change. The results provide a number of interesting insights into the mechanisms of long-term climate change.

The climate model used in this study is GENESIS (Global Environmental and Ecological Simulation of Interactive Systems) developed at the National Center for Atmospheric Research (NCAR) by Thompson and Pollard (1995). We consider 13 time slices from the Late Permian to the Present which are spaced approximately 20–30 million years apart and encompass large changes in geography and in atmospheric CO_2. The evolution of global climate in response to these parameters is considered first, followed by a discussion of the climates of three end-member geographies within the 260-million-year timespan. Finally, we discuss the limitations of the model results and the relative importance of other climate forcing factors (e.g., ocean heat transport, vegetation).

Long-Term Geographic Variations

A number of paleogeographic reconstructions have been produced for times back to the Cambrian (510–535 Ma) and even into the Precambrian. The uncertainties in continental locations, both relative to each other and to the Earth's spin axis, become larger further back in time. Although discrepancies exist between different reconstructions for the Paleozoic, they are generally consistent for the Mesozoic and Cenozoic. We focus on these later intervals.

The evolution of global geography from the Late Permian to the Present can be divided into three major paleogeographic regimes, based on relative continental positions: Pangean, Tethyan, and Transition to Modern (after Barron, 1989). From the Carboniferous to the Middle Jurassic, the major continents were assembled into the supercontinent Pangea (Fig. 2.1a), which extended nearly from pole to pole. A single ocean, Panthalassa, dominated the globe with the equatorial Tethys Sea forming an embayment between Gondwana to the south and Laurussia to the north. Over the Pangean time frame, the supercontinent as a whole moved northward by some 30° of latitude. Southern Gondwana moved off the south pole as northern Laurussia moved over the north pole (e.g., Scotese and Golonka, 1992). Smaller-scale changes included (but were not limited to) the northward movement of continental fragments across the Tethys Sea (South China block, Lhasa block), widespread uplift in southern Africa and eastern South America in the Triassic, and the gradual lowering of the Appalachian Mountains from the Permian to Jurassic.

Starting in the Middle Jurassic, several rifting events began to split apart the Pangean supercontinent. Rifting between Africa and Laurussia opened the central Atlantic seaway. Further rifting of the Pangean remnants during the Jurassic and Cretaceous opened the northern South Atlantic Ocean. India separated from Antarctica in the Early Cretaceous and began its rapid northward movement across the Tethys Sea. This paleogeographic regime is characterized by smaller landmasses and widespread seaways (Fig. 2.1b).

The Transition to Modern geographic regime is characterized by the continued opening of the South and North Atlantic oceans and the closing of the Tethys Sea as a result of the northward motion of Africa, India and Australia (Fig. 2.1c). Widespread tectonic uplift and mountain building events during the latter part of this regime include the Alpine–Himalayan orogeny and the uplift of the Tibetan Plateau, the Andean orogeny, and uplift of western North America. These widespread increases in topography may have contributed to climate change over this interval.

Long-Term Atmospheric Carbon Dioxide Changes

Atmospheric CO_2 variations over the last glacial cycle are recorded in deep ice cores retrieved from Antarctica (Barnola et al., 1987) and range from 200 ppmv to 280 ppmv (preindustrial). The long-term variability of atmospheric CO_2 is much more uncertain because we are unable to measure paleoconcentrations directly. Geochemical models (e.g., Berner et al., 1983; Berner, 1991) suggest that CO_2

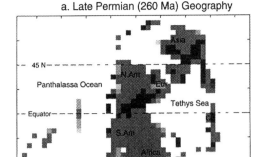

a. Late Permian (260 Ma) Geography

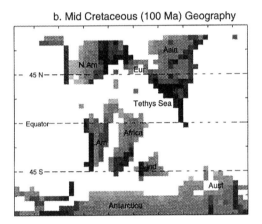

b. Mid Cretaceous (100 Ma) Geography

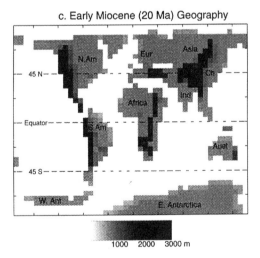

c. Early Miocene (20 Ma) Geography

Fig. 2.1. The major paleogeographic regimes since the late Paleozoic shown at model resolution: (a) Pangean regime – Late Permian, modified from Scotese and Golonka (1992); (b) Tethyan regime – mid-Cretaceous, modified from Barron et al. (1981); and (c) Transition to Modern regime – Miocene, modified from Barron et al. (1981).

might have varied by as much as ten times present-day levels over the past 260 million years as a function of plate spreading rates, a proxy for volcanic CO_2 outgassing rates, and continental silicate chemical weathering rates which draw down CO_2 levels. A best estimate curve of atmospheric CO_2 through the Phanerozoic from a carbon cycle model (Berner, 1991) is shown in Fig. 2.2. The error bars on this curve are based on a series of sensitivity analyses that vary outgassing rates and weathering rates within accepted estimates. These error bars indicate that CO_2 may seldom have exceeded the present level by more than a factor of four.

A variety of proxy methods of estimating paleo-atmospheric CO_2 levels are in broad agreement with this theoretical geochemical model. Measurement of $\delta^{13}C$ content of $CaCO_3$ (Cerling, 1991; Mora et al., 1991, 1996) and of $FeCO_3$ in goethite (Yapp and Poths, 1992) in ancient paleosols give past CO_2 values that fall close to the Berner (1991)

estimate. The marked decrease in pCO_2 from 400–300 Ma is especially evident (see Berner et al., 1993). Fractionation of $\delta^{13}C$ between carbonates and organic particulate matter in marine sediments (Arthur et al., 1991; Freeman and Hayes, 1992) and lacustrine sediments (Hollander and McKenzie, 1991) give pCO_2 values in the range of 2–8 times present values over the last 120 Ma.

A problem in using the theoretical estimates of Berner (1991) in climate model studies arises because the atmospheric CO_2 sink term, chemical weathering rates, is a complex function of surficial lithologies and climate (precipitation/runoff amounts and location) (e.g., Bluth, 1990). Thus, climate is a partial factor in the calculation of these CO_2 values (Berner, 1991). We address this problem with the series of sensitivity tests that isolate the effects of geography on climate and then show how climate sensitivity to atmospheric CO_2 varies with geography.

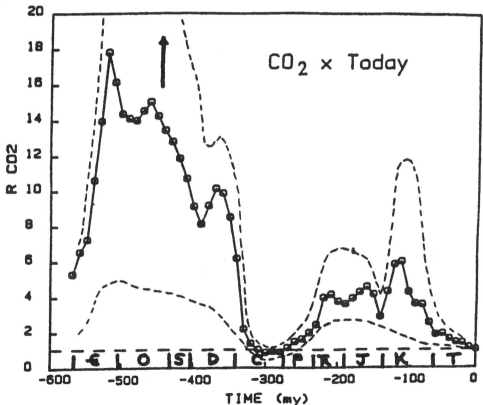

Fig. 2.2. A best estimate curve of atmospheric CO_2 versus time, calculated from a carbon cycle model (*Berner*, 1991). RCO_2 = mass of atmospheric $CO_2(t)$/mass of atmospheric CO_2 (0 Ma, 340 ppm). The dashed lines are a rough estimate of errors based on sensitivity analyses. See Berner (1991) for discussion. [From Berner, 1991] *Reprinted by permission of American Journal of Science.*

Model Description

The model used in this study is GENESIS (version 1.02) (Pollard and Thompson, 1994; Thompson and Pollard, 1995) which is an extensively modified version of the NCAR Community Climate Model CCM1. The model resolution is R15 (4.5° latitude by 7.5° longitude) with 12 layers in the vertical. Modifications from CCM1 include semi-Lagrangian transport of water vapor, a subgrid plume convection scheme, planetary boundary layer mixing, a new cloud parameterization, and a diurnal cycle. Also included is a land-surface transfer model that accounts for the physical effects of vegetation and multilayer models of soil, snow, and sea ice. The atmospheric model is coupled to a 50-m slab mixed-layer ocean that includes a mechanism for ocean heat transport following Covey and Thompson (1989).

The present-day performance of this model is comparable to that of previous coarse-grid models with a mixed-layer ocean. Reasonable values are predicted for surface temperatures and the diurnal range of temperatures, energy fluxes in the atmosphere, jet stream maxima strengths, and the locations of precipitation maxima. The largest errors are found in overly warm surface temperatures over high-latitude Northern Hemisphere land areas in summer due to low cloudiness values, and a larger than observed global precipitation value, caused in part by an overly large prescribed ocean surface aerodynamic roughness length (Thompson and Pollard, 1995).

GENESIS surface air temperatures (2 m above the surface) do contain some regional biases that can affect the results of this study. The global annual mean temperature is about 0.8°C warmer than the observed temperature dataset of Legates and Willmott (1990a), primarily due to average land temperatures being warmer than observed by 1.9°C. The largest error occurs over Antarctica which is about 15°C too warm in winter because the relative coarseness of the model allows too much lateral mixing of warmer air from the midlatitudes (Thompson and Pollard, 1995). High northern latitude areas above 60°N are generally too warm by a few degrees except for north central Asia which is too cold in winter. These high latitude temperature errors appear to result from a poor representation of

clouds in these regions; areas that are too warm have too few clouds in summer and too many clouds in winter (Thompson and Pollard, 1995).

GENESIS precipitation predictions compared to the observations of Legates and Willmott (1990b) show that the broad scale patterns are simulated quite well, including the seasonal movement of the Intertropical Convergence Zone (ITCZ), the dry subtropics and the relative dryness of the northern continents (Thompson and Pollard, 1995). The magnitude of global mean precipitation is much higher than observed: 4.5 mm/day or 164.3 cm/yr versus 3.1 mm/day or 113.1 cm/yr. (Our present-day control value is 160.4 cm/yr.) This overestimate is mostly due to the overly large prescription of the ocean roughness length used in the evaporation calculation. The major regional biases are overly intense rainfall in patches over land within the ITCZ and widespread areas of excessive convective precipitation in the southern oceans (Thompson and Pollard, 1995). The other major continental bias occurs near mountains where the coarse resolution of the model results in a broadening of orographic precipitation belts, as over the Great Basin of the western United States. This type of error occurs for all time slices.

The series of experiments used in this study were based on the reconstructed continental geographies and topographies of Barron et al. (1981) and of Scotese and Golonka (1992). A list of the times selected is given in table 2.1. The solar constant was kept at the present-day value of 1370 W/m^2 for all time slices, even though some stellar evolution models (e.g., Endal, 1981) suggest that the sun's luminosity may have been as much as 2% lower in the Permian. Earth's present orbital configuration was also used in all experiments. The motivation for this approach was to keep boundary conditions between time slices the same, as much as possible, so that differences in climate due to geography could be isolated and identified. The present-day "control" value for specified ocean heat transport was used for all experiments, as was a globally uniform set of land-surface characteristics (intermediate values for soil color and texture, and a savannah-type, or mixed canopy and groundcover, vegetation). The uniform vegetation and soil parameter specifications were set for the Present, rather than using the realistic values that are available. The only other surface boundary condition that was

Table 2.1. Global (Gb.) and Continental (Ct.) Averages for Control Experiments (340 ppm CO$_2$)

Time (Ma)	Land Area (x10^8 km^2)	Gb. Temp. (°C)	Ct. Temp. (°C)	Gb. Prec. (cm/yr)	Ct. Prec. (cm/yr)	Runoff (cm/yr)
0	1.50	13.7	8.7	160.4	122.8	39.2
20	1.64	16.0	12.2	164.4	132.3	42.0
40	1.50	15.9	11.9	168.0	138.6	44.9
60	1.56	15.8	11.5	167.0	138.0	45.5
70	1.48	16.2	11.4	169.0	132.8	45.1
80	1.52	15.9	10.2	170.4	135.7	44.9
90	1.26	16.4	11.2	174.3	129.5	42.2
100	1.25	15.4	9.4	173.0	132.9	45.7
130	1.41	14.6	8.1	168.4	113.7	31.4
155	1.53	14.5	10.8	159.9	113.7	32.7
200	1.42	14.5	10.8	160.8	96.5	24.0
240	1.50	15.1	10.9	161.6	104.2	28.1
260	1.39	15.5	14.0	161.3	110.5	30.3

changed was the specification of the Greenland and Antarctic ice sheets for the Present.

Model Results: Evolution of Climate Through Thirteen Time Slices

A total of 25 experiments were carried out for the 13 time slices considered here to determine the model sensitivity to changes in geography and atmospheric CO$_2$. Two cases were considered for each time slice, a control experiment with a "realistic" geography for that time slice and present-day CO$_2$ levels, and an elevated CO$_2$ experiment where the value of CO$_2$ was approximated from the Berner (1991) Phanerozoic curve. A number of annual average climatic parameters were calculated for the entire globe and for continental areas, including surface temperature and components of the hydrologic cycle. The results are shown in Tables 2.1 and 2.2.

Global Surface Temperature

The globally averaged annual surface temperatures for all time slices are shown in Fig. 2.3. Temperature trends with changes in geography alone and with changes in geography and CO$_2$ together are both illustrated to highlight the differences in climate sensitivity.

Table 2.2. Global and Continental Averages for Elevated CO$_2$ Experiments

Time (Ma)	CO$_2$ Level (x340 ppm)	Gb. Temp. (°C)	Ct. Temp. (°C)	Gb. Prec. (cm/yr)	Ct. Prec. (cm/yr)	Runoff (cm/yr)
0	1	13.7	8.7	160.4	122.8	39.2
20	1	16.0	12.2	164.4	132.3	42.0
40	2	18.6	15.4	174.8	147.2	49.8
60	2	18.4	14.8	173.3	146.3	49.4
70	3	20.3	17.1	178.7	146.9	52.8
80	3.5	20.4	16.6	180.4	148.2	52.0
90	3	21.4	18.4	185.6	141.9	49.6
100	6	21.7	18.1	187.1	149.0	54.0
130	3	19.9	15.2	181.2	125.3	36.8
155	4.5	21.3	19.0	175.9	129.6	40.3
200	4	20.4	18.3	175.5	113.2	29.9
240	4	20.4	17.7	173.8	115.5	34.0
260	1	15.5	14.0	161.3	110.5	30.3

The temperature trend for changes in geography alone displays remarkably little variability given the large changes in specified land-sea distribution and topography. Excluding the present-day time slice with specified ice sheets, temperatures range from a low of 14.5°C in the Jurassic to 16.4°C in the Turonian. Pangean geographies (Triassic—Jurassic) are generally cooler and Tethyan geographies (Late Cretaceous—Early Tertiary) are warmer. The largest change in temperature is the steep cooling from the Miocene to the Present that resulted from the realistic specification of ice sheets. Higher topography may have also contributed to this cooling. The largest rise in temperature occurs between 100 and 90 Ma and probably is the result of a rise in sea level and reduction of land area in the Turonian (90 Ma).

Globally averaged annual surface temperatures for the elevated CO_2 cases show much more variability through time. The major features of this trend include a steep warming from the Late Permian into the Mesozoic, consistently warm temperatures through the Triassic and Jurassic, a slight cooling in the Early Cretaceous followed by the warmest temperatures in the mid- to Late Cretaceous. From the Late Cretaceous, temperatures follow a long-term cooling trend through the Tertiary to the Present. Most of the variation in this temperature trend is due to the different levels of specified CO_2 (see Table 2.2). For example, the cooling in the Early Cretaceous (130 Ma) relative to the Late Jurassic (155 Ma) is a result of lower specified CO_2 (3x present atmospheric levels (PAL), versus 4.5 x PAL) as there is no significant change in temperature between these time slices when geography alone is varied. Comparison of the Tertiary trends (control versus elevated CO_2) also shows that the long-term cooling trend in the elevated CO_2 cases must be due to the progressively smaller amounts of specified CO_2.

Continental Surface Temperature

The trend of annual average continental surface temperature through time (Fig. 2.4) shows substantial variability with changes in geography. The Late Permian time slice (260 Ma) has the warmest continental temperatures (14.0°C) and the Early Cretaceous the coolest (8.1°C). Mid- and Late Cretaceous continental temperatures are slightly warmer than the Early Cretaceous but cooler than any time slice in the early Mesozoic or the Permian. From 90–20 Ma, temperatures increase slightly and are followed by a cooling into the Present where the Greenland and Antarctic ice sheets are included in the boundary conditions.

This difference in continental surface temperatures between the Permian and Cretaceous is of par-

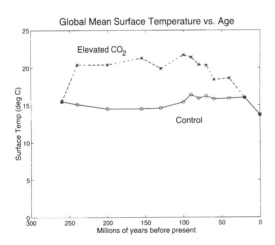

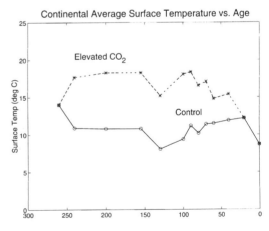

Fig. 2.3. Globally averaged, annual surface temperature as a function of age for control experiments (340 ppm CO_2; solid line) and elevated CO_2 experiments (values given in Table 2.2; dashed line).

Fig. 2.4. Average annual continental surface temperature as a function of age for control experiments (solid line) and elevated CO_2 experiments (dashed line).

ticular interest because it is apparently at odds with paleoclimatic reconstructions from the geologic record. The Late Permian followed the termination of the Permo-Carboniferous ice age and the mid- to Late Cretaceous is regarded as one of the warmest intervals during the Phanerozoic (Barron, 1983). These trends are due in part to changes in seasonality that result from the very different continental configurations through time. The Permian (260 Ma) geography featured large subtropical landmasses which experienced hot summers and relatively mild winters. The Cretaceous (130–70 Ma) geographies had larger landmasses at higher latitudes which experienced cooler summers and much colder winters than for the Permian geography. The mismatch between climate model predictions for changes in geography alone and the geologic record implies that additional climate forcing factors may be necessary to explain the time evolution of continental surface temperatures.

As in the globally averaged case, the elevated CO$_2$ levels result in higher continental surface temperatures, and the degree of warming relative to the control experiment for each time slice is a function of the specified CO$_2$. This continental surface temperature trend shows a warming through the early Mesozoic, a dramatic cooling in the Early Cretaceous followed by another warming in the mid-Cretaceous. A step-wise cooling through the Tertiary ends at the present-day value of 8.7°C. The time evolution of continental surface temperature is strongly affected by changes in both geography and in atmospheric CO$_2$ levels.

Globally Averaged Precipitation

Variations in globally averaged, annual precipitation rates are relatively small when geography alone is changed (Fig. 2.5a) because the proportion of ocean surface (the source of atmospheric water vapor) to land surface has varied within a small range (about 70%–75%). The driest geographies are Pangean (155–260 Ma). There is a slight increase in global precipitation in the Cretaceous time slices, coinciding with the breakup of Pangea and the development of the subtropical Tethys Seaway. A gradual decline in global precipitation through the Tertiary to the Present follows. The total variation between the wettest and driest time slices is 10.7 cm/yr. The elevated CO$_2$ globally

averaged precipitation values are consistently wetter than the control values (Fig. 2.5a). The trend of this curve is similar to the control, with a few important differences. Higher CO$_2$ is specified for the Pangean time slices than for the Transition to Modern time slices with the result that for global precipitation values, the Mesozoic is on average wetter than the Cenozoic. The Cretaceous as a whole is the wettest Period, peaking at the 100 and 90 Ma time slices. Following this time, the elevated CO$_2$ experiments show a more pronounced decline in global precipitation through the Present than do the control experiments. The total variation, 20.2 cm/yr, is about double that of the control.

Continental Average Precipitation

The continental hydrologic cycle is extremely sensitive to the boundary condition changes in these experiments. Continental average annual precipitation rates show large, systematic changes through time that are very dependent on the systematic changes in paleogeography through time (Fig. 2.5b). The Pangean geographies are the driest and the Tethyan geographies the wettest. A large increase in continental precipitation from the Late Jurassic and Early Cretaceous into the mid-Cretaceous coincides with the breakup of the Pangean supercontinent. Precipitation rates remain high through the early Tertiary and then decline from the Eocene to the Present. The total variation in this curve is 42 cm/yr from the driest time slice (200 Ma) to the wettest (40 Ma).

The trend of the elevated CO$_2$ continental precipitation curve (Fig. 2.5b) is essentially the same as for the control experiments, although with higher values. The Pangean time slices are still very dry, the mid-Cretaceous (100 Ma) with 6x CO$_2$ specified is the wettest experiment and a more pronounced drying trend occurs over the past 40 m.y. Although CO$_2$ variations do have some impact, changes in geography clearly dominate the trend in continental precipitation rates through time.

Average Continental Runoff

To fully characterize the continental hydrologies of different time periods, both evaporation and soil moisture storage must be considered. Runoff is a function of P-E and soil moisture storage, so tem-

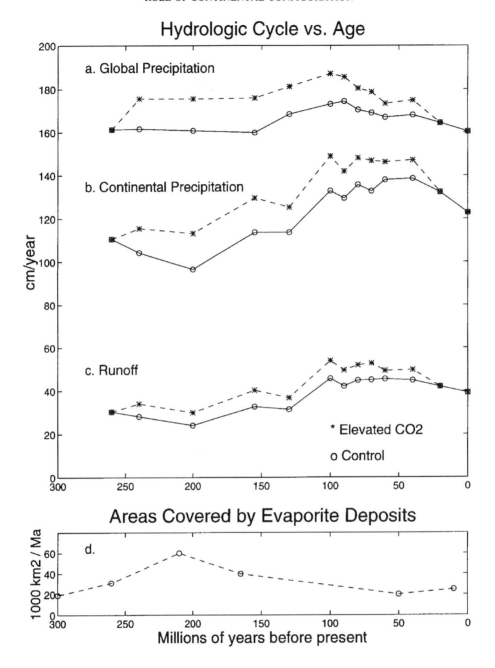

Fig. 2.5. (a) Globally averaged, annual precipitation rate, (b) average annual continental precipitation, (c) average annual continental runoff as a function of age for control experiments (solid lines) and elevated CO_2 experiments (dashed lines), and (d) areas covered by evaporite deposits through time, data from Gordon (1975).

perature becomes a factor in this aspect of the hydrologic cycle. The control experiments show changes in runoff through time (Fig. 2.5c) that are quite similar to the continental precipitation trends.

The magnitude of variability, however, is much larger for runoff. Again, the Pangean time slices have the lowest runoff rates and the mid-Cretaceous through Early Tertiary time slices have the highest runoff rates. The wettest time slice (100 Ma) has almost twice the average runoff rate as the driest time slice (200 Ma), 45.7 cm/yr versus 24.0 cm/yr. Specification of higher CO$_2$ levels raises average runoff rates for each time slice, but does not change the form of the trend significantly. Changes in paleogeography are clearly the dominant control of continental runoff rates through time.

Geologic Trends of Climate

An enormous amount of work has been done on the reconstruction of Late Paleozoic to Present climates from the geologic record and good summaries can be found in Crowley and North (1991) and Frakes et al. (1992). Here we give a very broad overview of the trends in climate through this timespan for qualitative comparison with the global and continental climate model results. More detailed model and reconstructed climate comparisons are given elsewhere (e.g. Fawcett, 1994; Barron et al., 1995). We exclude the higher frequency Quaternary glacial fluctuations from this analysis.

The Late Permian (260 Ma) experienced a transitional climate at the end of the long Permo-Carboniferous glaciation. Many parts of the globe, and especially Gondwanaland, warmed considerably through the Late Permian into the Triassic. Although not well quantified, global temperatures are generally regarded as having been warm through the Triassic and Jurassic from paleofloral distributions and a lack of evidence of permanent ice. These intervals were characterized by widespread continental aridity, as indicated by numerous evaporite deposits and eolian sand deposits in western North America, South America, and Europe. The areal extent of evaporite deposits during the past 300 million years reached a peak during this time period (Fig. 2.5d, data from Gordon (1975)). A cooling episode has been suggested for the latest Jurassic and Early Cretaceous based on the occurrence of possible high-latitude ice rafted deposits (Frakes and Francis, 1988).

The mid-Cretaceous is considered to be the warmest interval during the Phanerozoic, and Late Cretaceous climates were also quite warm but experienced a series of fluctuations to cooler temperatures. This interval has abundant evidence for humid continental climates and reduced areas of continental aridity.

Cenozoic climates are characterized by long-term cooling and drying trends from the Late Cretaceous (70 Ma) to the Quaternary glacial state (2 Ma). The cooling trend was punctuated by warm intervals in the early Eocene (50 Ma), the mid-Miocene (12 Ma), and the mid-Pliocene (4 Ma), each of which was followed by cooling episodes. The drying trend accelerated in the past 25 Ma as shown by expansion of dry vegetation types (savannah, grasses), by an increase in wind-blown eolian particles (Rea et al., 1985) and by a reduction in kaolinite (Chamley, 1989) in deep-sea sediments.

Discussion

This study has examined the roles of geography and atmospheric CO$_2$ in the evolution of global climate through the past 260 million years. The time series of global and continental averages of important climatic parameters shows that some aspects of simulated climate, notably continental hydrology, are quite sensitive to changes in geography. Other aspects of simulated climate including global mean temperature and precipitation are relatively insensitive to changes in geography alone, but are sensitive to additional climate forcing factors such as atmospheric CO$_2$.

Predicted and Reconstructed Climate Variations

Predicted global and continental surface temperature trends for changes in geography alone clearly do not match the geologic record of temperature through time. Surface temperature predictions for the higher CO$_2$ levels, adopted from Berner (1991), compare qualitatively well with the geologic record as they include features such as the warming from the Late Permian into the Triassic, the cooling in the Early Cretaceous, and the very warm mid-Cretaceous with the subsequent Cenozoic cooling trend. Elevated CO$_2$ may not be sufficient to explain all features of reconstructed temperatures, however, especially equator-to-pole temperature gradients for warm intervals like the mid-Cretaceous (100 Ma) and the early Eocene (50 Ma), and the warming and cooling events of the later part of the Cenozoic cooling trend. Barron et al. (1995) show that for mid-Cretaceous GENESIS experiments, increased CO$_2$ and increased oceanic heat flux that prevents the tropics from heating up too much provide the best match to the reconstructed temperatures. For the early Eocene, Sloan et al. (1995) suggest that enhanced oceanic heat transport could have contributed to high-latitude warming but consider it a secondary effect with the primary forcing in the atmosphere. Rind and Chandler (1991) and Chandler et al. (1992) have also suggested increased ocean heat transport as a means of removing sea ice and producing warmer climates, especially at higher latitudes. Chandler et al. (1992) considered the role of ocean heat transport in contributing to the warmth of the Early Jurassic climate. Another contribution to the late Cenozoic cooling could be the evolution of grasses and tundra-type vegetation in the Miocene (22–5 Ma), which increased albedos at high latitudes (Dutton and Barron, 1996).

The continental aspects of the hydrologic cycle, which are best suited for comparison with the geologic record of climate, show a very good match with proxy indicators of humidity/aridity through time. Areas of evaporite (salt) deposits compiled by Gordon (1975) (Fig. 2.5d) show the Triassic to have one of the driest continental climates of the Phanerozoic and the early Cenozoic to have one of the wettest. The differences between the control experiment trend and the elevated CO_2 trend are small but there is more variability in the latter. The Pangean time slice experiments are predicted to have had very dry continents, consistent with the numerous evaporite and eolian sand deposits of this interval. Predicted spatial patterns of low P-E are compatible with the distribution of these deposits, and regions of higher predicted continental P-E correspond to humid climate deposits such as thick coal sequences (Chandler et al., 1992; Fawcett, 1994; Fawcett et al., 1994; Wilson et al., 1994). Widespread humid climate indicators for the mid- and Late Cretaceous (100–65 Ma) are consistent with model predictions of very wet continents at these times. For the interval from 40 Ma to the Present, there is a significant decrease in predicted continental precipitation and runoff which is consistent with the late Cenozoic trend of increasing aridity.

The predicted, long-term elevated CO_2 trends for both surface temperature and continental hydrological balance match qualitatively the geologic record of climate. This suggests that the combined effects of geographic evolution and estimated CO_2 varia-

tions can explain much of the climatic variability over the Late Paleozoic to Present time span. However, detailed model to geologic record comparisons for individual time slices have shown that realistic geography and elevated CO_2, while necessary, are often not sufficient to fully reproduce the reconstructed climates. Other forcing factors must be considered, although inappropriate model sensitivity to the various forcing factors cannot be discounted.

Paleogeographic Control of the Hydrologic Cycle

These model results suggest that the hydrologic cycle is highly sensitive to the paleogeographic changes that have occurred since the Late Permian. In the geologic record, water plays a number of critical roles. It is the primary agent of weathering, erosion and sediment transport, it is a determining factor for ecosystems, and it impacts the climate system as a greenhouse gas, by transporting latent heat and by modifying albedos (snow and clouds). An understanding of how changes in geography control variability in the hydrologic cycle is essential to fully characterize past climates.

Early model studies using NCAR CCM version 0A suggested that the hydrologic cycle is highly sensitive to changes in climate and to climate forcing factors including paleogeography, atmospheric CO_2, and orbital variations (Barron et al., 1989). The intensity of the global hydrologic cycle increased with global surface temperature, reflecting the relationship between saturation vapor pressure and surface temperature. Land-sea distribution was found to modify the intensity of the hydrologic cycle by influencing the availability of moisture to the atmosphere. The amount of land area in the tropics where evaporation rates are highest is especially important (Barron et al., 1989). Otto-Bliesner (1995) also found a significant relationship between land-sea configuration through time and precipitation and continental runoff.

The number of GENESIS climate model experiments for different time slices allows a fuller examination of factors important to the hydrologic cycle. A plot of globally averaged, annual mean precipitation versus globally averaged, annual mean surface temperature for the 13 control experiments (Fig. 2.6) shows no clear correlation of these

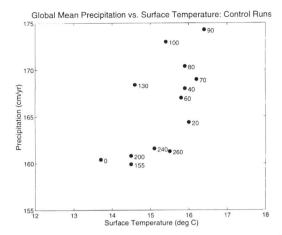

Fig. 2.6. Globally averaged precipitation rates as a function of globally averaged surface temperature for the control experiments of all time slices. Numbers give the age of each time slice in Ma.

on global precipitation rates. As these model results suggest, paleogeography is the most likely candidate.

A number of sensitivity tests beyond the single elevated CO_2 experiments have been carried out for three representative end-member geographies of the paleogeographic regimes, Late Permian (260 Ma), mid-Cretaceous (100 Ma), and Present (0 Ma) (Fig. 2.7). These additional sensitivity experiments include different levels of enhanced equator-to-pole ocean heat transport, different levels of CO_2 specified for each time slice, and combinations of the two (see Fawcett, 1994 and Barron et al., 1995 for more details). Globally averaged temperatures compared to globally averaged precipitation rates for these sensitivity tests show a clear division into the three geographic groups and within each geography there is a strong positive correlation between these parameters. The slope of this relationship is does control the intensity of the global hydrologic cycle (globally averaged precipitation) but geography first sets a "base level". This relationship also holds true for continental average precipitation and runoff (Fawcett, 1994).

two variables through time, although the range in temperature in this plot is only 2°C and the precipitation range is 10 cm/yr. This suggests that some factor moderates the influence of global temperature

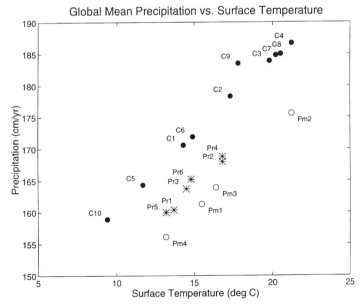

Fig. 2.7. Globally averaged precipitation rates as a function of globally averaged surface temperature for all simulations of three end-member geographies. Labels for present-day experiments: Pr1 = Uniform Ocean Heat Flux (OHF), Pr2 = 2xCO_2 + Uniform OHF, Pr3 = Control, Pr4 = 2xCO_2, Pr5 = Zero OHF, Pr6 =2xOHF. Mid-Cretaceous Experiments: C1 = Control, C2 = 2xCO_2, C3 = 4xCO_2, C4 = 6xCO_2, C5 = Zero OHF, C6 = 2xOHF, C7 = 4xCO_2 + 2xOHF, C8 = 4xCO_2 + 4xOHF, C9 = Mixed Forest Vegetation, C10 = Groundcover Only Vegetation. Late Permian Experiments: Pm1 = Control, Pm2 = 4xCO_2, Pm3 = 2xOHF, Pm4 = Groundcover Only Vegetation. Data from Fawcett (1994).

The geographic "base level" is a function of total land area and land area in the tropics, confirming the results of Barron et al., (1989). Of the three end-member geographies, the wettest (mid-Cretaceous) has the lowest amount of total land area (Table 2.1) and the least amount of land area in the tropics (16% land between 22.5°N and S, vs. 24% for the Present and the Late Permian). The increased ocean surface area in the mid-Cretaceous (and Tethyan time slices by extension) leads to higher evaporation rates, especially in the Northern Hemisphere tropics as shown by a comparison of zonally averaged annual evaporation rates (Fig. 2.8).

Sensitivity of Climate to Geography and Atmospheric CO_2

Globally averaged surface temperature, as predicted by the GENESIS climate model, is quite insensitive to changes in geography alone. This conclusion differs from earlier mean annual simulations where geography was found to play a significant role in explaining mid-Cretaceous warmth (e.g., Barron and Washington, 1984). The difference appears to result from the addition of seasonally varying insolation to the model. A more continental climate for Northern Hemisphere landmasses in the mid-Cretaceous GENESIS control experiment occurs because of the role of seasonally simulated sea ice (Barron et al., 1993). Globally averaged surface temperature is much more sensitive to variations in atmospheric CO_2. Comparison with the geologic record has suggested that a four to six times increase in CO_2 over

present day levels for the mid-Cretaceous can explain much of the warmth of that interval (Barron et al., 1993). Additional sensitivity tests show that both enhanced equator-to-pole ocean heat transport and elevated CO_2 give the best model fit to reconstructed latitudinal temperature gradients (Barron et al., 1995). This study found that specifying progressively smaller amounts of atmospheric CO_2 from the mid-Cretaceous to the Present (as in the Berner, 1991, CO_2 calculation) produced a long term global cooling that is qualitatively similar to the cooling trend reconstructed from the geologic record (e.g., Savin, 1977). However, the step-wise nature of the reconstructed trend with abrupt cooling events is not reproduced by this series of experiments, suggesting that CO_2 alone is insufficient to fully explain the trend. We note that CO_2 changes may have been more abrupt than the Berner (1991) model can resolve, and that feedbacks to the CO_2 change that are not incorporated into the model (e.g., vegetation changes, ocean circulation changes, growth of ice sheets) could also have had an impact.

Continental average surface temperature is very sensitive to changes in geography; the Late Permian is the "warmest" geography and the mid-Cretaceous is the "coldest" in terms of land area. The Late Permian has a larger amount of tropical and subtropical land exceeding 30°C than the mid-Cretaceous and that the mid-Cretaceous has larger amounts of high latitude land with subfreezing winter temperatures. The present-day experiment is also relatively cold, even though it has a significant amount of warm subtropical land, because it has the Greenland and Antarctic ice sheets specified as surface boundary conditions. The geologic record shows that the mid-Cretaceous was much warmer than the Late Permian. Additional climate forcing factors are needed to explain this fact, and increased amounts of atmospheric CO_2 help account for the warm time periods. The elevated CO_2 continental temperature trend through time is the product of complex interplay between geography and atmospheric CO_2.

Several aspects of the hydrologic cycle are strongly affected by changes in geography; the Pangean geographies are the driest and the Tethyan geographies (especially the mid-Cretaceous) are the wettest. There are several reasons for this. The amount of water available to the atmosphere is a

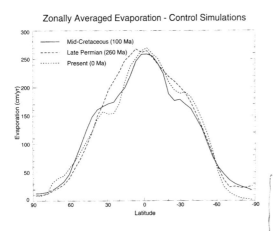

Fig. 2.8. Zonally averaged evaporation rates for the control experiments of the three end-member geographies.

function of the surface area of ocean, particularly in the tropics where evaporation rates are highest. The mid-Cretaceous has the lowest amount of land both globally and in the tropics, and the Late Permian has the highest, so the mid-Cretaceous has the highest globally averaged precipitation. The difference between wettest and driest time slices is relatively small, however, because the size of the world oceans acts as a buffer to differences in evaporation rates. The continental aspects of the hydrologic cycle are much more affected by geography and result from supercontinent versus distributed continent geometries. The size of the supercontinent Pangea restricts precipitation in the interior because of the distance from oceanic moisture sources. In the Late Permian, areas of high continental precipitation are restricted to the margins. The mid-Cretaceous continents are relatively dispersed and have available many moisture sources, and so little land area is dry. The position of the northern Tethys margin in the Northern Hemisphere mid-latitudes serves to focus precipitation, a feature noted in earlier model studies (Barron and Frakes, 1989). All of these factors combine to explain differences in precipitation rates between the different geographies.

Increases in atmospheric CO$_2$ for all geographies raise global temperatures and therefore raise global evaporation rates and the intensity of the global hydrologic cycle. The largest changes in temperature occur at high latitudes during winter and so they do not affect evaporation rates much at all. Temperature increases in the tropics, however, strongly affect the hydrologic cycle by increasing evaporation rates. Wetter tropics and generally wetter mid-latitudes result in a net increase in the continental water balance under higher CO$_2$ conditions (Fawcett, 1994). Changes in other boundary conditions (ocean heat transport, vegetation) that cause global warming or cooling also cause changes in the hydrologic cycle in the same direction as the temperature change.

Limitations to This Study

A variety of factors contribute to uncertainties in the model experiments described here. These include uncertainties in the boundary conditions specified; land-sea distribution, topography, and the amount of CO$_2$ for individual time slices. Although variations in these factors alone produce significant climatic variation that qualitatively match the geologic record, a limitation to this study is that we have not considered other potentially important factors including ocean heat transport, land-surface characteristics, and variations in solar output. Uncertainties in the specified atmospheric CO$_2$ values are relatively high because there are no direct methods for measuring ancient CO$_2$ levels for these older time periods.

The model's climate sensitivity to changes in boundary conditions is also a source of uncertainty in this study. The coarse resolution of the model requires parameterization of sub-grid-scale processes such as precipitation, which are physically based on present-day relationships. We assume that the same parameterizations are valid for other time periods although there is no way to be certain of this. Ocean processes are represented very crudely in the model and their potential role in climate change should not be underestimated.

The errors in the present-day GENESIS (version 1.02) simulation (such as the overprediction of global precipitation) should not affect the results of this study because we focus on the relative differences in climate between time slices rather than on the absolute values. However, the regional biases in temperature and precipitation outlined in section 2.1 could have an impact on the predicted time-evolution of climate, especially as the land-sea and topographic distributions change. The overprediction of rainfall over land in the ITCZ is potentially the largest problem because changes in land area near the equator will affect the continental precipitation averages. Also, the temperature biases at high latitudes will differ through time as the amount of high-latitude land changes. The largest error in the present-day simulation, the too cold winter temperatures over Antarctica, will not be as important for the older time slice predictions because the Antarctic ice sheet is only specified for the Present, and is the main reason for the cold Antarctic temperatures. At this time, we cannot directly quantify the effect of these regional biases in the simulation, we can only consider them to be caveats to the results of this study.

Conclusions

The main objective of this study was to evaluate the relative sensitivity of climate to realistic changes in paleogeography and atmospheric CO_2 over a geologically long time span. The major conclusions arising from this work are as follows:

1. The model results suggest large and well-defined changes as a function of changing geography and atmospheric CO_2 which can be compared with the geologic record.

2. Globally averaged parameters including surface temperature and precipitation are relatively insensitive to changes in geography alone.

3. Globally averaged surface temperature is much more sensitive to variations in atmospheric CO_2, as defined by the Berner (1991) geochemical model. The predicted temperature trend through time captures the major features of the reconstructed temperature trend, although it does not explain all features of the paleoclimatic record.

4. Continental average surface temperature is very sensitive to changes in geography alone, and the Late Permian control experiment is warmer than the mid-Cretaceous control experiment. The geologic record shows that the mid-Cretaceous was much warmer than the Late Permian, therefore, additional climate forcing factors are required to explain these climate differences.

5. Continental surface temperature is also sensitive to changes in CO_2, and its predicted trend through time is the result of complex interplay between geography and atmospheric CO_2 levels.

6. Global temperature is a primary control on the intensity of the hydrologic cycle, but geography significantly modifies this relationship by influencing the amount of moisture available to the atmosphere. The amount of land in the tropics is key to this relationship.

7. Continental precipitation and runoff are both very sensitive to changes in geography: the Pangean geographies are the driest and the Tethyan geographies are the wettest. These parameters are also sensitive to changes in atmospheric CO_2; however geography (supercontinent vs. distributed continent) controls the precipitation trends through time

Acknowledgments

Reviews by Mark Chandler and Bette Otto-Bliesner helped to improve the manuscript considerably. We also thank the editors, Kevin Burke and Tom Crowley, for helpful comments. Computer time was provided by the Earth System Science Center at The Pennsylvania State University.

References

Arthur, M. A., K. R Hinga, M. E. Q. Pilson, E. Whitaker, and D. Allard, 1991, Estimates of pCO_2 for the last 120 Ma based on the $\delta^{13}C$ of marine phytoplanktic organic matter (abstract), *Eos Trans, AGU*, 72, 166.

Barnola, J. M., D. Raynaud, Y. S. Korotkevich, and C. Lorius, 1987, Vostok ice core provides 160,000-year record of atmospheric CO_2, *Nature*, 329, 408–414.

Barron, E. J., 1983, A warm, equable Cretaceous: The nature of the problem, *Earth Sci. Rev.*, 19, 305–338.

Barron, E. J., 1989, Climate variations and the Appalachians from the Late Paleozoic to the Present: Results from model simulations, *Geomorphol.*, 2, 99–118.

Barron, E. J., and L. A. Frakes, 1989, Climate model evidence for variable continental precipitation and its significance for phosphorite formation, in *Phosphate Deposits of the World —Vol. 3*, W. C. Burnett and S. R. Riggs (eds.), 260–272, Cambridge University Press, Cambridge, U.K.

Barron, E. J. and W. M. Washington, 1984, The role of geographic variables in explaining paleoclimates: Results from Cretaceous climate model sensitivity studies, *J. Geophys. Res.*, 89, 1267–1279.

Barron, E. J., W. W. Hay, and S. L. Thompson, 1989, The hydrologic cycle: A major variable during Earth history, *Palaeogeog., Palaeoclimatol., Palaeoecol.*, 75, 157–174.

Barron, E. J., P. J. Fawcett, D. Pollard, and S. L. Thompson, 1993, Model simulations of Cretaceous climates: The role of geography and carbon dioxide, *Phil. Trans. Royal Soc. London*, B341, 307–316.

Barron, E. J., P. J. Fawcett, W. M. Peterson, D. Pollard, and S. L. Thompson, 1995, A "simulation" of mid-Cretaceous climate, *Paleoceanography*, 10, 953–962.

Barron, E. J., C. G. A. Harrison, J. L. Sloan, and W. W. Hay, 1981, Paleogeography, 180 million years ago to the present, *Ecol. Geol. Helv.*, 74, 443–470.

Berner, R. A., 1991, A model for atmospheric CO$_2$ over Phanerozoic time, *Am. J. Sci.*, *291*, 339–376.

Berner, R. A., A. C. Lasaga, and R. M. Garrels, 1983, The carbonate-silicate geochemical cycle and its effect on atmospheric carbon dioxide over the last 100 million years, *Am. J. Sci*, *283*, 641–683.

Bluth, G. J. S., 1990, Effects of paleogeology, chemical weathering, and climate on the global geochemical cycle of carbon dioxide: Ph.D. Thesis, 130 pp., Department of Geosciences, The Pennsylvania State University, University Park, Pa.

Cerling, T. E., 1991, Carbon dioxide in the atmosphere: Evidence from Cenozoic and Mesozoic paleosols, *Am. J. Sci.*, *291*, 377–400.

Chandler, M. A., D. Rind, and R. Ruedy, 1992, Pangean climate during the Early Jurassic: GCM simulations and the sedimentary record of paleoclimate, *Geol. Soc. Am. Bull.*, *104*, 543–559.

Covey, C., and S. L. Thompson, 1989, Testing the effects of ocean heat transport on climate, *Global and Planetary Change*, *1*, 331–341.

Crowell, J., and L. A. Frakes, 1970, Phanerozoic glaciation and the causes of the ice ages, *Am. J. Sci.*, *268*, 193–224.

Crowley, T. J., and G. R. North, 1991, *Paleoclimatology*, Oxford Monographs on Geology and Geophysics, Oxford University Press, New York.

Donn, W., and D. Shaw, 1977, Model of climate evolution based on continental drift and polar wandering, *Geol. Soc. Am. Bull*, *88*, 390–396.

Dutton, J. F., and E. J. Barron, 1997, Miocene to present vegetation changes: A possible piece of the Cenozoic cooling puzzle, *Geology*, *25*, 39–41.

Endal, A. S., 1981, Evolutionary variations of solar luminosity, in *Variations of the Solar Constant, NASA Conference Publication 2191*, 175–183.

Fawcett, P. J., 1994, Simulation of climate-sedimentary evolution: A comparison of climate model results to the geologic record for India and Australia, Ph.D. thesis, 327 pp., Department of Geosciences, The Pennsylvania State University, University Park, Pa.

Fawcett, P. J., E. J. Barron, V. D. Robison, and B. J. Katz, 1994, The climatic evolution of India and Australia from the Late Permian to mid-Juras-
sic: A comparison of climate model results with the geologic record, in: *Pangea: Paleoclimate, Tectonics, and Sedimentation During Accretion, Zenith, and Breakup of a Supercontinent*, G. D. Klein (ed.), pp. 139–157, Special Paper 288, Geological Society of America, Boulder, Colo.

Fischer, A. G., 1982, Long-term climatic oscillations recorded in stratigraphy, in *Climate in Earth History*, W. H. Berger and J. C. Crowell (eds.), pp. 97–104, National Academy, Washington, D.C.

Frakes, L. A., and J. E. Francis, 1988, A guide to Phanerozoic cold polar climates from high-latitude ice-rafting in the Cretaceous, *Nature*, *333*, 547–549.

Frakes, L. A., J. E. Francis, and J. I. Sytkus, 1992, *Climate Modes of the Phanerozoic*, Cambridge University Press, N.Y.

Freeman, K. H., and J. M. Hayes, 1992, Fractionation of carbon isotopes by phytoplankton and estimates of ancient CO$_2$ levels, *Global Biogeochemical Cycles*, *6*, 185–198.

Gordon, W., 1975, Distribution by latitude of Phanerozoic evaporite deposits, *J. Geol.*, *83*, 671–684.

Hollander, D. J., and J. A. McKenzie, 1991, CO$_2$ control on carbon-isotope fractionation during aqueous photosynthesis: A paleo-pCO$_2$ barometer, *Geology*, *19*, 929–932.

Legates, D. R., and C. J. Willmott, 1990a, Mean seasonal and spatial variability in global surface air temperature, *Theor. Appl. Climatol.*, *41*, 11–21.

Legates, D. R., and C. J. Willmott, 1990b, Mean seasonal and spatial variability in gauge-corrected, global mean precipitation, *Int. J. Climatol.*, *10*, 111–127.

Mora, C. I., S. G. Driese, and L. A. Colarusso, 1996, Middle to Late Paleozoic atmospheric CO$_2$ levels from soil carbonate and organic matter, *Science*, *271*, 1105–1107.

Mora, C. I., S. G. Driese, and P. G. Seager, 1991, Carbon dioxide in the Paleozoic atmosphere: Evidence from carbon isotope compositions of pedogenic carbonate, *Geology*, *19*, 1017–1020.

Otto-Bliesner, B. L., 1995, Continental drift, runoff, and weathering feedbacks: Implications from climate model experiments, *J. Geophys. Res.*, *100*, 11,537–11,548.

Pollard, D., and S. L. Thompson, 1994, Sea ice dynamics and CO_2 sensitivity in a global climate model, *Atmospheres and Oceans*, *32*, 449–467.

Rea, D. K., M. Leinen, and T. R. Janecek, 1985, Geologic approach to the long-term history of atmospheric circulation, *Science*, *227*, 721–725.

Rind, D., and M. A. Chandler, 1991, Increased ocean heat transports and warmer climate, *J. Geophys. Res.*, *96*, 7437–7461.

Savin, S. M., 1977, The history of the earth's surface temperature during the past 100 million years, *Ann. Rev. Earth Planet. Sci.*, *5*, 319–355.

Scotese, C. R., and J. Golonka, 1992, PALEOMAP Paleogeographic Atlas, PALEOMAP Progress Report #20, Department of Geology, University of Texas at Arlington, Tex.

Sloan, L. C., J. C. G. Walker, and T. C. Moore, Jr., 1995, Possible role of oceanic heat transport in early Eocene climate, *Paleoceanography*, *10*, 347–356.

Thompson, S. L., and Pollard, D., 1995, A global climate model (GENESIS) with a land-surface transfer scheme (LSX). Part I: Present climate simulation, *J. Climate*, *8*, 732–761.

Wilson, K. M, D. Pollard, W. W. Hay, S. L Thompson, and C. N. Wold, 1994, General circulation model simulations of Triassic climates: Preliminary results, in *Pangea: Paleoclimate, Tectonics, and Sedimentation During Accretion, Zenith, and Breakup of a Supercontinent*, G. D. Klein (ed.), pp. 91–116, Special Paper 288, Geological Society of America, Boulder, Colo.

PART III

ROLE OF CONTINENTAL ELEVATION

CHAPTER 3

Orographic Evolution of the Himalaya and Tibetan Plateau

T. Mark Harrison, An Yin, and Frederick J. Ryerson

The effect of orography on climate has been evaluated using several atmospheric circulation models (Manabe and Terpstra, 1974; Hahn and Manabe, 1975; Barron and Washington, 1984; Kutzbach et al., 1989; 1993). These simulations generally indicate that major climate changes can be triggered by the appearance of a mountain belt or the uplift of a large region which underscore the fundamental importance of understanding the orographic evolution of mountain belts in order to place constraints on reconstructions of past climate over time scales of 10^6 to 10^7 years (Crowley, this volume, Chapter 1). For example, there is accumulating evidence that the appearance of a high and extensive Himalayan-Tibetan mountain system significantly influenced Neogene climate (Hahn and Manabe, 1975; Ruddiman and Kutzbach, 1989; Quade et al., 1989; Kutzbach et al., 1993). In general, however, the complex and spatially variable uplift history of the plateau that is emerging from recent geologic studies (Harrison et al., 1992; Murphy et al., 1997) have not been considered in these models, and those that have attempted to make a link (e.g., Raymo et al., 1988) may have been misled by spurious inferences from paleobotanical studies (see Appendix).

The surface of the earth is being continuously altered under the influence of weathering, erosion, tectonics, sediment deposition, and magmatic processes. Factors which contribute to increasing surface elevation include thickening of the continental crust, injection of the ductile lower crust, and temperature increases or density decreases in both the lithosphere and underlying circulating mantle. Likewise, those that may reduce surface elevation include erosion, tectonic denudation, lithospheric thinning, and decreases in temperature and increases in density of the mantle and crust. Our first challenge is to understand the interplay between these processes in natural settings, and then interpret the observed behavior in terms of physical models.

Although long recognized as the best example of an active continental collision orogen (Argand, 1924), the Indo-Asian collision zone remains one of the most extraordinary and puzzling tectonic features on Earth (Fig. 3.1). Heightened interest in the evolution of this region developed over the past 25 years as its key role in documenting the failure of plate tectonics to describe the diffuse intraplate deformation characteristic of continental tectonics became clear (e.g., Dewey and Burke, 1973; Molnar and Tapponnier, 1975). As a consequence, the region has become the principal natural laboratory in which approaches for estimating the thickening and uplift histories of the continental lithosphere are tested.

Assessing the influence of topography on climate is complicated by the still relatively blunt instruments we have at our disposal for determining elevation histories (see Appendix). These approaches are in general better suited to determining when shifts in elevation occurred rather than the magnitude of such topographic changes. In this paper, we first outline models of possible uplift histories of southern Asia. We then review the methodologies that are drawn upon to assess the history of crustal thickening in the Indo-Asian collision zone and describe how this record can be translated into an uplift history. Finally we document our current understanding of the orographic history of this region, present a summary model, and briefly address its implications to climate reconstructions. In particular, we wish to emphasize that the plateau has grown in size over time in a spatially complex fashion. Although only approximate estimates of the elevation history of the Himalaya and Tibet can be made at the present time, this information may well

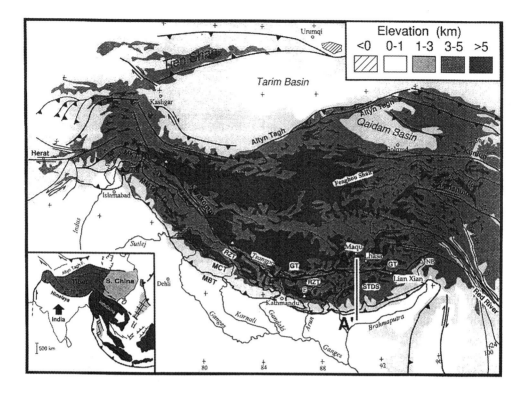

Fig. 3.1. Map of the Indo-Asian collision zone showing the principal elevated terranes (Himalaya, Tibet, Tien Shan), basins (Qaidam, Tarim), and faults (see Table 3.1). MBT, Main Boundary Thrust; MCT, Main Central Thrust; STDS, Southern Tibetan Detachment System; RZT, Renbu Zedong Thrust; GT, Gangdese Thrust; MKT, Main Karakorum Thrust; MMT, Main Mantle Thrust. E, Mt. Everest; K, Mt. Kailas; NB, Namche Barwa; NP, Nanga Parbat. The cross section AA' is shown in Fig. 3.3.

be adequate to evaluate simple models that introduce the effect of an evolving Asian topography on regional and global climate patterns.

Models for the Elevation History of the Himalaya-Tibet Orogenic System

Introduction

The exceptional extent and elevation of the Tibetan Plateau and Himalayan range, a mountain system that may not have been equaled on this planet for over one billion years, has generally been attributed to the convergence between India and southern Asia over the past 55 Ma. However, the geological history of the Tibetan interior has not yet been convincingly established, leaving open the degree to which this generalization is true. In this section, we briefly comment on the timing and spatial variability of uplift as predicted by a variety

of tectonic models for the Himalaya and Tibetan Plateau. These models can be classified into three categories; those that predict wholesale uplift, progressive growth, or in which an elevated terrane was inherited.

Wholesale Uplift

Mantle Lithosphere Delamination A suggestion arising from modeling lithospheric evolution as an isoviscous continuum is that thickening leads to conditions in which a thick mantle boundary layer can be convectively thinned in a relatively brief period of time (e.g., England and Houseman, 1989). The change in potential energy in the remaining lithosphere would then lead to a sudden uplift of the plateau, long after the initiation of thickening. This model has been called upon to explain an apparent sudden increase in the elevation of the Tibetan Plateau during the Late Miocene (11–5 Ma)

Table 3.1. Timing and magnitude of slip on major faults in southern Asia.

Structure	Magnitude of Slip	Age and Duration	Reference
Himalayan Thrust System	140–~500 km		(1,2)
Main Central Thrust		24–19, 9–4 Ma	(3,23)
Main Boundary Thrust		11–1 Ma	(4)
Main Frontal Thrust		Pleistocene to Present	(4)
Tethyan Thrust System	200–400 km	~55–30 Ma	(5)
Renbu–Zedong Thrust	>20 km	19–11 Ma	(6)
Gangdese Thrust System	~80 km	27–23 Ma	(7)
Fenghuo Shan Thrust System	>100 km	~50–35 Ma	(8,9)
Red River Fault			
Left–lateral	550 ± 150 km	35–17 Ma	(10,11)
Right–lateral	40–6 km	5 Ma to Present	(12)
Altyn Tagh Fault (left–lateral)	~500 km	~15 Ma to Present	(13,14*)
Tien Shan Thrust System	100–200 km	~21 Ma to Present	(15–17)
Kunlun Fault	unknown	active in Quaternary	(18)
Qinling Fault	~150 km	active in Quaternary	(19)
Haiyuan Fault	<15 km	active since Pliocene	(20,21)
Karakorum–Jiali Fault System	>160 km	8 Ma to Present	(18,22)

* see (15) for a different estimate

References: (1) Lyon–Caen and Molnar (1985), (2) Schelling and Arita (1991), (3) Harrison et al. (1995b), (4) Burbank et al. (1996), (5) Ratschbacher et al. (1994), (6) Quidelleur et al. (1997), (7) Yin et al. (1994), (8) Leeder et al. (1988), (9) Qinghai BGMR (1989), (10) Leloup et al. (1995), (11) Harrison et al. (1996b), (12) Leloup et al. (1993), (13) Peltzer and Tapponnier (1988), (14) Yin and Nie (1996), (15) Burchfiel and Royden (1991), (16) Avouac et al. (1993), (17) McKnight (1993), (18) Peltzer et al. (1985), (19) Tapponnier and Molnar (1977), (20) Burchfiel et al. (1991), (21) Zhang et al. (1991), (22) Armijo et al. (1989), (23) Harrison et al. (1997a).

(Harrison et al., 1992; Molnar et al., 1993). However, when similar calculations are carried out using a temperature-dependent rheology, the mantle lithosphere tends to resist detachment due to its high strength (Buck and Toksoz, 1983; Lenardic and Kaula, 1995). Thus it is possible that the strength of the mantle lithosphere may stabilize it against catastrophic or even gradual detachment. Lenardic and Kaula (1995) suggest that mantle lithosphere at the margin of the thickened region could be thermally eroded until the base of the crust is exposed to the asthenosphere. This in turn becomes a weak zone that permits the thickened mantle lithosphere to be rapidly delaminated, in a manner similar to that first proposed by Bird (1978), producing rapid uplift.

Delayed Continental Underplating Powell (1986) proposed that from 20–5 Ma, the Indian lithosphere subducted into the mantle beneath Tibet. At 5 Ma, the Indian continental crust is hypothesized to have broken off from the still downgoing slab and had rapidly risen to the base of the Tibetan continental crust. Uplift results from iso-static adjustment of this newly thickened Tibetan crust. This model predicts that uniform plateau uplift occurred simultaneously with crustal thickening over the last 5 Ma but requires an unusually fluid mantle lithosphere overlying the subducted continental crust.

Progressive Growth

Underthrusting of Indian Lithosphere Originally suggested more than 70 years ago, this model proposes that virtually the entire Tibetan Plateau is underlain by Indian lithosphere resulting from a continent scale thrust fault (Argand, 1924). However, the low seismic velocities (and thus high implied temperatures) in the upper mantle underlying Tibet are inconsistent with it being the cold, underthrusted lithosphere of India (McNamara et al., 1994). Although subject to relatively large uncertainties, paleomagnetic studies suggest the total convergence between the Indian plate and southern Tibet since collision is substantially less that the ~2500 km of convergence between India and Siberia that occurred over the past 55 Ma (Besse and Courtillot, 1988). Another prediction of this model, one that we will return to later in this Chapter, is that the locus of uplift of the Tibetan Plateau has advanced from the suture zone northward at a rate approximately equivalent to the motion of the Indian plate.

Underthrusting of Asian Lithosphere Willett and Beaumont (1994) proposed that Asiatic continental subduction occurred along the northern boundary of Tibet. As a consequence, roll back of the south dipping mantle slab beneath northern Tibet would have resulted in upwelling of mantle asthenosphere and attendant uplift.

Distributed Shortening In this model, folding and faulting shortens and thickens the lithosphere throughout Tibet with about 50% shortening producing the roughly double-normal-thickness crust (Dewey and Burke, 1973; England and Houseman, 1986). Uplift is distributed throughout the plateau and may or may not exhibit a regular temporal or spatial progression. However, as discussed in section Cenozoic Crustal Shortening in the Himalaya, Tibet and Tien Shan, p. 46, the surface expression of this deformation is not easily recognized in Tibet today.

Continental Injection Zhao and Morgan (1985) proposed that the lower crust of Tibet behaves as a relatively low-viscosity fluid into which additions from India are assimilated, the additional volume acting to raise the plateau hydraulically. Their model predicts that the entire plateau was uplifted as a single unit with significant topography not appearing until about 10 Ma ago. However, as described later, it appears that much of the uplift and unroofing of Tibet has occurred in distinct episodes and that southern Tibet has uplifted independently of the rest of the plateau.

Inherited Plateau

Multiple Collision The question of how much crustal thickening in southern Asia is due to the collision with India and how much was inherited from prior orogenic episodes remains open in the absence of detailed geological investigation throughout the elevated regions. Recent geological mapping of the Tibetan interior (Yin et al., 1994; Murphy et al., 1997) suggests that a significant portion of southern Tibet was elevated due to continued convergence between the Lhasa and Qiangtang

blocks and remained a topographic highstand until initiation of the Indo-Asian collision.

Intra-Arc Thickening Because southern Asia was the site of an Andean-type subduction margin for ~100 Ma prior to the onset of the Indo-Asian collision, it has been speculated that southern Tibet may have been a high standing arc during that period (e.g., England and Searle, 1986). The deformation of the Late Cretaceous (100-65 Ma) Takena Formation prior to the eruption of the Linzizong volcanics (ca. 60 Ma) suggests a significant crustal thickening event occurred in southern Tibet before the Indo-Asian collision began (Coulon et al., 1986; Pan, 1992). This event may be analogous to the development of the sub-Andean thrust belt (Isacks, 1988) which contributed to the formation of the Altiplano plateau.

With the above framework in mind, we proceed to describe the results of geological and geochemical investigations that bear on the elevation history of the Tibetan Plateau and surrounding regions. We find that no one of the above described models can adequately explain the present distribution of topography. Rather we find that a spatially and temporally complex mix of these mechanisms is required to explain the orographic evolution of southern Asia.

Relationship among Uplift, Denudation, and Lithospheric Shortening

Overview

In general, the techniques we have available to address the thickening history of Asia (see Appendix) are best suited to determining when deformation initiates rather than assessing the magnitude of uplift. In this section we describe how knowledge of the magnitude of lithospheric shortening and denudation may be translated into estimates of the amount of surface uplift. In the best possible circumstances (e.g., in terms of exposure, availability of marker beds, barometry, depth control of structural variations, etc.), this approach can yield estimates of shortening that are accurate to ± 25%. The translation of this information into uplift history requires knowledge of the density structure of the lithosphere which introduces an additional 10%–15% uncertainty into the calculations.

Although the most useful information for determining elevation histories resides at the Earth's surface, there are relatively few strategies that permit use of rocks that originate there, and these approaches can only be used in special cases (see Appendix). Because of more general applicability, one of the most useful ways of assessing crustal thickening and thinning histories comes from the thermal imprint on rocks at depth that is recorded in minerals in the form of isotopic variations (e.g., McDougall and Harrison, 1988). Thermal histories derived in this way can relate to either denudation at the Earth's surface or to thermal effects of tectonic processes that either thicken or thin the crust and, by the requirement of isostasy, raise or lower elevation. Other physical processes, such as stable isotopic variations and cosmic ray spallogenic reactions, have application in assessing elevation histories as do inferences from paleogeography and paleoecology. For readers wishing a brief introduction to these methodologies, capsule reviews are provided in the Appendix.

Because of its general applicability, thermochronometry has played a useful role in providing clues to the orographic evolution of the Himalaya and Tibetan Plateau. Thermochronometry can be utilized in two ways. First, because drainage throughout southern Tibet and the Himalaya largely predates the establishment of the present topography, abrupt increases in rates of denudation along the southern margin of the collision zone can, in certain cases, be confidently ascribed to uplift of the Earth's surface (Harrison et al., 1992). Second, the thermal effect of normal and reverse faulting on adjacent rocks that are recorded by thermochronometers can be interpreted, using physical models, to yield slip histories (e.g., Harrison et al., 1995a, 1996, 1997a). These in turn can be used to constrain paleoelevation by the assumption of isostasy.

Denudation

Relating cooling at depth to surface uplift requires special circumstances. First, the observed cooling must be ascribable to motion towards the Earth's surface and not as a consequence of a change in basal heat flow, hydrothermal and magmatic effects, or tilting of the crust. Second, because the motion of hot rock at mid crustal depths towards the Earth's

surface tends to upwardly perturb the equilibrium geotherm, care must be taken to account for this effect. The effect that heat advection has in complicating the interpretation of geochronology in terms of denudation rate was documented by Clark and Jäger (1969) and has been periodically reviewed since then (see Harrison and Clarke, 1979; Parrish, 1983; Zeitler, 1985; England and Molnar, 1990a). The magnitude of this effect can be estimated from the Peclet number, $Pe = Uz/\kappa$, where U is velocity, z is initial depth below the surface, and κ is the thermal diffusivity (typically $\sim10^{-6}$ m^2/sec). A Pe number of order ~0.01 marks the approximate boundary where the upward displacement of isotherms becomes significant. For values considerably below that threshold, advection is a minor effect. Third, surface conditions must be appropriate (e.g., weathering agents, pre-existing drainage, steep gradient to base level) to immediately translate an elevation increase into increased denudation rates.

Pre-thickening Elevation

In cases where lithospheric thickening and subsequent denudation can be documented, a first approximation of the pre-shortening elevation, e_1, can be obtained from a simple model based on conservation of mass and Airy isostasy (Yin et al., 1997):

$$e_1 = [(\rho_{ml} - \rho_c)/S_v\rho_{ma}]\{(\rho_{ml} - \rho_c)\,(e_2 + D + T_{cr}$$
$$+ L_{c2} - T_{cr}\,S_v) + (\rho_{ma} - \rho_{ml})\,(T_L\,(1 - S_v) + e_2$$
$$+ D + (e_2\rho_c - L_{c2}(\rho_{ml} - \rho_c))/(\rho_{ma} - \rho_{ml}))\} \qquad (3.1)$$

where e_2 is the elevation following thickening, ρ_c is the density of the crust, ρ_{ml} is the mantle lithosphere density, ρ_{ma} is the mantle asthenosphere density, ρ_{c2} the length of the crustal root, T_{cr} is the thickness of continental crust when its surface is at the sea level, L_{c2} is the length of the crustal root, T_L is the thickness of the continental lithosphere when its surface is at the sea level, S_v is the stretching strain (original length vs. final length) in the vertical direction, and D is the magnitude of denudation during lithospheric shortening. The vertical stretching is inversely proportional to the horizontal shortening strain, h, that is, $S_v = 1/Sh$.

This model is a very simple approximation and ignores the effect of density change resulting from metamorphic reactions and changes in pressure and temperature as a function of depth. It also neglects any thermal erosion of the lithosphere as it thickens. Thus an estimate based on this model provides only a lower bound for the elevation of the crust prior to thickening. If we choose the isostatic compensation depth at the Moho, the paleoelevation, e_1, is given by:

$$e_1 = \{e_2 - ((T_c \, S_H - D) \, (\rho_m - \rho_c)/\rho_m)\}/(1+S_H) \qquad (3.2)$$

where T_c is crustal thickness when the surface is at sea level, and S_H is horizontal shortening strain (final length/original length). This equation approximates the case in which the mantle lithosphere is thermally eroded during thickening so that no lithospheric root develops. Equation (3.2) yields an upper limit for elevation prior to lithospheric shortening. Thus the most probable value of elevation prior to the onset of thickening lies between the results estimated from equations (3.1) and (3.2). The two most important parameters in equations (3.1) and (3.2) can be observed directly: denudation (D) via thermochronometry, and crustal shortening (S_H) via field mapping and construction of balanced cross sections.

Elevation During Thrust Formation

Recognizing when and where a thrust fault is initiated provides a second constraint on paleoelevation. As we will later argue, the thrust systems in southern Tibet and the Himalaya developed when the hinterland was already substantially elevated. Simple mechanical models suggest that the most likely location for a fracture to form is the boundary where a thin plate impinges upon a thicker one.

The stress concentration in an elastic medium under compression containing a step function thickness variation is singular at the elevation discontinuity. This suggests that the trace of a thrust fault that develops in response to this stress distribution will be located at low elevation adjacent to the discontinuity. Results of a numerical model of indentation into a layered lithosphere capable of flow (Lenardic and Kaula, 1995) also show that the maximum compressive stress occurs at the break in slope of the topography. Although the surface break of hinterland thrusts developing as part of a transfer fault system need not occur at low elevation

(e.g., Tapponnier et al., 1990), the general expectation along the collision front is that thrusts will surface near the topographic boundary between the thin Indian indenter and the thickened Asian crust. Indeed, the current traces of all Himalayan thrusts developed over the past 25 Ma are at relatively low (<2 km) elevation (Masek et al., 1994). We thus argue that, at least along the collision front, the surface rupture of a thrust fault is associated with a paleoelevation of <2 km. Thermochronometric determination of when thrust faulting is initiated provides the related timing information.

The Orographic Evolution of the IndoAsian Collision Zone

Introduction

As would be expected from the interaction of two continental masses of irregular shape and contrasting size, the transition between the first contact of India with southern Asia and their final suturing over a great lateral distance was protracted. Sedimentological evidence indicates that the northwest tip of India first collided with Asia at ~55 Ma (e.g., Le Fort, 1996). By 40 Ma, the two continents appear to have met along the full length of a ~3000 km long suture zone (Dewey et al., 1988). Paleomagnetic evidence indicates that the Indo-Australian plate moved northward by 2600 ± 900 km relative to the Eurasian plate after the initiation of collision (Dewey et al., 1989; Le Pichon et al., 1992). In that same interval, southern Tibet moved north with respect to Eurasia by 2000 ± 600 km (Besse and Courtillot, 1988). These two values suggest <1000 km underthrusting of India beneath Asia.

The tectonic evolution of Asia in response to the Cenozoic Indo-Asian collision has been the subject of numerous syntheses (e.g., Tapponnier and Molnar, 1979; Peltzer and Tapponnier, 1988; Burchfiel and Royden, 1991; Harrison et al., 1992; Yin and Nie, 1996). Despite significant recent progress in the examination of specific tectonic elements (e.g., Burchfiel et al., 1992; Chinese State Bureau of Seismology (CSBS), 1992; Molnar et al., 1994; Yin et al., 1994; Leloup et al., 1995; Harrison et al., 1996; Quidelleur et al., 1997), the age and magnitude of deformation in the Indo-Asian collisional system as a whole remains incompletely understood. Here we describe the salient geological features of the three regions of pronounced late Cenozoic crustal thickening (the Himalaya, Tibet, and Tien Shan) and also the three major strike-slip fault systems (Red River, Altyn Tagh, and Karakorum) (Fig. 3.1). The development of large basins (the Tarim and Qaidam) is an integral aspect of the orographic evolution of the collisional system and is described briefly as well (Fig. 3.1). In summarizing the evolution of the southern part of the collisional system, we have focused primarily on the central Himalayan arc in preference to the western or eastern terminations of the arc, or syntaxes (Fig. 3.1). The geology of western syntaxes in particular is complicated by both an island arc collision prior to contact of India with Asia and a tectonic configuration that appears unrepresentative of the overall collision zone (e.g., Searle et al., 1988). At the outset, we draw the reader's attention to the emphasis we place on the growth and uplift history of the Tibetan Plateau *before* the initiation of the Indo-Asian collision, which is merely the last episode in the long history of Tethyan ocean closure (Sengor and Natalin, 1996).

Geologic History

Crustal Shortening in Tibet Prior to the Indo-Asian Collision   The starting point for mass balance calculations aimed at assessing the relative contributions of various tectonic accommodation mechanisms (e.g., England and Houseman, 1986; Le Pichon et al., 1992) has typically been to assume that the altitude of Tibet was at or near sea level (Norin, 1946) immediately prior to collision. However, recent geological investigations in the Maqu and Choqin regions (Fig. 3.2) of the central and northern Lhasa Block suggest that significant crustal thickening (50%) occurred in those regions between 150 Ma and 60 Ma (Coulon et al., 1986; Pan, 1992; Murphy et al., 1997). This crustal thickening has been interpreted to be related either to intra-arc shortening during the subduction of Indian oceanic lithosphere (Coulon et al., 1986; Pan, 1992) or to continued collision between the Lhasa and Qiangtang blocks (Yin et al., 1994; Murphy et al., 1997) (Fig. 3.1). The lithology of clasts in the middle Cretaceous-early Tertiary Xigaze forearc deposits are consistent with a relatively low topogra-

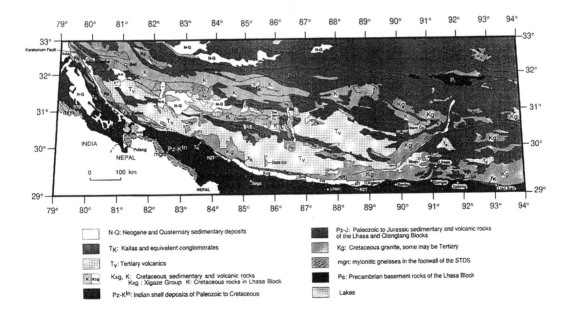

Fig. 3.2. Geological map of southern Tibet modified from Liu et al. (1988) and Yin et al. (1994). The Gangdese batholith consists of Cretaceous to Tertiary granites and Tertiary volcanics. Indian shelf deposits are synonymous with Tethyan metasedimentary rocks. GT, Gangdese Thrust; RZT, Renbu Zedong Thrust.

phy along the southernmost margin of Asia throughout this interval and an elevated hinterland (Durr, 1996). This suggests that a significant portion of the uplift created during Early Cretaceous (144-100 Ma) crustal shortening was likely preserved until the Indo-Asian collision began and can be considered as an integral part of the total uplift of the plateau.

Cenozoic Crustal Shortening in the Himalaya, Tibet and Tien Shan Several large Cenozoic thrust systems have been recognized in the Lesser Himalaya (Gansser 1964; Lyon-Caen and Molnar, 1984, 1985), Tethyan Himalaya (Burg, 1983; Liu et al., 1988; Ratschbacher et al., 1994), Gangdese Shan in southern Tibet (Yin et al., 1994), northern Tibet (Leeder et al., 1988; Liu et al., 1988), and Tien Shan (Avouac et al., 1993) (Fig. 3.1). From south to north, these major thrust systems are the Main Himalayan Thrust, Tethyan thrust system, Renbu Zedong thrust system, Gangdese thrust system, Fenghuo Shan thrust system, Nan Shan thrust system, and Tien Shan thrust system (Figs. 3.1 and 3.2). Their ages and extent of shortening along these thrust systems are summarized in Table 3.1. The total amount of shortening between the Indian shield and southern Tibet is approximately 1000 km (Patriat and Achache, 1984; Dewey et al., 1989; Chen et al., 1993; Patzelt et al., 1996), and 100–200 km across the Tien Shan (Burchfiel and Royden, 1991; Avouac et al., 1993; McKnight, 1993). The amount of Tertiary shortening across the interior of the Tibetan Plateau is not well constrained, but relatively little has been securely documented (Chang et al., 1986; Murphy et al., 1997).

When the left-lateral slip that occurred along the Red River fault (see section Large-scale Strike-slip Faulting, p. 46) is restored (Leloup et al., 1995), the Eocene thrusting in northern Tibet and southern Qaidam becomes located close to the northern tip of the Indochina block (Fig. 3.1). If, as seems probable, the activity in the Fenghuo Shan thrust belt and left slip on the Red River fault overlapped both in time and space, the development of the Fenghuo Shan thrust belt may be related to the eastward extrusion of the Indochina block (Yin and Nie, 1996).

Large-scale Strike-slip Faulting Although strike-slip faulting itself does not itself contribute

to elevation changes, the development of these features appears to have been generally related to crustal thickening events in the Indo-Asian collisional system, either as transfer faults (Burchfiel et al., 1989) or as a termination mechanism for the large extrusion of continental blocks (Yin and Nie, 1996). In either case, the timing and magnitude of strike-slip faulting are important, and in some cases contribute unique information in determining the timing of crustal thickening and thus the uplift history of Asia during the Indo-Asian collision.

Red River Fault System The Red River fault can be traced from eastern Tibet, through the Yunnan province of China, to the South China Sea for a distance of more than 1000 km (Peltzer and Tapponnier, 1988; Leloup et al., 1995). The fault system has had a complex history during the Indo-Asian collision (Fig. 3.1). It was a left-slip fault between 35 and 17 Ma with a displacement greater than 80 km, and possibly as large as 550 km, as suggested by both the opening of the South China Sea (Briais et al., 1993) and finite-strain analysis across the fault zone (Lacassin et al., 1993). The sense of motion on the Red River fault system reversed at about 5 Ma (Leloup et al., 1993; Harrison et al., 1996) becoming right-slip with a normal-slip component in places (Allen et al., 1984; Leloup et al., 1995). The later stage of the left-slip history of the Red River fault system was associated with transtensional tectonics, which migrated diachronously from Vietnam in the southeast to the Ailao Shan of Yunnan in the northwest between 30 and 17 Ma (Harrison et al., 1996). The change in slip direction along the Red River fault likely reflects a second phase of continental extrusion which carried Tibet eastward between the Altyn Tagh, Karakorum, and Red River faults.

Alternatively, slip along the Red River fault has been attributed to accommodation of rotation of rigid blocks about the eastern syntaxes of the Himalaya orogenic belt (Cobbold and Davy, 1988; England and Molnar, 1990b). In this view, the magnitude of slip is much smaller than that suggested for eastward extrusion of the Indochina Block during the collision of India and Asia (e.g., Peltzer and Tapponnier, 1988). However, the timing of initiation of transtensional tectonics along the Red River fault and the successful prediction of the kinematic history of the Red River fault from the seafloor magnetic-reversal history in the South China Sea are consistent with the extrusion hypothesis of the Indochina Block in the middle Cenozoic (Harrison et al., 1996).

Altyn Tagh Fault System The Quaternary slip rate along the Altyn Tagh fault (Fig. 3.1) has been inferred to be as much as 2–3 cm/yr (Peltzer et al., 1989) and as little as ~0.5 cm/yr (CSBS, 1992). Estimates of total left-slip vary from 1200 km, on the basis of offset of the Qilian Shan suture zone and the western Kunlun Shan (CSBS, 1992), ~500 km, by matching the offset of the late Paleozoic magmatic belts in the western and eastern Kunlun Shan (Peltzer and Tapponnier, 1988) and by matching the Qilian Shan suture zone with the central Tarim uplift zone, and ~200 km, based on inferences from the measured Pliocene-Quaternary slip rate as a constant along the Altyn Tagh fault and the duration of the Indo-Asian collision (Burchfiel and Royden, 1991; CSBS, 1992) (Fig. 3.1). As noted by both Tapponnier et al. (1990) and Burchfiel et al. (1989), the Altyn Tagh fault is currently a transfer fault linking thrusts in the Tibetan Plateau with the Nan Shan. This kinematic model requires that the slip history along the Altyn Tagh fault be related to the sequence of thrusting in Tibet and the Nan Shan. Because the magnitude of deformation in Tarim is much less than that in Tibet and the Nan Shan, the total left-slip along the Altyn Tagh fault should decrease northeastward along strike. If thrusting in the plateau propagated from south to north (e.g., England and Houseman, 1986; Molnar et al., 1987), then the Altyn Tagh fault has progressively lengthened since its initiation. Although the initiation age of the Altyn Tagh system would decrease northeastward in this model, each segment of the fault would have been continuously active once formed. Alternatively, thrusting could have propagated from north to south in Tibet (Harrison et al., 1992) which implies that the active length of the Altyn Tagh fault has decreased since its initiation and that slip along the Altyn Tagh fault would have ended progressively from northeast to southwest. If the thrusting sequence was not unidirectional, then both the length and the active portion of the Altyn Tagh fault would have varied with time.

The junction point of the eastern Kunlun and Altyn Tagh faults divides the Altyn Tagh system into eastern and western segments (Fig. 3.1). The eastern segment is transpressional, as indicated by a series of folds and thrusts that involve Neogene strata and are parallel to the Altyn Tagh fault (Burchfiel et al., 1989). The western segment is transtensional, as expressed by Altyn-Tagh parallel normal faults both north (Avouac and Peltzer, 1993) and south (CSBS, 1992) of the Altyn Tagh fault. As reported by CSBS (1992), late Neogene volcanic centers in western Tibet (e.g., Arnaud et al., 1992; Turner et al., 1993) are mostly distributed within pull-apart basins along the western Altyn Tagh fault system. The change from strike-perpendicular contraction to strike-perpendicular extension along the Altyn Tagh system implies that the Tarim basin has rotated clockwise with respect to the Tibetan Plateau in the late Cenozoic. We therefore propose that the Neogene volcanism in western Tibet may have been related to the late Cenozoic clockwise rotation of the Tarim basin with respect to Tibet (Yin et al., 1995). Such a clockwise rotation of Tarim has been proposed to explain a decrease in shortening from west to east in the Tien Shan (Chen et al., 1991; Avouac et al., 1993). This interpretation is in contrast with the earlier models that the Neogene volcanism in Tibet was related to convective removal of a thickened lithospheric mantle (England and Houseman, 1989; Turner et al., 1993) or subduction of the Tarim basin beneath the Tibetan Plateau (Arnaud et al., 1992). Neither interpretation directly predicts a spatial and temporal association between surface structures and the location of the volcanic centers.

Karakorum Fault The Karakorum fault is an active, right-slip fault that is over 800 km long (Fig. 3.1). Offset estimates vary from 120 km to ~1000 km (Peltzer and Tapponnier, 1988; Ratschbacher et al., 1994; Searle, 1996). Previous workers have proposed the Karakorum fault accommodates eastward extrusion of Tibet by serving as the conjugate to the left-slip Altyn Tagh fault (Peltzer and Tapponnier, 1988). At its western end, the Karakorum fault has been interpreted as terminating in the W-dipping Konggar detachment (Brunel et al., 1994; Ratschbacher et al., 1994). Its eastern segment has been proposed to extend across Tibet, forming the Karakorum-Jiali fault system (Armijo et al., 1989). South of the Karakorum fault are the Kohistan and Ladakh batholiths and the western syntaxes where, at Nanga Parbat, very rapid denudation and uplift has occurred over the last 10 Ma (e.g., Chamberlain and Zeitler, 1996).

Sedimentary Basins and Denudation

Tarim Basin The present Tarim basin is a late Cenozoic feature, as it is mostly covered by Neogene and Quaternary sediments (Zhou and Chen, 1990; Jia et al., 1991) (Fig. 3.1). Recent drilling, seismic-reflection profiles, and field studies in the basin suggest that it has a protracted history of deformation since the late Precambrian (Jia et al., 1991; Hendrix et al., 1992). Two main regions in the Tarim basin have received Cenozoic sediments: the Kuche and Hotan foredeeps. Their development has been interpreted as being related to the southern Tien Shan and western Kunlun thrust belts, respectively (Zhou and Chen, 1990; Zhou and Zheng, 1990). The age of the initiation of the foredeep basins can constrain the age of initial thrusting in both thrust belts. An abrupt jump in sedimentation rate suggests that the southern Tien Shan and Hotan thrust belts were initiated in the mid-Miocene (~15 Ma), or about 35 m.y. after the collision between India and Asia began. As the Hotan thrust belt is interpreted to be geometrically and kinematically linked with the Altyn Tagh strike-slip system to the east and the Pamir thrust system-Chaman strike-slip fault to the west (Yin et al., 1993), the Altyn Tagh fault can also be inferred to have initiated in the mid-Miocene. This conclusion is similar to that of Avouac et al. (1993) based on extrapolation of the current slip rate and the total magnitude of shortening in the Tien Shan. Note that the lower part of the biostratigraphy of the Hotan section is well constrained based on foraminifera and other marine fossils with ages from the Upper Cretaceous (~75 Ma) to Lower Oligocene (~30 Ma) (Hao, 1983; Tang, 1990). The Kuche section consists entirely of alluvial and fluvial deposits, and thus its age is more problematic. Results of a recent magnetostratigraphic study (Craig et al., 1994), fission-track cooling ages of the Tien Shan (Hendrix et al., 1994), and inferences about the duration of Cenozoic shortening in the Tien Shan by the current slip rate and total amount of shortening (Avouac et al., 1993) are all consistent with age assignments for the Kuche section by Ye and Huang (1992) and the inference that thrusting in the southern Tien Shan began at about 20 Ma.

Qaidam Basin The Qaidam basin lies between the Qilian Shan and the Tibetan Plateau and its present configuration began to develop in the Cenozoic (e.g., Song and Wang, 1993) (Fig. 1). Because the basin is extensively covered by Quaternary deposits, its subsurface geology has been inferred from seismic-reflection profiles and drill-hole data (Song and Wang, 1993), as well as from the exposed rocks to the north along the southern flank of the Qilian Shan and to the south along the eastern Kunlun Shan (Qinghai Bureau of Geology and Mineral Resources, 1989). Seismic-reflection profiles obtained from the southwestern part of the basin suggest that the Mesozoic strata are no more than several-hundred meters thick. In contrast, the Cenozoic sequence is on average between 4 and 5 km thick, and locally exceeds 10 km (Song and Wang, 1993). Three major unconformities, Paleocene-Eocene (~53 Ma), early Pliocene (~4 Ma), and early Quaternary (~2 Ma), are recognized in the Cenozoic sequence of the Qaidam basin. An influx of conglomerates in the late Paleocene and early Eocene (~55–50 Ma) is consistent with observations in the Fenghuo Shan region in north-central Tibet, where Eocene foreland basin sedimentation related to thrusting has been reported by Leeder et al. (1988). We interpret the southern margin of the Qaidam and the northern Tibetan Plateau as having experienced late Paleocene–early Eocene (~50 Ma) N-S compression, possibly related to an early phase of Indo-Asian collision and extrusion of the Indochina block (Yin and Nie, 1996); the profound early Pliocene (~4 Ma) unconformity in the Qaidam basin may have been produced by the beginning of a second phase of thrusting, possibly linked with the eastward propagation of the Altyn Tagh fault. Because sedimentary facies of Mesozoic strata in the Tarim and Qaidam basins have not been investigated in detail, when and how the two basins were separated remains unknown.

Pre-collisional Orography (Cretaceous–Paleocene)

The Tibetan expeditions of Sven Hedin identified occurrences of Albian–Aptian (~120–100 Ma) marine limestones (Norin, 1946). This has been widely interpreted as indicating that much of Tibet was slightly below sea level during the Cretaceous (e.g., England and Houseman, 1986; Dewey et al., 1988). However, the locations of their sampling are difficult to place on a modern geological map and the broad conclusion that marine limestones were ubiquitous during the Cretaceous is contradicted by results of a recent geotraverse across the western Lhasa block through Choqin (Yin et al., 1994; Murphy et al., 1997) (Fig. 3.2). The northern Lhasa block was undergoing significant thickening between 145 and 110 Ma and cannot be characterized as being near or below sea level throughout that time. The Norin fossils may, however, reflect a back-arc marine seaway present during part of the Cretaceous (Yin et al., 1994).

In the Lhasa Block, the Upper Cretaceous Takena Formation (dominantly red beds) is folded and unconformably overlain by the Paleocene-Eocene Linzizong Formation (mostly intermediate-felsic volcanic rocks). At Maqu (Fig. 3.2), the Linzizong Formation is mildly folded and cut by at least one south-directed thrust fault with a total shortening over 50 km of ~15% (Pan, 1992). Near Choqin (Fig. 3.2), the equivalent of the Linzizong Formation, the Choqin tuff, is undeformed (Murphy et al., 1997). England and Searle (1986) argued that the Lhasa block resisted Tertiary deformation because it was substantially thickened and elevated in the Gangdese arc prior to collision. However, the clastic sediments arriving in the Xigaze forearc basin were derived from the northern Lhasa block (Willems et al., 1995; Dürr, 1996) indicating that the Gangdese arc was, at least in places, topographically subdued during the Late Cretaceous.

Early-collisional Orography (Middle Eocene–Early Oligocene) (~52–30 Ma)

Leeder et al. (1988) documented a fold and thrust belt that developed in central-northern Tibet shortly after collision between India and Asia began. Eocene clastic sedimentary rocks in the Fenghuo Shan are shortened by folding and thrusting by at least 50% over a distance of 100 km (Kidd et al., 1988) (Fig. 3.1). Surprisingly, this is the only existing documentation of Eocene–Early Oligocene crustal thickening on the Tibetan Plateau. Thin skinned thrusting was almost surely occurring within the Tethyan Himalaya during the Paleogene and as much as 1500 km of shortening may have been taken up between the Indian shield and shelf sequence (Patzelt et al., 1996).

Several lines of evidence suggest the existence of an early thickening event in the proto Himalaya, referred to as the Eo-himalayan (58–36 Ma) orogeny (Le Fort, 1996). This phase of metamorphism is largely inferred from high pressure, prograde assemblages in the Greater Himalayan Crystallines that are overprinted by Neogene (24–2 Ma) metamorphism (Pêcher, 1989; Hodges et al., 1994) related to slip on the Main Central Thrust (see Fig. 3.3 and section *Evolution of the Himalayan and Southern Tibetan Thrusts*, p. 50). Restitic monazite ages between ~30 and 45 Ma are commonly found in Neogene leucogranites (e.g., Edwards and Harrison, 1997) and likely represent products of Eocene prograde metamorphism of the Indian basement. Igneous rocks with a similar range of crystallization ages have also been documented (e.g., Parrish and Hodges, 1993; Hodges et al., 1996). These results indicate that a pure shear deformation was occurring within the Indian continental basement as the Tethyan shelf sequence was being tectonically telescoped following onset of the collision.

Scattered Eocene to Oligocene (58-24 Ma) molasse deposits, interpreted to result from uplift of this proto Himalaya, have been recognized in northern India and Pakistan. Eocene rocks of the Murree Formation (Bossart and Ottiger, 1989) have been biostratigraphically dated in a reentrant feature in northern Pakistan. Although the paleomagnetism study of Najman et al. (1994) suggested an age of the unfossiliferous Dagshai Formation of northern India of 36 ± 7 Ma, a subsequent $^{40}Ar/^{39}Ar$ investigation of detrital muscovite places a maximum depositional age of this formation at <24 Ma (Najman et al., 1994). This requires that the Dagshai and overlying Kasuli Formations be early Miocene in age or younger.

The comparative lack of evidence for Eocene to Oligocene thickening in Tibet and the Himalaya may reflect the fact that Indochina began to be extruded to the southeast along the Red River fault beginning at about 35 Ma (Tapponnier et al., 1986; Leloup et al., 1995; Harrison et al., 1996). Perhaps as much as one half of the convergence between India and Asia during the Oligocene (~35–24 Ma) may have been taken up by extrusion along this and associated faults (Harrison et al., 1992). As a consequence, significant crustal shortening, and consequent uplift and erosion, in Tibet may have been delayed until Indochina had extruded clear of the indenter.

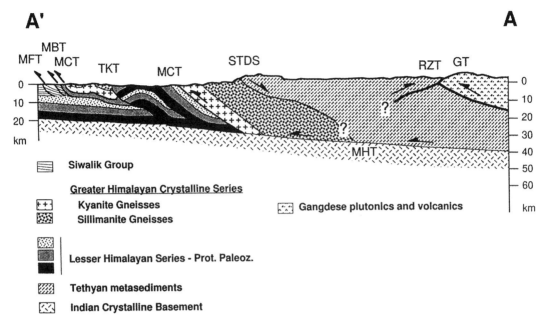

Fig. 3.3. N-S cross section across the collision front (section A-A' in Fig. 3.1) modified from Schelling and Arita (1991) and Zhao et al. (1993). MFT, Main Frontal Thrust; MBT, Main Boundary Thrust; TKT, Tamar Khola Thrust; MCT, Main Central Thrust; MHT, Main Himalayan Thrust; STDS, Southern Tibetan Detachment System; RZT, Renbu Zedong Thrust; GTS, Gangdese Thrust System. No vertical exaggeration.

Late-collisional Orography (Late Oligocene–
Recent) (30–0 Ma)

Evolution of the Himalayan and Southern Ti-
betan Thrusts Knowing the timing and sequence of
thrusting within the Himalaya and southern Tibet is
pivotal to understanding the evolution of the Indo-
Asian collision. It has long been suspected (e.g.,
Gansser, 1964) that thrusting within the Himalaya
is a relatively late stage response in the overall accommodation of India's collision with Asia, which
began at about 50 Ma (Burbank et al., 1996b). Even
today it remains unclear why crustal-scale thrusting
within the collision zone was delayed for perhaps
20–30 Ma following the onset of collision (e.g.,
Harrison et al., 1992).

North of the Indus-Tsangpo suture, the south-directed Gangdese Thrust (GT) placed the Cretaceous–
Early Oligocene (144–30 Ma) Gangdese batholith,
atop Tethyan metasediments (Yin et al., 1994)
(Figs. 3.2 and 3.3). The south-directed GT has been
recognized only recently because the younger,
north-directed Renbu Zedong Thrust (RZT) generally
obscures its exposure throughout its length except
in an ~100-km-long window near Zedong (Yin et

al., 1994) (Fig. 3.2). Throughout southeastern Tibet, the RZT is the dominant structural feature along
the suture zone (Burg, 1983; Liu et al., 1988; Kidd
et al., 1988; Yin et al., 1994). Within the Tethyan
Himalaya (Fig. 3.1), imbricate south-directed
thrusts, perhaps of the same generation as the GT,
generally juxtapose greenschist grade rocks on
both sides of the faults suggesting relatively small
displacements (Wang et al., 1983; Burg, 1983;
Burg et al., 1987) (Fig. 3.2).

South-directed thrusts within the Himalaya (Fig.
3.3), including the Main Central Thrust (MCT),
Main Boundary Thrust (MBT), and the Main Frontal
Thrust (MFT) (Gansser, 1964; Bouchez and Pêcher,
1981; Arita, 1983; Le Fort, 1986; Mattauer, 1986;
Burbank et al., 1996b) all appear to sole into a
common decollement, the Main Himalayan Thrust
(MHT) (Zhao et al., 1993; Nelson et al., 1996;
Brown et al., 1996) (Fig. 3.3). In general, the MCT
places high grade gneisses of Indian origin (the
Greater Himalayan Crystalline, or Tibetan slab) on
top of middle grade schists (Lesser Himalayan, or
Midland Formations). The MBT juxtaposes those
schists against the unmetamorphosed Miocene–
Pleistocene (24–0 Ma) molasse (Siwalik Group),

and the MFT is presently active within Quaternary sediments (Fig. 3.3).

Juxtaposition of the Greater Himalayan Crystallines on top of the Lesser Himalayan Formations along much of the Himalaya is associated with a zone of inverted metamorphism largely contained within a highly sheared, 4–8 km thick zone of distributed deformation immediately beneath the MCT fault (Fig. 3.4). Kinematic indicators within this broad region commonly referred to as the MCT zone, document, that the shear sense is uniformly top-to-the-south (e.g., Colchen et al., 1986). Proposed causes of the apparent increase in metamorphic grade with higher structural position (i.e., shallower depth) include: heating of the colder lower plate by emplacement of a hot nappe, either due to thermal relaxation alone or aided by strain heating (e.g., Arita, 1983; England et al., 1992), folding of pre-existing isograds (Searle and Rex, 1989), imbricate thrusting (Brunel and Keinast, 1986), mantle delamination (Bird, 1978), and the ductile shearing of an existing zone of right-way-up metamorphism (Hubbard, 1996; Harrison et al., 1997a).

Gangdese Thrust The first evidence of crustal scale thrusting in the collision zone is initiation of the south-directed Gangdese Thrust (GT) during the late Oligocene (30–24 Ma) (Figs. 3.2 and 3.3). West of Lhasa, the GT juxtaposes Late Cretaceous Xigaze Group forearc-basin deposits on top of Tethyan sedimentary rocks of the Indian continental margin. East of Lhasa, the fault juxtaposes the Late Cretaceous–Eocene Gangdese batholith over Tethyan sedimentary rocks (Yin et al., 1994) (Fig. 3.2). Near Zedong, 150 km southeast of Lhasa, the Gangdese thrust is marked by a mylonitic shear zone that consists of deformed granite and metasedimentary rocks. A lower age bound for the Gangdese Thrust of 18.3 ± 0.5 Ma is given by crosscutting relationships (Yin et al., 1994). GT-related thickening during the Late Oligocene–Early Miocene isalso inferred to have occurred in a structurally similar position in southwest Tibet (Ryerson et al., 1995). In south-central Tibet, the Gangdese Thrust System appears to have had at least two modes of accommodation; as opposing thrusts bounding a pop-up of the Xigaze Formation (Fig. 3.2) and a transfer system of imbricate thrusts within the Tethyan Himalaya.

In general, thrusting leads to the interplay between lateral heat flow across the fault and denudation at the rising upper boundary resulting in two broad thermal regimes within the hanging wall.

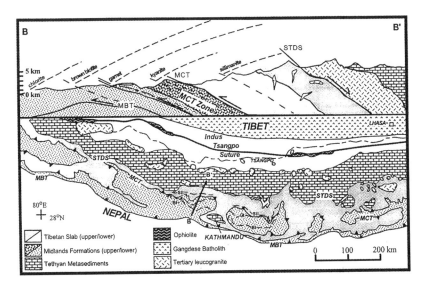

Fig. 3.4. Geological sketch map of the Himalaya and Southern Tibet after Le Fort et al. (1987) showing the relationship between the Tibetan Slab (Greater Himalyan Crystallines) and Midlands Formations (Lesser-Himalayan Formations) and reference isograds of the inverted metamorphic section. Main Central Thrust, MCT; Main Boundary Thrust, MBT; Southern Tibetan Detachment System, STDS. The cross section marked BB' is shown in the inset. Inset: Generalized cross section through the central Himalaya after Colchen et al. (1986) illustrating the pattern of inverted metamorphism beneath the MCT.

(1) A zone exists directly above the thrust surface whose thickness is a function of the thermal time constant for the inferred duration of thrusting. Within this zone, the thermal evolution is dominated by heat loss to the footwall. Because arrival of the footwall ramp is diachronous across the hanging wall in the direction of thrust transport, this conductive length scale (i.e., the zone refrigerated by the footwall) will vary from relatively thick directly above the thrust surface to thin further away. (2) Sufficiently far above the thrust surface, cooling is dominated by erosional denudation in response to thrust-related crustal thickening. Again, because thickening is diachronous, mid-crustal rocks adjacent to the thrust will be uplifted and denuded before rocks further away. A transition region exists between these two regimes in which cooling results from both processes.

Near the GT thrust surface, thermochronometry provides clear evidence of rapid cooling beginning at 27 Ma that is interpreted to result from refrigeration by the cold Tethyan sediments (Yin et al., 1994). This event, which is observed at several locations along strike (e.g., Copeland et al., 1995), is consistent with thrust initiation beginning at about 27 Ma. A $^{206}Pb/^{238}U$ zircon age of 30 ± 2 Ma for the hanging wall granitoid cut by the GT provides a firm upper bound for initiation of this thrust (T.M. Harrison, unpublished data).

Thermochronological results from the Gangdese batholith hanging wall well north of the thrust trace (Copeland et al., 1987; 1995; Richter et al., 1991; Harrison et al., 1992, 1993) indicate that rapid cooling of this belt was delayed until the Early Miocene (24–17 Ma). Because there is a several million year lag between the onset of rapid denudation and initiation of cooling in the mid-crust (Copeland et al., 1987; Richter et al., 1991), uplift of these locations in the hanging wall due to isostasy can be inferred to have occurred at or prior to ~22 Ma. The coincidence between this age and the timing of GT thickening across the Gangdese belt suggests diachronous arrival of the thrust tip across the northern Gangdese batholith and indicates the extent of crustal thickening. Accelerated erosion immediately following thickening-driven uplift is likely because the drainage from southern Tibet across the Himalaya is generally antecedent to elevated topography (Seeber and Gornitz, 1983).

The above mechanism explains the pattern of cooling ages throughout the Gangdese batholith. Assuming a flat-ramp geometry for the GT, with the flat following the brittle-ductile transition at ~15 km depth, the footwall ramp would propagate northward relative to the overriding plate at the same rate as the GT, estimated to be 12-24 mm/yr (Yin et al., 1994). At this rate, the Gangdese batholith near Lhasa would not have been transported up the GT ramp (and thus uplifted and subsequently denuded) until approximately 23 Ma when we infer fault activity ceased. The widespread preservation of precollisional Linzizong volcanic rocks along the northern margin of the Gangdese Shan (Liu et al., 1988) and the relatively modest amount of ca. 22 Ma denudation (and thus thickening) in the vicinity of Lhasa (<4 km) (Copeland et al., 1995) suggest that the northern tip of the footwall ramp did not extend beyond the northern margin of the Gangdese batholith. Seismic reflection profiling across southern Tibet detected moderately steep north-dipping reflectors beneath the northern flank of the Gangdese batholith that may mark the GT ramp (Brown et al., 1996; Alsdorf et al., 1995; Hauck et al., 1995; Mechie et al., 1995). The fault geometry described above implies a total GT displacement of ~80 km. Farther west, a minimum displacement of 46 ± 9 km along the GT is estimated (Yin et al., 1994).

Renbu Zedong Thrust For 500 km across southeast Tibet, the RZT marks the site of the suture between Asian rocks and those of Indian affinity (Fig. 3.2). The north-directed RZT truncates the GT, juxtaposing generally low grade Tethyan shelf rocks directly on top of the Gangdese batholith (Burg, 1983; Wang et al., 1983; Liu et al., 1988; Kidd et al., 1988). The RZT is clearly younger than the GT because its hanging wall strata locally thrust over the trace of the GT, putting both the hanging wall and footwall rocks of the GT below the RZT footwall (Yin et al., 1994). This relationship thus places an upper age limit on the RZT of ~23 Ma. Because the north-south trending Yadong-Gulu rift, which was initiated at 8 ± 1 Ma (Harrison et al., 1995a), clearly crosscuts the RZT, its activity is restricted to the interval 23–8 Ma.

Near Renbu (Fig. 3.2), Ratschbacher et al. (1994) obtained a 17.5 ± 0.9 Ma K-Ar age from white mica filling a tension gash in the RZT shear zone. Although possibly a cooling age, the low grade of metamorphism affecting these rocks was not likely to have exceeded the white mica closure temperature of ~350–400°C suggesting that this age may locally date the time of deformation. Between Renbu and the Zedong window, the nature of the RZT is obscured because of both structural complexity (e.g., thrust imbricates) and by burial beneath wide Tsangpo river channel deposits. East of the Zedong window, the geological relationships are straightforward and the RZT is very well exposed. Near Lang Xian (Fig. 3.2), thermochronological results for hanging wall schists and footwall granitoids are consistent with fault activity having occurring between 19 and 11 Ma (Quidelleur et al., 1997). A cooling episode recorded by all the K-feldspar samples beginning at ~10 Ma has been interpreted to reflect regional uplift of southern Tibet, perhaps as a result of displacement along a ramp of the Main Himalayan Thrust (Quidelleur et al., 1997).

Main Himalayan Thrust Results of recent seismic reflection profiling in southern Tibet have been interpreted as indicating that all the Himalayan thrusts (i.e., MCT, MBT, MFT) are all splays of the same decollement, termed the Main Himalayan Thrust (Zhao et al., 1993) (Fig. 3.3). This gently N-dipping feature extends well into southern Tibet (Hauck et al., 1995; Nelson et al., 1996).

Main Central Thrust Balanced cross sections and flexural modeling of gravity data suggest that the north-dipping MCT (Figs. 3.3 and 3.4) accommodated a minimum of 140 km of displacement and perhaps more than 300 km (Gansser, 1964; Arita, 1983; Lyon-Caen and Molnar, 1985, Pêcher, 1989; Schelling and Arita, 1991; Schelling, 1992; Srivastava and Mitra, 1994). Geochronology from the MCT hanging wall clearly indicates that anatexis and simple shear deformation was occurring at 22 ± 1 Ma (Parrish and Hodges, 1993; Coleman and Parrish, 1995; Harrison et al., 1995b; Hodges et al., 1996). Cooling ages in the upper portion of the MCT hanging wall suggest that slip had terminated by ~18 Ma (Hubbard and Harrison, 1989; Copeland et al., 1991). Although it remains unclear when the MCT was initiated, several lines of evidence suggest that it did not experience significant displacement prior to about 25 Ma. A world-wide increase in oceanic $^{87}Sr/^{86}Sr$ at ~20 Ma, if due to erosion of the

MCT hanging wall (Palmer and Edmonds, 1989; Richter et al., 1992), would be consistent with the MCT hanging wall not becoming exposed until the Early Miocene. The south-directed GT appears to have been active throughout the Late Oligocene, providing an accommodation mechanism prior to ~24 Ma (Yin et al., 1994). Lastly, no metamorphic or igneous products ascribable to the MCT older than ~24 Ma have yet been documented.

Although it has been widely assumed that the pattern of inverted metamorphism also developed during the Early Miocene (e.g., England et al., 1992), recent in situ Th-Pb dating of monazite included in garnet have revealed that the peak metamorphic recrystallization in the MCT zone of the central Himalaya occurred at 8–6 Ma (Harrison et al., 1997a). The apparent inverted metamorphism (Fig. 3.4) appears to have resulted from activation of a broad shear zone beneath the MCT zone which juxtaposed two right-way-up metamorphic sequences (e.g., Hubbard, 1996). Preliminary thermal/mechanical modeling suggest that the MCT fault was reactivated at ca. 10 Ma (following ~10 m.y. of inactivity) with a slip rate of ~20 mm/yr. At ~6 Ma, activity shifted progressively across the MCT zone from north to south and slip is inferred to have terminated at ~3-4 Ma. Recognition of this remarkably youthful phase of metamorphism resolves a number of outstanding problems in Himalayan tectonics, such as why the MCT (and not the more recently initiated thrusts) marks the break in slope of the present-day mountain range, and transcends others, such as the need for exceptional physical conditions (e.g., very high flow stress, mantle delamination, rapid decompression) to explain the generation of the Himalayan leucogranites (Harrison et al., 1997a,b).

Main Boundary Thrust The surface expression of the MBT is a generally steep north-dipping fault that marks the contact between the Lesser Himalayan Formations and the underlying Miocene–Pleistocene (24–0 Ma) Siwalik Formations (Johnson et al., 1982, 1985) (Fig. 3.3). Dramatic changes in sedimentation patterns recorded in the magnetostratigraphy of the Himalayan foreland led Burbank et al. (1996b) to conclude that slip on the MBT began at ~11 Ma and was active during the Pleistocene. Whether or not the MBT was active during the Late Miocene/Pliocene reactivation of the MCT and activation of the MCT zone is not presently clear. The geomorphology of the Himalayan front suggests that the MBT was active until recently (Nakata, 1989). Molnar (1984) estimated a minimum displacement on the MBT of ~100 km, and Schelling and Arita (1991) suggested about half that amount.

Main Frontal Thrust The frontal ramp of the Himalayan thrust is the presently active Main Frontal Thrust (e.g., Schelling and Arita, 1991) which places Siwalik Group sediments atop Quaternary deposits of the Ganga basin (Fig. 3.3).

The Sedimentary Record The Kailas/Dazhuka-type conglomerate along the Indus-Tsangpo suture zone in southern Tibet typically contains abundant cobbles of Gangdese plutonic and volcanic rocks (Fig. 3.2). Near Mt. Kailas (Fig. 3.2), the conglomerate is documented to have developed a thickness of at least 3 km (Gansser, 1964) but in most other locations is currently no more than several hundred meters thick. Because deposition of this rock records the timing of arc uplift, establishing its age is pivotal. However, age assessment of this unit (Late Eocene to Miocene?) has been problematic (Gansser, 1964). Using thermochronometry to establish an upper limit for derivation from the Gangdese basement, and dating cross-cutting dikes to constrain a lower bound, the Kailas conglomerate has been dated at locations in west, central, and eastern Tibet (Harrison et al., 1993; Ryerson et al., 1995). At all three sites, the results are consistent with an Early Miocene (24–17 Ma) age of deposition indicating that rapid denudation in southern Tibet was underway at ca. 20 Ma.

$^{40}Ar/^{39}Ar$ dating of individual detrital K-feldspars from modern Tsangpo River (Fig. 3.2) sediments near Lhasa yields a clustering of ages between 15 and 20 Ma (Copeland and Harrison, 1990) that is consistent with an episode of rapid unroofing at that time. Similar analyses of detritus from Siwalik Group samples from Pakistan and Nepal are dominated by Early–Middle Miocene and Late Eocene ages (Harrison et al., 1993). The stratigraphy of the Bengal fan (Stow et al., 1989) as well as $^{40}Ar/^{39}Ar$ dating of individual detrital grains (Copeland and Harrison, 1990) suggest a significant episode of Early Miocene denudation with rapid

erosion continuing at various places within the collision zone to the present. The thickness of Indus fan sediments accumulated on the western continental margin of India indicates that rapid subsidence began there at ~25 Ma (Whiting and Karner, 1991).

Summary Many aspects of the thrust activity in the Himalaya and southern Tibet remain poorly understood, but some details have emerged. The Gangdese Thrust, the earliest identified crustal thickening event, is believed to have been active from 28–23 Ma (Yin et al., 1994). The thickening effects from Early Miocene slip on the Main Central Thrust are thought to have been responsible for production of the High Himalayan leucogranites which range in age from 24–19 Ma (Coleman and Parrish, 1995; Harrison et al., 1995b). The north-directed Renbu Zedong Thrust was active in the interval 19–11 Ma placing it out-of-sequence with respect o the GT and MCT, but fills a gap in time that is not otherwise represented by current estimates of the timing of other recognized thrust faults. Activity on the Main Boundary Thrust initiated at ~11 Ma and continued, perhaps intermittently, into the Pleistocene (Burbank et al., 1996b). The MCT fault was reactivated during the late Miocene (11–5 Ma) in an event which led to formation of the underlying MCT zone and the apparent Himalayan inverted metamorphic sequences. Shearing within the MCT zone appears to have continued into the Pliocene (5–2 Ma). Note that out-of-sequence thrusting is also common within the molasse basins of Pakistan (Burbank et al., 1996b).

We can estimate the elevation prior to the development of the Gangdese thrust in southern Tibet using equation (3.1). Given the estimated amount of shortening of 30%–40% (Yin et al., 1994) and the estimated ~15 km of denudation (Copeland et al., 1995), an estimate of the elevation of the southern Tibetan Plateau in the location of the Gangdese plutonic belt can be taken from Fig. 3.5 to be approximately 3 km. In other words, about 2 km would have been gained by crustal shortening subsequent to ~28 Ma. Note that in the model, we assume a uniform lithospheric shortening on a vertical profile without thermal erosion of the thickened lithosphere. If we include this mechanism, equation (3.2) predicts that ~4 km elevation has been gained since 28 Ma. Note that different mass schemes would be justified if the mechanism of shortening were un-

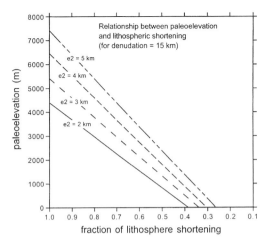

Fig. 3.5. Relationship between elevation prior to shortening and horizontal strain (defined as present length vs. original length) for a total denudation of 15 km (Eq. 3.2).

derplating or injection of India into the ductile lower crust.

A second constraint on the elevation at ca. 28 Ma comes from our expectation that the surface break of a thrust fault occurs at a break in topographic slope. For example, the surface exposure of the presently active Main Frontal Thrust (Fig. 3.3) is at an elevation of approximately 250 m, and the

Main Boundary Thrust, thought to have been active during the Pleistocene (Burbank et al., 1996b) is at <1 km elevation (Masek et al., 1994). By analogy, the surface rupture of the GT, then located approximately 50–100 km south of the suture, was very likely at an elevation of <1 km.

From the above calculations and geological constraints, we conclude that at ca. 30 Ma, the average elevation of the Himalaya was 1 km, and the Gangdese Shan was at an elevation of 3 ± 1 km.

Extensional Systems

Southern Tibetan Detachment System The Southern Tibetan Detachment System (STDS) is a down-to-the-north, low-angle normal fault that is traceable along the length of the Himalaya (Burg et al., 1984; Burchfiel et al., 1992) (Figs. 3.1 and 3.6). By dating crystallization ages of crosscutting leucogranite bodies, usually by U-Th-Pb dating of monazite (a light rare earth phosphate), it has been possible to constrain the timing of this feature at several locations. In the western Himalaya, a small leucogranite in the Zanskar region beneath and not cut by the STDS yields an age of 19 ± 1 Ma (Noble and Searle, 1995). Footwall $^{40}Ar/^{39}Ar$ cooling ages

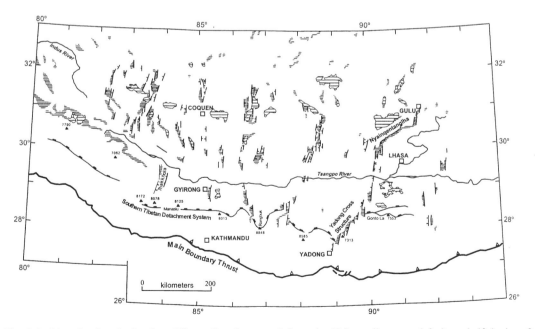

Fig. 3.6. Map showing the Southern Tibetan Detachment and the major N-S trending normal faults and rift basins of southern Tibet (modified from Armijo et al. (1986) and Burchfiel et al. (1992)). Normal faults are marked by thick solid lines with the dip direction indicated by the orthogonal ticks. Graben fill is represented by closely spaced horizontal lines.

between 18–15 Ma (T.M. Harrison, unpublished data) suggest that activity on the basal detachment was probably limited to the interval 20-17 Ma. In the central Himalaya, a phase of the Manaslu pluton which cuts the STDS has been dated at 23.1 ± 0.4 Ma (Harrison et al., 1995b) (Fig. 3.4). About 250 km east of Manaslu in the vicinity of Mt. Everest (Fig. 3.4), the Rongbuk granite cuts the STDS shear zone (Burchfiel et al., 1992). Harrison et al. (1995b) interpreted a crystallization age for this granite of 22 ± 1 Ma based on both Th-Pb ion microprobe dating of monazite and earlier U-Pb accessory mineral dating results. However, in the eastern Himalaya at Gonto La (Figs. 3.4 and 3.6), near the Tibet-Bhutan border, the detachment remained active until at least 12.5 ± 0.4 Ma (Edwards and Harrison, 1997). Note that the central and eastern High Himalaya are offset by ~70 km across the Yadong Cross Structure (Fig. 3.6). There are at least three possible explanations that explain this diachroneity.

The first is that the STDS represents a single, northward propagating normal fault system. If this were the case, then the contrast in age between the central and eastern Himalaya might simply reflect a rollback of the location of the basal detachment. Existing knowledge of the magnitude of offset on normal faults in the Tethyan metasediments (e.g., Burchfiel et al., 1992) is not sufficiently detailed for us to assess this hypothesis, leaving open the possibility that the STDS developed diachronously along strike. The second interpretation is that the topographic gradient required to permit STDS-type normal faulting (Burchfiel and Royden, 1985) occurred much later in the eastern Himalaya relative to the central Himalaya, possibly along the Kakhtang thrust which is documented only in Bhutan (Gansser, 1964). The third interpretation is that the north-directed RZT (see section Evolution of the Himalayan and Southern Tibetan Thrusts, p. 50) accommodated the displacement of the north-directed STDS in the eastern Himalaya (Fig. 3.3). Directly north of Bhutan, the RZT is known to have experienced >20 km of slip between 19–11 Ma (Quidelleur et al., 1997) permitting the possibility that synchronous movement occurred on the two structures. Results of seismic reflection profiling in southern Tibet are consistent with the two faults connecting at depth (Zhao et al., 1993; Nelson et al. 1996).

Burg et al. (1984) first suggested that the STDS was a gravitational collapse structure in response to the thickening of Tethyan metasediments. Burchfiel and Royden (1985) also proposed that the development of the STDS was due to gravitational collapse, but specifically in response to crustal thickening of the Himalaya along the MCT. Although the MCT may not have been active during the middle Miocene, other features such as the MBT and Kakhtang thrust may have accommodated that role. Yin (1993) found that the state of stress within a thrust hanging wall is critically dependent on pore-fluid pressure along the fault. He suggested that anatexis within the MCT hanging wall provided a sink for volatiles evolved from the lower plate during metamorphism, thereby reducing the pore-fluid pressure along the thrust and creating stress conditions in the hanging wall favorable for normal faulting along the STDS. Note that both the Burchfiel and Royden (1985) and Yin (1993) models propose a causal relationship between slip on the STDS and MCT.

Alternatively, Harris and Massey (1994) proposed that the leucogranites were produced under vapor-absent conditions by the dehydration melting reaction mus + qtz = melt + alsil + ksp, and became mobile only after decompression associated with activity on the STDS markedly increased the melt fraction in which case there is no causal link. The many leucogranites cut by the STDS suggest that decompression melting via tectonic denudation may not provide a general explanation for High Himalayan anatexis. Recently, Harrison et al. (1997b) have shown that dehydration melting can occur along the Himalayan thrust in the absence of rapid decompression.

Which accommodation mechanism ultimately proves to have been dominant is less relevant to our discussion than the salient point that all the models described above share the view that the High and Tethyan Himalaya were rapidly elevated during the Early Miocene (24–17 Ma) followed by, or coeval with, a phase of extension. Note that in the case of simultaneous slip of similar magnitude on the MCT and STDS, neither elevation reduction nor surface uplift need have occurred during this phase. While the existing timing constraints are broadly consistent with some degree of overlap in the timing of the MCT and STDS, they do not conclusively support such a link.

East-West Extension   Seismic evidence indicates that the dominant strain regime on the Tibetan Plateau today is extension in the east-west direction (Molnar, 1988). East-west extension on the southern Tibetan Plateau (e.g., Armijo et al., 1986) has been accommodated by a series of generally N-S trending rifts throughout southern Tibet (Fig. 3.6). Development of these graben marks a significant shift in the state of stress within the Tibetan crust, and determining the timing of these features is recognized as an important step in understanding the evolution of the plateau. There is some agreement that this extension reflects spreading following the attainment of maximum sustainable elevation for the conditions extant in southern Tibet (e.g., Tapponnier et al., 1986; England and Houseman, 1989; Dewey, 1988). Stratigraphic and geomorphological relationships indicate that the graben-defining normal faults have been active in southern Tibet throughout the Pleistocene (Armijo et al., 1986).

The Nyainqentanghla is a NE-SW trending mountain range of unusually prominent relief in southern Tibet, located approximately 100 km NW of Lhasa (Fig. 3.6). The Nyainqentanghla bounds the western margin of the Yangbajian graben, a segment of the Yadong-Gulu rift (Armijo et al., 1986; Pan and Kidd, 1992). The Yadong-Gulu rift, the most prominent graben system in southern Tibet, is unusual in that although most Quaternary graben in southern Tibet strike N-S, the central portion of this rift strikes NE-SW (Fig. 3.6), likely controlled by preexisting structures in the Nyainqentanghla basement (Armijo et al., 1986). Within the framework of the orogenic collapse hypothesis (e.g., Dewey, 1988), the time of normal fault initiation marks the attainment of maximum elevation of the southern Tibetan Plateau. Knowing when the Tibetan Plateau reached its present size and elevation is relevant not only to geodynamics but also to global atmospheric circulation, the onset of the Asian monsoon, and factors controlling ecological evolution in Asia.

The eastern edge of the Nyainqentanghla massif is marked by a low angle detachment fault shear zone of amphibolite grade mylonites. The $^{40}Ar/^{39}Ar$ thermochronometric results from samples collected along two deeply incised valleys within the massif reveal that a rapid cooling event propagated from ~8 Ma in the core of the range to ~4 Ma within the high strain zone at the eastern boundary (Harrison et al., 1995a). Numerical simulations of slip on a normal fault which fit the thermal histories indicate that fault initiation occurred at 8 ± 1 Ma with an average slip rate between 8–4 Ma of ~3 mm/yr. Because the extension direction of the Yangbajian graben is representative of most rifts on the southern Tibetan Plateau, these data suggest that crustal thickness and elevation had reached values close to their present dimensions by 8 ± 1 Ma. Preliminary results from other N-S graben suggest initiation ages of 9 ± 1 Ma (Harrison et al., 1995a). Furthermore, the observation that the Yadong-Gulu rift cross-cuts (Edwards et al., 1996) a portion of the STDS that was active at 12 Ma (Edwards and Harrison, 1997) provides that age as a firm upper bound on the initiation age of rifting.

At the same time as normal faulting was apparently getting underway on the plateau, the Indo-Australian oceanic lithosphere was beginning to deform. A region of deformation and anomalous seismicity south of India has been interpreted as a diffuse plate boundary that initiated at 8.0–7.5 Ma (e.g., Curray and Munasinghe, 1989). Harrison et al. (1992) suggested that onset of this deformation was a consequence of the increased resistance gained by Tibet once it had attained maximum elevation. In this interpretation, folding and faulting of the Indo-Australian plate developed in response to the increased north-south compressional stress that resulted from plateau uplift. Molnar et al. (1993) subsequently assessed this force balance and concluded that an increase in the average elevation of Tibet from around 4 to 5 km would result in a change of stress from values below that needed to deform oceanic lithosphere to values that exceeded the threshold for buckling.

Because the presence of a high and large Tibetan Plateau appears to be an important driving force for the Asian monsoon, it follows that the timing of the intensification of the monsoon also bears on plateau uplift. Carbon isotopic data from pedogenic carbonates in the Siwalik Formation in Pakistan and Nepal indicate a shift from C_3 to C_4 type vegetation (see Appendix section Stable Isotope Environmental Signatures, p. 61) in the Himalayan foothills beginning at about 7.5 Ma (Quade et al., 1989, 1995; Harrison et al., 1993; Stern et al., 1994). Quade et al. (1989) proposed that this shift from forest to grassland was the result of intensification of the monsoon at about 7.5 Ma, a conjecture supported by changing planktonic assemblages during the late Miocene (11–6 Ma) and other indicators in Arabian Sea sediments (e.g., Kroon et al., 1991; deMenocal, 1995). The global nature of the C_3 to C_4 shift could reflect the radiation of grasses outward from southern Asia, but could also reflect a decrease in atmospheric CO_2 during the late Miocene (Cerling et al., 1993; Quade et al., 1995). Although retreat of the Paratethys has a potentially similar influence to Tibetan uplift in driving changes in the Asian monsoon (Ramstein et al., 1997), this sea had essentially disappeared prior to the C_3 to C_4 shift and development of the graben.

The three independent consequences of plateau uplift, east-west extension on the Tibetan Plateau, intraplate deformation of the Indo-Australia plate, and the intensification of the Asian monsoon, all point to the plateau having achieved something approaching its present extent and elevation by ~9 Ma (Harrison et al., 1992, 1995a; Molnar et al., 1993). Only after the southern Tibetan Plateau reached its maximum elevation would the first phenomenon be triggered, and the remaining two would require both high elevation and possibly a larger threshold area.

Coleman and Hodges (1995) argued that evidence of east-west extension adjacent to the High Himalayan Thakkhola graben implies that the Tibetan Plateau had attained its high mean elevation by 14 Ma. There is, in fact, considerable evidence for a phase of Middle Miocene (17–11 Ma) east-west extension that was entirely restricted to the High and Tethyan Himalaya (Yin et al., 1994; Edwards et al., 1996). A high but narrow (~200 km) mountain range has certainly existed from the Himalaya to the southern Gangdese Shan since at least the Early Miocene (~24 Ma) (Copeland et al., 1987, 1995; Harrison et al., 1992; Yin et al., 1994) but is unlikely to have substantially perturbed climate beyond a local rain shadow. Features such as the Thakkhola graben, which clearly does not extend across the suture zone onto the Tibetan Plateau (Rothery and Drury, 1984; Armijo et al., 1986; Mercier et al., 1987), may reflect a response to this localized relief, or, more likely, to the spreading of the Himalayan arc (Armijo et al., 1986).

Although the entire plateau is presently in extension, it is only in the southern half that the rift valleys are well developed (England and Houseman, 1989). The southern Tibetan graben cannot be explained as a consequence of arc spreading as the syntaxes at ca. 8 Ma (Dewey et al., 1989) were located well to the south of the northern extent of rifting. Within the orogenic collapse hypothesis, there are several reasons why we might expect rifting to be restricted to southern Tibet. Kong (1995) found that the present orientation of the Indian indenter, the distribution of elevation within Tibet, and extrusion occurring at the eastern boundary predicts high extensional strain rates in southern Tibet and relatively low values in the north. In essence, the traction on the Himalayan thrust was found to be insufficient to prevent rifting in southern Tibet whereas the magnitude in northern Tibet could be taken up by continuum processes. Development of the south graben could reflect accretion of Indian material to southern Asia which would lead to higher extensional strains than farther north. Alternatively, the combined effects of internal drainage and a thin elastic upper crust in northern Tibet (due to higher heat flow) could effectively conceal the development of rift valleys there.

A balance between the current convergence rate of ~15 mm/yr and the ~1%/Ma extension rate has possibly resulted in the lowering of the average elevation of the Himalaya and Tibet from ~5500 m to the present value of 5020 m (Armijo et al., 1986; Fielding, 1996).

Summary

The thickened crust resulting from the Indo-Asian collision is our best natural laboratory in which to study the interaction of continental crust, the subjacent lithosphere, and asthenosphere during continent-continent collision. Although long appreciated as the best modern example of a continental collision orogen, and more recently as a contributing factor influencing global climate during the Neogene, our knowledge of the uplift history of the plateau is only now coming into focus.

Models advanced to explain the evolution of the present distribution of crust within the Indo-Asian orogen include those that predict wholesale uplift (mantle delamination, delayed under-plating), progressive growth (Indian under-thrusting, Asian under-thrusting, continental injection), lateral responses (orogenic collapse, horizontal extrusion), and inheritance of an elevated terrane (multiple collision, intra-arc thickening). Investigations of the timing and magnitude of deformation in Tibet and surrounding regions provide some constraints on the crustal thickening and elevation histories that help select among these models. However, the pattern of crustal displacements in response to the collision is inconsistent with most permutations of these single mechanisms but instead requires a specific, time-dependent transfer among several of these processes, often with multiple mechanisms operating simultaneously. The parameters that appear most important in dictating which mechanisms are dominant at any one time are: the location and geometry of pre-existing lithospheric weakness, the distribution of topography before and during the collision, the geometry of the indenter and extruded blocks, the magnitude of boundary stresses, and the age of the lithosphere (Kong et al., 1997).

Our summary of the results of the investigations we have reviewed in this Chapter are shown schematically in Figure 3.7, a map of the present configuration of southern Asia. The considerable uncertainties involved in tectonic reconstructions

of Asia, particularly the precollisional history, do not yet warrant our presenting these results palinspastically restored to their predeformational positions. However, our knowledge of the elevation history of southern Asia is probably sufficiently known over the past ~8 Ma, and the geographic locations sufficiently close to present values (within 4° latitude), to allow us to directly use Figure 3.7 to test simple climate evolution models. For earlier periods, it may be possible in certain cases to approximate paleogeographic positions from recent reconstructions of Asia (e.g., Sengor and Natalin, 1996; Yin and Nie, 1996).

The Tibetan Plateau began to form locally in response to the collision of the Lhasa block with southern Asia during the Early Cretaceous, particularly in the southern Tibet. This event is well-documented in the northern Lhasa block where a fold and thrust belt developed between 144–110 Ma and remained substantially elevated until the onset of the collision of India with Asia. The first evidence of uplift related to thickening due to the collision of India with Asia is deformation and clastic sediment

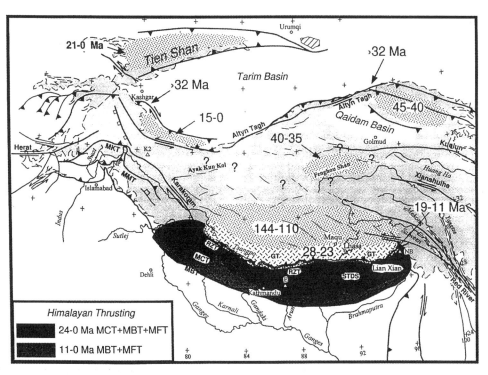

Fig. 3.7. Summary map showing the approximate timing and location of tectonic events resulting in crustal thickening and uplift in the Indo-Asian collision zone. MBT, Main Boundary Thrust; MCT, Main Central Thrust; STDS, Southern Tibetan Detachment System; RZT, Renbu Zedong Thrust; GT, Gangdese Thrust; MKT, Main Karakorum Thrust; MMT, Main Mantle Thrust; E, Mt. Everest; K, Mt. Kailas; NB, Namche Barwa; NP, Nanga Parbat.

production in the Nan Shan and Fenghuo Shan regions of northern Tibet which began at about 45–
32 Ma. Possibly in response to this thickening, the
left-lateral Red River fault was initiated at 35 Ma
permitting the eastward extrusion of Indochina until
~17 Ma. This had the effect of reducing the magnitude of crustal thickening in Tibet, perhaps by accommodating as much as one-half of Indo-Asian
convergence during the Oligocene (37–24 Ma).
Thin-skinned thrusting was occurring in the
Tethyan Himalaya throughout the Paleogene while
cover rocks on the northern margin of the Indian
shield experienced crustal thickening. Although little crustal thickening of Tibet within fold and thrust
belts can be documented for the Eo-oligocene (58–
24 Ma), mass balance considerations all but require
that channeled flow in the ductile lower crust (e.g.,
Zhao and Morgan, 1985; Bird, 1991) has continuously thickened the crust, uniformly raising much
of Tibet to between 1 and 3 km in elevation.

Thus by the beginning of the Late Oligocene (30
Ma), a relatively low (~2 km) Tibetan Plateau
(except, perhaps, along the collision zone between
the Lhasa and Qiangtang blocks) was likely in existence while the Tethyan and proto-Himalaya were
topographically subdued and the present Himalayan
range had not yet begun to develop. At ca. 28 Ma,
crustal thickening began in southern Tibet along
the Gangdese Thrust, moving southward to the Himalaya shortly thereafter in a series of south-directed thrusts which appear to be splays of the same
decollement. Recall that since we can estimate the
amount of crustal shortening and denudation, we can
infer the magnitude of surface uplift. The Main Central Thrust is thought responsible for generating the
24–9 Ma Himalayan leucogranites which suggests
an Early Miocene age for the initial phase of MCT
fault activity. This general pattern of propagation
toward the foreland was interrupted by the north-directed Renbu Zedong Thrust which was active in
southern Tibet between 19–11 Ma. Extension in the
High Himalaya along the South Tibetan Detachment
System occurred concurrently with slip on the MCT
and RZT. The Tien Shan thrusts and thickening in
the western Kun Lun were also initiated during the
Early Miocene (~20 Ma). By ca. 20 Ma, the
Gangdese Shan, Tethyan Himalaya, and High Himalaya were now likely a significant climatic barrier, with an average elevation of perhaps 4–5 km,
behind which stood a large but still relatively sub-

dued Tibetan Plateau. At this point, thickening deformation jumped the Tarim basin to the Tien Shan.

Thickening in northwest Tibet, apparently related to transtension along the Altyn Tagh fault,
began during the middle Miocene (~15 Ma). Subsequent to initiation of the Main Boundary Thrust at
~11 Ma, a broad zone of deformation beneath the
MCT fault was active at some time between 10–4
Ma producing the classic Himalayan inverted metamorphic sequences. Recognizing the juxtaposition
of hanging wall gneisses partially melted at ca. 22
Ma and young (ca. 7 Ma), lower grade footwall
rocks across this tectonically telescoped section
renders unnecessary appeals for high shear stress
during faulting in order to create the leucogranite
melts (e.g., England et al., 1992).

By ~9 Ma, the Tibetan Plateau had attained an
average elevation of approximately 5 km and began
to differentially extend E-W in a set of N-S trending
graben. Although the entire plateau is in extension,
these features are particularly well developed in
southern Tibet, perhaps because of accretion of India to southern Asia leading to higher extensional
strains.

The discrete changes we infer in the manner in
which continuous Indo-Asian convergence was accommodated over the past 55 Ma provides a cau-
tionary tale against interpreting episodic phenomena in the geological record in terms of discontinu-
ous processes, particularly in light of the growing
appreciation that complex physical systems driven
by structureless inputs can exhibit highly intermittent dynamics. Perhaps the clearest lesson emerging from our study of the Indo-Asian collision is
that the continental lithosphere's complex history
and geometry (Kong et al., 1997) exerts a powerful
control on continuous plate convergence being
manifested in the geological record as episodic
phenomena.

Appendix: Methodologies

Thermochronometry

Thermochronometry is the use of isotopic variations that are preserved by mineral chronometers
during cooling to reconstruct geological temperature histories. Because virtually all tectonic processes involve heat flow disturbance, thermal histories can shed light on a variety of geophysical pro-

cesses (e.g., denudation, thrusting, rifting, magmatism) that might otherwise be unobtainable.

The ^{40}K to ^{40}Ar decay system has proven to be a particularly useful tool for thermochronometry because of the incompatibility and mobility of the daughter product in silicate hosts. At very high temperatures in the crust, Ar diffuses rapidly from the host mineral. However, as the rock cools, the diffusivity (D) decreases according to the Arrhenius relationship, $D = D_0\,e^{-E/RT}$, where D_0 is a constant, E is the activation energy, R is the gas constant, and T is the absolute temperature (E and D_0 can be measured directly through experiment). At some point, the daughter product is moving so slowly that it is effectively trapped in the mineral structure. The transition between these two states is given by the closure temperature (Dodson, 1973) which is defined as:

$$T_c = E/R \ln ((A \cdot R \cdot T_c^2 \cdot D_0/r^2)/(E \cdot dT/dt)) \qquad (A.1)$$

where A is a constant related to the diffusion geometry, r is the mineral radius, and dT/dt is the cooling rate of the rock. A mineral age does not necessarily date the time of a rock forming event but rather tells us when the rock passed through the temperature given by the above expression. The simplest application of the closure concept is to adopt generalized kinetic parameters for minerals commonly dated by, say, the ^{40}Ar/^{39}Ar variant of K-Ar dating (see McDougall and Harrison, 1988) such as micas and amphiboles, and relate the bulk mineral age to T_c. By linking the ages and closure temperatures with a variety of minerals that have contrasting daughter product retentivities (including other systems such as the thermal stability of fission tracks in apatite), this method can yield estimates of geological cooling rates. The thermal degassing of Ar from K-feldspar represents a special case of the closure process from which more detailed thermal histories can be obtained.

There are two distinct sources of information available from a K-feldspar ^{40}Ar/^{39}Ar step-heating experiment: an age spectrum and an Arrhenius plot derived by plotting model diffusion coefficients (D/r^2) versus the temperature of discrete laboratory heating steps. Ages are calculated from the flux of ^{40}Ar relative to neutron irradiation produced ^{39}Ar (a proxy for the parent ^{40}K) permitting the spatial distribution of the ^{40}Ar to be determined. Most

^{40}Ar/^{39}Ar age spectra for slowly cooled K-feldspars are significantly different from that expected from a single diffusion domain size and yield nonlinear Arrhenius plots that are inconsistent with diffusion from domains of equal size. Assuming that natural K-feldspars contain a discrete distribution of domain sizes (Lovera et al., 1989), an internally consistent explanation for the commonly observed features of K-feldspar age spectra and their associated Arrhenius plots is obtained. This approach, termed the multi-diffusion domain (MDD) model, permits recovery of continuous thermal history segments potentially providing much higher resolution than can be obtained by simply linking bulk ages and closure temperatures.

Stable Isotope Environmental Signatures

Climate is the primary control on floral assemblages. Because plants synthesize carbohydrates from CO_2 using different photosynthetic pathways, we expect the variable isotopic fractionations produced by these pathways to be preserved in fossil soils permitting inferences of environmental change. The rate of carbon fixation is largely controlled by the activity of the enzyme ribulose biphosphate carboxylase and the concentration of CO_2 at the entry point to the photosynthetic cycle. Plants such as trees and shrubs are called C_3 plants because carbon fixation occurs using a three carbon molecule (the Calvin cycle). Most grasses are C_4 plants which utilize a more complex photosynthetic pathway in which the first product of carbon fixation is malate, a four carbon molecule. In this pathway, a mesophyll cell first fixes CO_2 from the atmosphere which is then passed on to the malate in a bundle-sheath cell. The effect of the mesophyll cell is to preconcentrate CO_2 giving the C_4 plant greater tolerance to low levels of atmospheric CO_2, but at the cost of being less energy efficient. The malate molecule reacts to form the C_3 compound plus pyruvate, releasing CO_2 which is then fixed by the Calvin cycle.

The carbon isotopic composition of buried soil carbonate provides information on the nature of vegetation growing during soil formation because the C_3 and C_4 chemical pathways fractionate carbon isotopes differently (Quade et al., 1995). The three-carbon molecule utilized by C_3 plants tends to favor the lighter isotope resulting in $\delta^{13}C_{PDB}$ values that average -27 ± 6‰. The four-carbon molecule used by C_4 plants is isotopically less selective yielding $\delta^{13}C_{PDB}$ values of about -13. The carbon isotopic composition of soil carbonate is set by the relative contribution of C_3 and C_4 plants to the soil organics and the ~15‰ $\delta^{13}C$ enrichment between carbonate and soil CO_2. Of that ~15‰, +4.4‰ is due to a kinetic effect during diffusion and about +11‰ is an equilibrium isotope fractionation between CO_2 and carbonate (Quade et al., 1995). Thus paleosols derived from soils hosting only C_3 and C_4 plants are characterized by $\delta^{13}C$ values of about -12‰ and +2‰, respectively.

A second approach involves the recognition that the oxygen isotope composition of meteoric waters is recorded by soil carbonates (Cerling and Quade, 1993) as the isotopic composition of soil water is largely determined by the isotopic composition of local meteoric water. The formation of mountain belts can potentially shift the $\delta^{18}O$ of leeward precipitation towards highly negative values as a result of the rain shadow effect and thus detection of such a shift potentially has implications to tectonics (e.g., Chamberlain et al., 1996). Furthermore, because the $\delta^{18}O$ and δD ratios of precipitation are strongly dependent on ambient temperature (Craig, 1961), inferences about elevation are possible through the adiabatic lapse rate.

Cosmic Ray Exposure Dating

Cosmic ray protons approaching the earth hit atomic nuclei in the upper atmosphere producing a variety of particles such as protons, pions, and muons. These generally relativistic secondary particles cascade toward the earth undergoing collisions with air nuclei that produce a particle shower moving in approximately the direction of the incident cosmic ray. Although most of these particles are eventually stopped, the surviving particles (mostly neutrons) that contact the surface strip nucleons in the upper 0.5–1 m of rock (see Cerling and Craig, 1994). These spallation reactions produce distinctive cosmogenic nuclides. For example, ^{10}Be atoms are produced near the Earth's surface by neutron spallation of oxygen, aluminum and silicon. Assuming that the cosmic ray flux is constant, or that its time-dependence is known, the production rate of cosmogenic nuclides varies with both rock composition (i.e., abundance of target nuclei)

and position. Position is important because production rates vary with geomagnetic latitude (the deflection of cosmic ray protons along magnetospheric equipotential lines results in partial shielding of latitudes <60°), the portion of sky that is exposed, depth from the surface, and altitude. This last parameter holds potential for the assessment of paleoelevation.

Cosmic ray collisions with atmospheric gases cause cosmogenic production rates to be attenuated at greater atmospheric depths. Thus rocks exposed on mountain tops receive a higher flux than those at sea level. For both ^{10}Be and ^{26}Al, a 250 m increase in altitude raises production rates in rocks by ~15%. However weathering and erosion can complicate interpretation of surface exposure ages. The greatest potential of this method may be realized when two isotopic systems are used in concert, allowing both exposure age and erosion rate to be constrained (Lal, 1991). To use cosmogenic isotope abundance as a paleoaltimeter, the variation in the altitude dependent parameter, which is the production rate, must be constrained along with the effects of erosion. The sample age must be determined independently. Even if such a setting can be defined, the actual increase in production rate during uplift is likely to be of the same magnitude as the current uncertainties in known production rates (~10%). For instance, if a surface of known age is uplifted at a rate of 2 mm/yr for 100,000 yr, the resulting uplift is 200 m. This corresponds to an integrated (over the uplift path) increase in production rate of approximately 6%.

A potentially more tractable application of cosmogenic surface dating to topographic change is in dating geomorphic surfaces where uplift can be determined by an independent measurement. For instance, the deformation of fluvial terraces developed along the piedmont of active mountain ranges can be measured to yield cumulative uplift and/or shortening. If this deformation can be associated with a specific datable surface, a rate (potentially comprised of both incision and uplift) may be inferred (Molnar et al., 1994; Burbank et al., 1996a).

Paleoecology

Numerous paleobotanical and paleontological studies throughout Tibet have been conducted by Chinese investigators (e.g., Li et al., 1981, 1986; Xu,

1981). By comparing nonmarine fossil floral and faunal associations with those in contemporary settings, these studies derive apparently quantitative estimates of surface elevations. The results of these studies generally suggest that uplift of the Tibetan Plateau is an essentially Pleistocene phenomenon. At least in the Himalaya and southern Tibet, this conclusion is inconsistent with our knowledge of the lithospheric thickening history. Although useful as a general indicator of prevailing climate, translating fossil remains into quantitative estimates of elevation is fraught with considerable uncertainty (England and Molnar, 1990a). A potentially superior approach is to calibrate paleotemperature from leaf physiognomy (Wolfe, 1990,1992). For an assumed sea level temperature and lapse rate, paleotemperature derived from this method can be translated into quantitative estimates of paleoelevations (Gregory and Chase, 1992). This approach has not yet been applied to the elevated terranes of the Indo-Asian collision.

Geological Relationships

Several geological relationships provide clear constraints on the elevation history of continental rocks. For example, marine shelf carbonate rocks are both directly dateable by paleontological means and must have been close to sea level when they formed. The sedimentological and deformational records of orogeny provide broad evidence for both positive topography and subsident basins. Knowledge of the present elevation and the magnitude of crustal/lithospheric shortening and postorogenic denudation (from petrological or thermochronological methods) permits an estimate of crustal thickness and elevation histories (see section Pre-thickening Elevation, p.43).

Geomorphology

Geomorphological investigations, such as segmentation and uplift of dated surfaces (e.g., Molnar et al., 1994; Avouac et al., 1993; Summerfield, 1991; Matthews, 1974) can yield quantitative constraints on uplift histories. This approach, however, requires the conjunction of several special circumstances, including the preservation of a datable surface, which limits the general application of this approach. For example, Shackleton and Chang

(1988) argued for the existence of a peneplanation surface in the Gangdese Shan near Lhasa, but problems with age assignment makes it difficult to use this information in reconstructing the elevation history of that region. As a consequence, this approach has not provided a great deal of information regarding the uplift history of the collision zone aside from the obvious fact that the present topography can be explained by virtually all evolutionary models (Fielding, 1996).

Acknowledgments

We thank our colleagues Peter Copeland, Mike Edwards, Marty Grove, Bill Kidd, Patrick Le Fort, Herve Leloup, Gilles Peltzer, Jay Quade, and Paul Tapponnier for contributing to the body of work that led to this synthesis. We thank Tom Crowley and Kevin Burke for inviting our participation in this book, and to them and three anonymous reviewers for offering constructive reviews of the manuscript. This work was supported by grants from the National Science Foundation and Department of Energy.

References

Allen, C. R., A. R. Gillespie, Y. Han, K. E. Sieh, B. Zhang, and C. Zhu, 1984, Red River and associated faults, Yunnan province, China: Quaternary geology, slip rates, and seismic hazard, *Geol. Soc. Am. Bull., 95*, 686–700.

Alsdorf, D., et al., 1995, Crustal deformation of the Lhasa block, Tibet Plateau, from INDEPTH seismic reflection profiling, *Geol. Soc. Am. Abs. w. Progs., 27*, A334.

Argand, E., 1924, La Tectonique de L' Asie, *Comptes Rendus, Proc. 13th Int. Geol. Congress, 1*, 171–372.

Arita, K., 1983, Origin of the inverted metamorphism of the Lower Himalaya, central Nepal, *Tectonophysics, 95*, 43–60.

Armijo, R., P. Tapponnier, J. L. Mercier, and T.-L. Han, 1986, Quaternary extension in southern Tibet: Field observations and tectonic implications, *J. Geophys. Res., 91*, 13,803–13,872.

Armijo, R., P. Tapponnier, and T. Han, 1989, Late Cenozoic right-lateral strike-slip faulting in southern Tibet, *J. Geophys. Res., 94*, 2787–2838.

Arnaud, N. O., P. Vidal, P. Tapponnier, P. Mate, and W. M. Deng, 1992, The high K_2O volcanism of northwestern Tibet: Geochemistry and tectonic implications, *Earth Planet. Sci. Lett.*, *111*, 351–367.

Avouac, J.-P. and G. Peltzer, 1993, Active tectonics in southern Xinjiang, China: Analysis of terrace rises and normal fault scarp degradation along the Hotan-Qira fault system, *J. Geophys. Res.*, *98*, 21,773–21,807.

Avouac, J.-P., P. Tapponnier, M. Bai, H. You, and G. Wang, 1993, Active thrusting and folding along the northern Tien-Shan and late Cenozoic rotation of the Tarim relative to Dzungaria and Kazakhstan, *J. Geophys. Res.*, *98*, 6755–6804.

Barron, E. J., and W. M. Washington, 1984, The role of geographic variable in explaining paleoclimates: Results form Cretaceous climate model sensitively studies, *J. Geophys. Res.*, *89*, 1267–1279.

Besse, J., and V. Courtillot, 1988, Paleogeographic maps of the continents bordering the Indian Ocean since the Early Jurassic, *J. Geophys. Res.*, *93*, 11791–11808.

Bird, P., 1978, Initiation of intracontinental subduction in the Himalaya, *J. Geophys. Res.*, *83*, 4975–4987.

Bird, P., 1991, Lateral extrusion of lower crust from under high topography in the isostatic limit, *J. Geophys. Res.*, *96*, 10,275–10,286.

Bosart, P., and R. Ottiger, 1989, Rocks of the Murree Formation in northern Pakistan: Indicators of a descending foreland basin of late Paleocene to middle Eocene age, *Eclogae Geol. Helv.*, *82*, 133–165.

Bouchez, J. L., and A. Pêcher, 1981, Himalayan Main Central Thrust pile and its quartz-rich tectonites in central Nepal, *Tectonophysics*, *78*, 23–50.

Briais, A., P. Patriat, and P. Tapponnier, 1993, Updated interpretation of magnetic anomalies and seafloor spreading stages in the South China Sea: Implications for the Tertiary tectonics of SE Asia, *J. Geophys. Res. 98*, 6299–6328.

Brown, L. D., et al., 1996, Bright spots, structure, and magmatism in Southern Tibet from INDEPTH seismic reflection profiling, *Science, 274*, 1688–1691.

Brunel, M., N. Arnaud, P. Tapponnier, Y. Pan and Y. Wang, 1994, Kongur Shan normal fault: Type example of mountain building assisted by extension (Karakorum fault, eastern Pamir), *Geology, 22*, 707–710.

Brunel, M., and J. R. Kienast, 1986, Étude pétro–structurale des chevauchments ductiles himalayens sur la transersale de L'Everest–Makalu (Népal oriental), *Can. J. Earth Sci.*, *23*, 1117–1137.

Buck, W. R., and M. N. Toksoz, 1983, Thermal effects of continental collision: Thickening a variable viscosity layer, *Tectonophysics, 100*, 53–69.

Burbank, D. W., J. Leland, E. Fielding, R. S. Anderson, N. Brozovic, M. R. Reid, and C. Duncan, 1996a, Bedrock incision, rock uplift and threshold hillslopes in the northwestern Himalayas, *Nature, 379*, 505–510.

Burbank, D.W., R. A. Beck, and T. Mulder, 1996b, The Himalayan foreland basin, in *The Tectonics of Asia*, A. Yin and T.M. Harrison (eds.), pp. 149-188, Cambridge University Press, New York.

Burchfiel, B. C., and L. H. Royden, 1985, North–south extension within the convergent Himalayan region, *Geology, 13*, 679–682.

Burchfiel, B. C., and L. H. Royden, 1991, Tectonics of Asia 50 years after the death of Emile Argand, *Ecologae Geol. Helv.*, *84*, 599–629.

Burchfiel, B. C., Q. Deng, P. Molnar, L. H. Royden, Y. Wang, P. Zhang, and W. Zhang, 1989, Intracrustal detachment with zones of continental deformation, *Geology, 17*, 748–752.

Burchfiel, B. C., Z. Chen, K. V., Hodges, Y. Liu, L. H., Royden, C. Deng, and J. Xu, 1992, The South Tibetan Detachment System, Himalayan orogen: Extension contemporaneous with and parallel to shortening in a collisional mountain belt, *Geol. Soc. Am. Spec. Pap.*, *269*, 1–41.

Burg, J. P. (compiler), 1983, *Carte geologique du sud du Tibet, scale, 1:500,000*, Cont. Natl. de la Rech. Sci., Paris.

Burg, J. P., M. Brunel, D. Gapais, G. M. Chen, and G. H. Liu, 1984, Deformation of leucogranites of the crystalline Main Central Sheet in southern Tibet (China), *J. Struct. Geol.*, *6*, 535–542.

Burg, J. P., A. Leyreloup, J. Giaradeau, and G. M. Chen, 1987, Structure and metamorphism of a tectonically thickened continental crust: The Yalu Tsangpo suture zone (Tibet), *Phil. Trans. R. Soc. Lond.*, *A321*, 67–86.

Cerling, T. E., and H. Craig, 1994, Geomorphology and in situ cosmogenic isotopes, *Annu. Rev. Earth Planet. Sci., 22*, 273–317.

Cerling, T. E., and J. Quade, 1993, Stable carbon and oxygen isotopes in soil carbonates, *Geophys. Monogr., 78*, 217–231.

Cerling, T. E., Y. Wang, and J. Quade, 1993, Global ecologic changes in the Miocene: Expansion of the C_4 ecosystem, *Nature, 361*, 344–345.

Chamberlain, C. P., and P. K. Zeitler, 1996, Assembly of the crystalline terranes of the northwestern Himalaya and Karakorum, northwestern Pakistan, in *The Tectonics of Asia*, A. Yin and T. M. Harrison (eds.), pp. 138–148, Cambridge University Press, New York.

Chamberlain, C. P., M. A. Poage, R. C. Reynolds, and D. Craw, 1996, The topographic evolution of the southern Alps, New Zealand, from the isotopic analysis of clay, *Geol. Soc. Am. Abs. w. Progs., 28*, A249.

Chang C., et al., 1986, Preliminary conclusions of the Royal Society and Academia Sinica 1985 geotraverse of Tibet, *Nature, 323*, 501–507.

Chen, Y., et al., 1991, Paleomagnetic study of Mesozoic continental sediments along the northern Tien Shan (China) and heterogeneous strain in central Asia, *J. Geophys. Res., 96*, 4065–4082.

Chen, Y., V. Courtillot, J. P. Cogne, J. Besse, Z. Yang, and R. Enkin, 1993, The configuration of Asia prior to the collision of India: Cretaceous paleomagnetic constraints, *J. Geophys. Res., 98*, 21,927–21,941.

Chinese State Bureau of Seismology (CSBS), 1992, *The Altyn Tagh Active Fault System*, Seismology Publishing House, Beijing.

Clark, S. P., and E. Jäger, 1969, Denudation rate in the Alps from geochronologic and heat flow data, *Am. J. Sci., 267*, 1143–1160.

Cobbold, P. R., and P. Davy, 1988, Indentation tectonics in nature and experiment, 2. Central Asia, *Bull. Geol. Inst. Uppsala, 14*, 143–162.

Colchen, M., P. Le Fort, and A. Pêcher, 1986, *Annapurna-Manaslu-Ganesh Himal notice de la carte geologiquie au 1/200.00e Bilingual edition: French–English*. Centre National de la Recherche Scientifique Paris.

Coleman, M. E., and R. R. Parrish, 1995, Constraints on Miocene high-temperature deformation and anatexis within the Greater Himalaya

from U–Pb geochronology, *EOS, 76*, F708.

Coleman, M., and K. V. Hodges, 1995, Evidence for Tibetan Plateau uplift before 14 Myr ago from a new minimum estimate for east–west extension, *Nature, 374*, 49–52.

Copeland, P., and T. M. Harrison, 1990, Episodic rapid uplift in the Himalaya revealed by $^{40}Ar/^{39}Ar$ analysis of detrital K–feldspar and muscovite, Bengal Fan, *Geology, 18*, 354–357.

Copeland, P., T. M. Harrison, W.S.F. Kidd, X. Ronghua and Z. Yuquan, 1987, Rapid early Miocene acceleration of uplift in the Gandese Belt, Xizang-southern Tibet, and its bearing on accommodation mechanisms of the India–Asia collision, *Earth Planet. Sci. Lett., 86*, 240–252.

Copeland, P., T. M. Harrison, K. V. Hodges, P. Maréujol, P. LeFort and A. Pêcher, 1991, An Early Pliocene thermal perturbation of the Main Central Thrust, Central Nepal: Implications for Himalayan tectonic, *J. Geophys. Res., 96*, 8475–8500.

Copeland, P., T. M. Harrison, P. Yun, W.S.F. Kidd, M. K. Roden, and Y. Zhang, 1995, Thermal evolution of the Gangdese batholith, southern Tibet: A history of episodic unroofing, *Tectonics, 14*, 223–236.

Coulon, C., H. Maluski, C. Bollinger, and S. Wang, 1986, Mesozoic and Cenozoic volcanic rocks from central and southern Tibet: $^{39}Ar/^{40}Ar$ dating, petrological characteristics and geodynamical significance, *Earth Planet. Sci. Lett., 79*, 281–302.

Craig, H., 1961, Isotopic variations in meteoric waters, *Science, 133*, 1702–1705.

Craig, P., A. Yin, and S. Nie, 1994, Reconstruction of Cenozoic depositional systems in the southern Tian Shan, NW China: Implications for timing of deformation, *Geol. Soc. Am. Abstr. w. Progr., 26*, 462.

Curray, J. R., and T. Munasinghe, 1989, Timing of intraplate deformation, northeastern Indian Ocean, *Earth Planet. Sci. Lett., 94*, 71–77.

DeMenocal, P. B., 1995, Plio–Pleistocene African climate, *Science, 270*, 53–59.

Dewey, J. F., 1988, Extensional collapse of orogens, *Tectonics, 7*, 1123–1139.

Dewey, J. F., and K. Burke, 1973, Tibetan, Variscan and Precambrian basement reactivation: products of continental collision, *J. Geol., 81*, 683–692.

Dewey, J. F., R. M. Shackelton, C. Chang, and Y.

Sun, 1988, The tectonic evolution of the Tibetan Plateau, *Phil. Trans. R. Soc. Lond., A327*, 379–413.

Dewey, J. F., S. Cande, and W. C. Pitman, 1989, Tectonic evolution of the India–Eurasia collision zone, *Eclogae Geol. Helv., 82*, 717–734.

Dodson, M. H., 1973, Closure temperature in cooling geochronological and petrological systems, *Cont. Min. Petrol., 40*, 259–274.

Dürr, S. B., 1996, Provenance of Xigaze fore–arc basin clastic rocks (Cretaceous, south Tibet), *Geol. Soc. Am. Bull., 108*, 669–691.

Edwards, M. A., and T. M. Harrison, 1997, When did the roof collapse? Late Miocene N–S extension in the High Himalaya revealed by Th–Pb monazite dating of the Khula Kangri granite, *Geology, 25*, 543–546.

Edwards, M. A., W.S.F. Kidd, J. Li, Y. Yue, and M. Clark, 1996, Multi stage development of the southern Tibet detachment system near Khula Kangri. New data from Gonto La, *Tectonophysics, 260*, 1–20.

England, P., and G. Houseman, 1986, Finite strain calculations of continental deformation 2. Comparison with the India–Asia collision zone, *J. Geophys. Res., 91*, 3664–3676.

England, P., and G. Houseman, 1989, Extension during continental convergence, with application to the Tibetan Plateau, *J. Geophys. Res., 94*, 17,561–17,569.

England, P., and P. Molnar, 1990a, Surface uplift, uplift of rocks, and exhumation of rocks, *Geology, 18*, 1173–1177.

England, P., and P. Molnar, 1990b, Right-lateral shear and rotation as the explanation for strike-slip faulting in eastern Tibet, *Nature, 344*, 140–142.

England, P., P. Le Fort, P. Molnar, and A. Pêcher, 1992, Heat sources for Tertiary metamorphism and anatexis in the Annapurna–Manaslu region, Central Nepal, *J. Geophys. Res.,* 97, 2107–2128.

England, P., and M. P. Searle, 1986, The Cretaceous–Tertiary deformation of the Lhasa block and its implications for crustal thickening in Tibet, *Tectonics, 5*, 1–14.

Fielding, E. J., 1996, Tibet uplift and erosion, *Tectonophysics, 260*, 55–84.

Gansser, A., 1964, *The Geology of the Himalayas*, Wiley Interscience, New York.

Gregory, K. M., and C. Chase, 1992, Tectonic significance of paleobotanically estimated climate and altitude of the late Eocene erosion surface, Colorado, *Geology, 20*, 581–585.

Hahn, D. G., and S. Manabe, 1975, The role of mountains in the south Asia monsoon circulation, *J. Atmos. Sci., 32*, 1515–1541.

Hao, Y. C. (Chief Editor), 1983, *The Cretaceous System of China: The Stratigraphy of China, Series No. 12*, Geol. Publishing House, Beijing.

Harris, N., and J. Massey, 1994, Decompression and anatexis of Himalayan metapelites, *Tectonics, 13*, 1537–1546.

Harrison, T. M., and G.K.C. Clarke, 1979, A model of the thermal effects of igneous intrusion and uplift as applied to the Quottoon Pluton, British Columbia, *Can. J. Earth Sci., 16*, 411–420.

Harrison, T. M., P. Copeland, W.S.F. Kidd, and A. Yin, 1992, Raising Tibet, *Science, 255*, 1663–1670.

Harrison, T. M., P. Copeland, S. Hall, J. Quade, S. Burner, T.P. Ojha, and W.S.F. Kidd, 1993, Isotopic preservation of Himalayan/Tibetan uplift, denudation, and climatic histories in two molasse deposits, *J. Geol., 101*, 159–177.

Harrison, T. M., P. Copeland, W.S.F. Kidd, and O. M. Lovera, 1995a, Activation of the Nyainqentanghla shear zone: implications for uplift of the southern Tibetan Plateau, *Tectonics, 14*, 658–676.

Harrison, T. M., K. D. McKeegan and P. LeFort, 1995b, Detection of inherited monazite in the Manaslu leucogranite by [208]Pb/[232]Th ion microprobe dating: Crystallization age and tectonic significance, *Earth Planet. Sci. Lett., 133*, 271–282.

Harrison, T. M., P. H. Leloup, F. J. Ryerson, P. Tapponnier, R. Lacassin, and W. Chen, 1996, Diachronous initiation of transtension along the Ailao Shan–Red River Shear Zone, Yunnan and Vietnam, in *The Tectonics of Asia*, A. Yin and T. M. Harrison (eds.), pp. 205-226, Cambridge University Press, New York.

Harrison, T. M., F. J. Ryerson, P. Le Fort, A. Yin and O. M. Lovera, 1997a, A Late Miocene–Pliocene origin for the Central Himalayan inverted metamorphism, *Earth Planet. Sci. Lett., 146*, E1–E8.

Harrison, T. M., O. M. Lovera, and M. Grove, 1997b, New insights into the origin of two contrasting Himalayan granite belts, *Geology*, in press.

Hauck, M. L., et al., 1995, Ramping of the Main Himalayan Thrust and development of the South Tibetan Detachment and Kangmar basement dome: INDEPTH reflection profiles in Southern Tibet, *Geol. Soc. Am. Abs. w. Progs.*, 27, A336.

Hendrix, M. S., S. A. Graham, A. R. Carroll, E. R. Sobel, C. L. McKnight, B. S. Schulein, and Z. Wang, 1992, Sedimentary record and climatic implications of recurrent deformation in the Tian Shan: Evidence from Mesozoic strata of the north Tarim, south Junggar and Turpan basins, northwest China, *Geol. Soc. Am. Bull.*, 104, 53–79.

Hendrix, M., T. Dumitru, and S. A. Graham, 1994, Late Oligocene-early Miocene unroofing in the Chinese Tian Shan: An early effect of the India–Asia collision, *Geology*, 22, 487–490.

Hodges, K. V., W. E. Hames, W. J. Olszewski, B. C. Burchfiel, L. H. Royden, and Z. Chen, 1994, Thermobarometric and ^{40}Ar/^{39}Ar geochronologic constraints on Eohimalayan metamorphism in the Dinggy area, southern Tibet, *Contrib. Mineral. Petrol.*, 117, 151–163.

Hodges, K. V., R. R. Parrish, and M. P. Searle, 1996, Tectonic evolution of the central Annapurna Range, Nepalese Himalayas, *Tectonics*, 15, 1264–1291.

Hubbard, M. S., 1996, Ductile shear as a cause of inverted metamorphism: example from the Nepal Himalaya, *J. Geol.*, 104, 493–499.

Hubbard, M. S., and T. M. Harrison, 1989, ^{40}Ar/^{39}Ar age constraints on deformation and metamorphism in the MCT Zone and Tibetan Slab, eastern Nepal Himalaya, *Tectonics*, 8, 865–880.

Isacks, B. L., 1988, Uplift of the central Andean Plateau and bending of the Bolivian orocline, *J. Geophys. Res.*, 93, 3211–3231.

Jia, C., H. Yao, G. Wi, and L. Li, 1991, Plate tectonic evolution and characteristics of major tectonic units of the Tarim basin, in *The Tarim Basin*, X. Tong and D. Liang (eds.), pp. 207–225, Xinjiang Scientific Publ. House, Urumuqi.

Johnson, N. M., N. D. Opdyke, G. D. Johnson, E. H. Lindsay, and R. A. K. Tahirkheli, 1982, Magnetic polarity stratigraphy and ages of Siwalik Group rocks of the Potwar Plateau, Pakistan, *Paleogeog. Paleoclimat. Paleoecol.*, 37, 17–42.

Johnson, N. M., J. Stix, L. Tauxe, P. F. Cerveny, and R. A. K. Tahirkheli, 1985, Paleomagnetic chronology, fluvial processes, and tectonic implications of the Siwalik deposits near Chinji Village, Pakistan, *J. Geol.*, 93, 27–40.

Kidd, W.S.F., Y. Pan, C. Chang, M. P. Coward, J. F. Dewey, A. Gansser, P. Molnar, R. M. Shackelton, and Y. Sun, 1988, Geological mapping of the 1985 Chinese–British Tibetan (Xizang–Qinghai) Plateau Geotraverse route, *Phil. Trans. R. Soc. Lond.*, A327, 287–305.

Kong, X., A. Yin, and T. M. Harrison, 1997, Evaluating the role of pre–existing weakness and topographic distributions in the Indo–Asian collision by use of a thin–shell numerical model, *Geology*, 25, 527–530.

Kong, X., 1995, Numerical modelling of the neo tectonics of Asia: A new spherical shell, finite–element method with faults, Ph.D. thesis, 227 pp., UCLA.

Kroon, D., T. Steens, and S.R. Troelstra, 1991, Onset of the monsoonal related upwelling in the western Arabian Sea as revealed by planktonic foraminifers, in W.L. Prell et al. (eds.), *Proceedings of the Ocean Drilling Project, Sci. Results*, 117, pp. 257–263, College Station, Texas (Ocean Drilling Program).

Kutzbach, J. E., P. J. Guetter, W. M. Ruddiman and W. L. Prell, 1989, Sensitivity of climate to late Cenozoic uplift in southern Asia and the American west: Numerical experiments, *J. Geophys. Res.*, 94, 18,393–18,407.

Kutzbach, J. E., W. L. Prell, and W. M. Ruddiman, 1993, Sensitivity of Eurasian climate to surface uplift of the Tibetan Plateau, *J. Geol.*, 101, 177–190.

Lal, D., 1991, Cosmic ray labeling of erosion surfaces; in situ nuclide production rates and erosion models, *Earth Planet. Sci. Lett.*, 104, 424–439.

Lacassin, R., P. H. Leloup, and P. Tapponnier, 1993, Bounds on strain in large Tertiary shear–zones of SE Asia from boudinage restoration, *J. Structural Geol.*, 15, 677–692.

Leeder, M. R., A. B. Smith, and J. Yin, 1988, Sedimentology, palaeoecology and palaeoenvironmental evolution of the 1985 Lhasa to Golmud Geotraverse, *Phil. Trans. R. Soc. Lond.*, A327, 107–143.

Le Fort, P., 1986, Metamorphism and magmatism during the Himalayan collision, in *Collision*

Tectonics, M.P. Coward and A.C. Ries (eds.), pp. 159–172, Geol. Soc. Spec. Publ., 19, London.

Le Fort, P., 1996, Evolution of the Himalaya, in *The Tectonics of Asia*, A. Yin and T.M. Harrison (eds.), pp. 95-106, Cambridge University Press, New York.

Le Fort, P., M. Cuney, C. Deniel, C. France–Lanord, S.M.F. Sheppard, B.N. Upreti, and P. Vidal, 1987, Crustal generation of the Himalayan leucogranites, *Tectonophysics, 134*, 39–57.

Lenardic, A., and W. M. Kaula, 1995, More thoughts on convergent crustal plateau formation and mantle dynamics with regard to Tibet, *J. Geophys. Res., 100*, 15,193–15,203.

Leloup, P. H., T. M. Harrison, F. J. Ryerson, C. Wenji, L. Qi, P. Tapponnier, and R. Lacassin, 1993, Structural, petrological and thermal evolution of a Tertiary ductile strike-slip shear zone, Diancang Shan (Yunnan, PRC), *J. Geophys. Res., 98*, 6715–6743.

Leloup, P. H., R. Lacassin, P. Tapponnier, D. Zhong, X. Lui, L. Zhang, and S. Ji, 1995, Kinematics of Tertiary left–lateral shearing at the lithospheric–scale in the Ailao Shan-Red River shear zone (Yunnan, China), *Tectonophysics, 251*, 3–84.

Le Pichon, X., M. Fournier, and L. Jolivet, 1992, Kinematics, topography, shortening, and extrusion in the India–Eurasia collision, *Tectonics, 11*, 1085–1098.

Li, J., B. Li, F. Wang, Q. Zhang, X. Wen, and B. Zhang, 1981, The process of the uplift of the Qinghai–Xizang plateau, in *Geological and Ecological Studies of Qinghai–Xizang (Tibet) Plateau 1, Geology, Geologic History and Origin of the Qinghai–Xizang plateau*, pp. 111–118, Beijing.

Li, T., X. Xiao, G. Li, Y. Gao, and W. Zhou, 1986, The crustal evolution and uplift of the Qinghai–Tibet plateau, *Tectonophysics, 127*, 279–289.

Liu, Z.Q., et al., 1988, *Geologic Map of Qinghai–Xizang Plateau and its Neighboring Regions (1:500,000)*, Chengdu Institute of Geology and Mineral Resources, Academic Sinica, Geology Publisher, Beijing (in Chinese).

Lovera, O. M., F. M. Richter, and T. M. Harrison, 1989, ^{40}Ar/^{39}Ar geothermometry for slowly cooled samples having a distribution of diffusion domain sizes, *J. Geophys. Res., 94*, 17,917–17,935.

Lyon–Caen, H., and P. Molnar, 1984, Gravity anomalies and the structure of western Tibet and the southern Tarim Basin, *Geophys. Res. Lett., 11*, 1251–1254.

Lyon–Caen, H., and P. Molnar, 1985, Gravity anomalies, flexure of the Indian plate, and the structure, support and evolution of the Himalaya and the Ganga Basin, *Tectonics, 4*, 513–538.

Manabe, S., and T. B. Terpstra, 1974, The effects of mountains on the general circulation of the atmosphere as identified by numerical experiments, *J. Atmos. Sci., 31*, 3–42.

Masek, J. G., B. L. Isacks, T. L. Gubbels, and E. J. Fielding, 1994, Erosion and tectonics at the margins of continental plateaus, *J. Geophys. Res., 99*, 13,941–13,956.

Mattauer, M., 1986, Intracontinental subduction, crustal stacking wedge and crust–mantle decollement, in *Collision Tectonics*, M. P. Coward and A. C. Ries (eds.), pp. 35-60, Geological Society Special Publication, 19, London.

Matthews, W. H., 1974, The structural setting of Cenozoic volcanic and sedimentary rocks at the British Columbia mainland, Cordilleran Section, Geol. Soc. Canada Ann. Meet., Vancouver, B.C.

McDougall, I., and T. M. Harrison, 1988, *Geochronology and Thermochronology by the* ^{40}Ar/^{39}Ar *Method*, Oxford University Press, New York.

McKnight, C. L., 1993, Structural styles and tectonic significance of Tian Shan Foothill fold and thrust belts, NW China, Ph.D. thesis, Stanford University, California., 207 pp.

McNamara, D. E., T. J. Owens, P. G. Silver, and F. T. Wu, 1994, Shear wave anisotropy beneath the Tibetan Plateau, *J. Geophys. Res., 99*, 13,655–13,665.

Mechie, J., et al., 1995, A deep crustal seismic section beneath the Indus-Yarlung Suture, Tibet, from wide–angle seismic data, *EOS, 76*, F566.

Mercier, J.-L., R. Armijo, P. Tapponnier, E. Carey–Gailhardis, and T. L. Han, 1987, Change from Tertiary compression to Quaternary extension in southern Tibet during the India–Asia collision, *Tectonics, 6*, 275–304.

Molnar, P., 1984, Structure and tectonics of the Himalaya: Constraints and implications of geophysical data, *Annu. Rev. Earth Planet. Sci., 12,* 489–518.

Molnar, P., 1988, A review of geophysical constraints on the deep structure of the Tibetan Plateau, the Himalaya and the Karakoram, and their tectonic implications, *Phil. Trans. R. Soc. Lond., A326,* 33–88.

Molnar, P., and P. Tapponnier, 1975, Cenozoic tectonics of Asia; effects of a continental collision, *Science, 189,* 419–426.

Molnar, P., B.C. Burchfiel, Z. Zhao, K. Liang, S. Wang, and M. Huang, 1987, Geologic evolution of Northern Tibet: Results of an expedition to Ulugh Muztagh, *Science, 235,* 299–305.

Molnar, P., P. England, and J. Martinod, 1993, Mantle dynamics, the uplift of the Tibetan Plateau, and the Indian monsoon, *Rev. Geophys., 31,* 357–396.

Molnar, P., E. T. Brown, B. C. Burchfiel, Q. Deng, X. Feng, J. Li, G. M. Raisbeck, J. Shi. Z. Wu, F. Yiou, and H. You, 1994, Quaternary climate change and the formation of river terraces along growing anticlines on the north flank of the Tien Shan, China, *J. Geol., 102,* 583–602.

Murphy, M. A., A. Yin, T. M. Harrison, S. B. Durr, W.S.F. Kidd, Chen Z., F. J. Ryerson, Wang X., and Zhou X., 1997, Significant crustal shortening in south–central Tibet prior to the Indo–Asian collision, *Geology, 25, 713–722.*

Najman, Y.M.R., M. S. Pringle, M. R. W. Johnson, A. H. F. Robertson, and J. R. Wijbrans, 1997, Laser [40]Ar/[39]Ar dating of singel detrital muscovite grains from early foreland–basin sedimentary deposits in India: Implications for early Himalayan evolution, *Geology, 25,* 535–538.

Najman, Y.M.R., R. J. Enkin, M.R.W. Johnson, A.H.F. Robertson, and J. Baker, 1994, Paleomagnetic dating of the earliest continental Himalayan foredeep sediments: Implications for Himalayan evolution, *Earth Planet. Sci. Lett., 128,* 713–718.

Nakata, T., 1989, Active faults of the Himalaya in India and Nepal, in *Tectonics of the Western Himalaya,* L. L. Malinconico and R. J. Lillie (eds.), pp. 243–264, Spec. Paper 232 Geological Society of America, Boulder, Colo.

Nelson, K. D., and 27 others, 1996, Partially molten middle crust beneath Southern Tibet: Synthesis of Project INDEPTH results, *Science, 274,* 1684–1696.

Noble, S. R., and M. P. Searle, 1995, Age of crustal melting and leucogranite formation from U–Pb zircon and monazite dating in the western Himalaya, Zanskar, India, *Geology, 23,* 1135–1138.

Norin, E., 1946, Geological explorations in western Tibet, reports from the scientific expedition to the northwestern provinces of China under the leadership of Dr. Sven Hedin, Publ. 29 (III), Geology, 7, Tryckeri Aktiebolaget, Thule, Stockholm.

Palmer, M. R., and J. M. Edmonds, 1989, The strontium isotope budget of the modern ocean, *Earth Planet. Sci. Lett., 92,* 11–26.

Pan, Y., 1992, Unroofing history and structural evolution of the southern Lhasa Terrane, Tibetan Plateau: Implications for the continental collision between India and Asia, 395 pp., Ph.D. thesis, State University of New York at Albany.

Pan, Y., and W.S.F. Kidd, 1992, Nyainqentanglha shear zone: A late Miocene extensional detachment in the southern Tibetan Plateau, *Geology, 20,* 775–778.

Parrish, R. R., 1983, Cenozoic thermal evolution and tectonics of the Coast Mountains of British Columbia 1, fission track dating, apparent uplift rates, and patterns of uplift, *Tectonics, 2,* 601–632.

Parrish, R. R., and K. V. Hodges, 1993, Miocene (22 ± 1) metamorphism and two stage thrusting in the Greater Himalayan sequence, Annapurna Santuary, Nepal, *Geol. Soc. Am. Abs. w. Progs., 25,* A174.

Parrish, R. R., and K. V. Hodges, 1996, Isotopic constraints on the age and provenance of the Lesser and Greater Himalayan sequences, Nepalese Himalaya, *Geol. Soc. Amer. Bull., 108,* 904–911.

Patriat, P., and J. Achache, 1984, India–Eurasia collision chronology has implications for crustal shortening and driving mechanism of plates, *Nature, 311,* 615–621.

Patzelt, A., H. Li, J. Wang, and E. Appel, 1996, Paleomagnetism of Cretaceous to Tertiary sediments from southern Tibet: Evidence for the extent of the northern margin of India prior to the collision with Eurasia, *Tectonophysics, 259,* 259–284.

Pêcher, A., 1989, Metamorphism in the central Himalaya, *J. Metamorphic Petrol., 7,* 31–41.

Peltzer, G., and P. Tapponnier, 1988, Formation and evolution of strike–slip faults, rifts, and basins during the India–Asia collision: An experimental approach, *J. Geophys. Res., 93*, 15,085–15,117.

Peltzer, G., P. Tapponnier, and R. Amijio, 1989, Magnitude of late Quaternary left–lateral displacement along the north edge of Tibet, *Science, 246*, 1285–1289.

Powell, C. M., 1986, Continental underplating model for the rise of the Tibetan Plateau, *Earth Planet. Sci. Lett., 81*, 79–94.

Qinghai Bureau of Geology and Mineral Resources (BGMR), 1989, *Geologic History of the Qinghai Region*, Geological Publishing House, Beijing.

Quade, J., T. E. Cerling, and J. R. Bowman, 1989, Development of Asian monsoon revealed by marked ecological shift during the latest Miocene in the northern Pakistan, *Nature, 342*, 163–166.

Quade, J., J.M.L. Cater, T. P. Ojha, J. Adam, and T. M. Harrison, 1995, Late Miocene environmental change in Nepal and the northern Indian subcontinent: stable isotope evidence from paleosols, *Geol. Soc. Am. Bull., 107*, 1381–1397.

Quidelleur, X., M. Grove, O. M. Lovera, T. M. Harrison, A. Yin and F. J. Ryerson, 1997, The thermal evolution and slip history of the Renbu Zedong Thrust, southeastern Tibet, *J. Geophys. Res., 102*, 2659–2679.

Ramstein, G., F. F. Fluteau, J. Besse, and S. Joussaume, 1997, Effect of orogeny, plate motion, and land-sea distribution on Eurasian climate change over the past 30 million years, *Nature, 386*, 788–795.

Ratschbacher, L., W. Frisch, G. Lui, and C. Chen, 1994, Distributed deformation in southern and western Tibet during and after the India–Asia collision, *J. Geophys. Res, 99*, 19,817–19,945.

Raymo, M. E., W. F. Ruddiman, and P. N. Froelich, 1988, Influence of late Cenozoic mountain building on ocean geochemical cycles, *Geology, 16*, 649–653.

Richter, F. M., D. B. Rowley, and D. J. DePaolo, 1992, Sr isotope evolution of seawater: the role of tectonics, *Earth Planet. Sci. Lett., 109*, 11–23.

Richter, F. M., O. M. Lovera, T. M. Harrison, and P. Copeland, 1991, Tibetan tectonics from a single feldspar sample: An application of the $^{40}Ar/^{39}Ar$ method, *Earth Planet. Sci. Lett., 105*, 266–276.

Rothery, D. A., and S. A. Drury, 1984, The neotectonics of the Tibetan Plateau, *Tectonics, 3*, 19–26.

Ruddiman, W. F., and J. E. Kutzbach, 1989, Forcing of Late Cenozoic northern hemisphere climate by plateau uplift in southern Asia and the American west, *J. Geophys. Res., 94*, 18,409–18,427.

Ryerson, F. J., A. Yin, T. M. Harrison, and M. Murphy, 1995, The Gangdese and Renbu Zedong thrust systems: Westward extension to Mt. Kailas, *Geol. Soc. Am. Progs. w. Abs., 27*, A335.

Schelling, D., 1992, The tectonostratigraphy and structure of the eastern Nepal Himalaya, *Tectonics, 11*, 925–943.

Schelling, D., and K. Arita, 1991, Thrust tectonics, crustal shortening, and the structure of the far–eastern Nepal, Himalaya, *Tectonics, 10*, 851–862.

Searle, M. P., 1996, Geological evidence against large-scale offsets along the Karakoram fault: Implications for timing and amounts of uplift, exhumation and extrusion of thickened crust, in A. M. Macfarland, R. B. Sorkabi, and J. Quade (eds.), p. 131, *Abstracts of the 11th Himalayan–Karakorum–Tibet Workshop*, Flagstaff, Arizona.

Searle, M. P., and D. C. Rex, 1989, Thermal model for the Zanskar Himalaya, *J. Metamorph. Geol. 7*, 127–134.

Searle, M. P., D. J. W. Cooper, and A. J. Rex, 1988, Collision tectonics of the Ladakh–Zanskar Himalaya, *Phil. Trans. R. Soc. Lond., A326*, 117–150.

Seeber, L., and V. Gornitz, 1983, River profiles along the Himalayan arc as indicators of active tectonics, *Tectonophysics, 92*, 335–367.

Sengor, A. M. C., and B. A. Natalin, 1996, Paleotectonics of Asia: fragments of a synthesis, in *The Tectonics of Asia*, A. Yin and T.M. Harrison (eds.), pp. 486–640, Cambridge University Press, New York.

Shackleton, R. M., and C. Chang, 1988, Cenozoic uplift and deformation of the Tibetan Plateau: the geomorphological evidence, *Phil. Trans. R. Soc. Lond., A327*, 365–377.

Song, T., and X. Wang, 1993, Structural styles and stratigraphic patterns of syndepositional faults in a contractional setting: Example from

Quaidam basin, northwestern China, *AAPG Bull., 77*, 102–117.

Srivastava, P., and G. Mitra, 1994, Thrust geometries and deep structure of the outer and lesser Himalaya, Kumaon and Garwal (India): Implications for evolution of the Himalayan fold-and-thrust belt, *Tectonics, 13*, 89–109.

Stern, L.A., G. D. Johnson, and C. P. Chamberlain, 1994, Carbon isotope signature of environmental change found in fossil ratite eggshells from a South Asian Neogene sequence, *Geology, 22*, 419–422.

Stow, D. A. V., J. R. Cochran and ODP Leg 116 Shipboard Party, 1989, The Bengal Fan: Some preliminary results from ODP drilling, *Geomarine Lett., 9*, 1–10.

Summerfield, M. A., 1991, Global Geomorphology, Longman Scientific and Technical, Essex.

Tang, T. F. (chief editor), 1990, *Late Cretaceous–early Tertiary Sedimentary Evolution of Tarim*, Science Press, Beijing.

Tapponnier, P., and P. Molnar, 1979, Active faulting and Cenozoic tectonics of Tien Shan, Mongolia, and Baikal regions, *J. Geophys. Res., 84*, 3425–3459.

Tapponnier, P., G. Peltzer, and R. Armijo, 1986, On the mechanics of the collision between India and Asia, in *Collision Tectonics*, M.P. Coward and A.C. Ries (eds.), pp. 115–157, Geol. Soc. Spec. Publ., 19.

Tapponnier, P., B. Meyer, J. P. Avouac, G. Pelzer, Y. Gaudemer, G. Sunmin, X. Hongfa, Y. Kelun, C. Zhitai, C. Shuahua, and D. Huagang, 1990, Active thrusting and folding in the Qilian Shan, and decoupling between upper crust and mantle in northeast Tibet, *Earth Planet. Sci. Lett., 97*, 382–403.

Turner, S., C. Hawkesworth, J. Liu, N. Rogers, S. Kelley, and P. van Calsteren, 1993, Timing of Tibetan uplift constrained by analysis of volcanic rocks, *Nature, 364*, 50–54.

Wang, X., Z. M. Li, and Q. Xiyiao, 1983, *Geologic Map of the Xigaze–Zedong Region (1:100,000), Xizang (Tibet) with Report*, Xizang Bureau of Geology, Lhasa.

Whiting, B. M., and G. D. Karner, 1991, 25 Ma flexural modification of western India continental margin: effects of Indus Fan loading, *EOS, 72*, 472.

Willems, H., X. Wan, J. Yin, L. Dongdui, G. Liu, S. Durr, and K. Gräfe, 1995, *The Mesozoic development of the N–Indian passive margin and of the Xigaze Forearc Basin in southern Tibet, China, Excursion Guide for IGCP 362 Working–Group Meeting "Integrated Stratigraphy"*, Berichte, Fachbereich Geowissenschaften, Universität Bremen, 64, 113 p.

Willett, S. D., and C. Beaumont, 1994, Subduction of Asian lithospheric mantle beneath Tibet inferred from models of continental collision, *Nature, 369*, 642–645.

Wolfe, J. A., 1990, Estimates of Pliocene precipitation and temperature base don multivariate analysis of leaf physiognomy, *U.S. Geol. Surv. Open–File Reports, 90–64*, 39–42.

Wolfe, J. A., 1992, An analysis of present–day terrestrial lapse rates in the western conterminous United States and their significance to paleoaltitudinal estimates, *U.S. Geol. Surv. Bull., 1964*, 1–14.

Xu, S., 1981, The evolution of the paleogeographic environments on the Tanggula Mountains in the Pliocene–Quaternary, in *Geological and Ecological Studies of Qinghai–Xizang (Tibet) Plateau 1, Geology, Geologic History and Origin of the Qinghai–Xizang Plateau*, pp. 247–255, Beijing.

Ye, C. H., and R. J. Huang, 1992, The Tertiary (chapter 13), in *Stratigraphy of Tarim*, pp. 308–363, Science Press, Beijing (in Chinese).

Yin, A., 1993, Mechanics of wedge-shaped blocks I: An elastic solution for compressional wedges, *J. Geophys. Res. 98*, 14,245–14,256.

Yin, A., and S. Nie, 1996, A Phanerozoic palinspastic reconstruction of China and its neighboring regions, in *The Tectonics of Asia*, A. Yin and T.M. Harrison (eds.), pp. 442–485, Cambridge University Press, New York.

Yin, A., S. Nie, T. M. Harrison, J. Fillipone, X. Qian, and M. Li, 1993, A kinematic model for the Tertiary development of the Tian Shan, Altyn Tagh, Chaman, and Karakorum fault systems in central Asia during the India–Asia collision, *Geol. Soc. Am. Abst. w. Prog., 25*, 121.

Yin, A., T. M. Harrison, F. J. Ryerson, W. Chen, W. S. F. Kidd, and P. Copeland, 1994, Tertiary structural evolution of the Gangdese thrust system, southeastern Tibet, *J. Geophys. Res., 99*, 18,175–18,201.

Yin, A., T. M. Harrison, and F.J . Ryerson, 1995, Transtension along the left-slip Altyn Tagh and

Kunlun faults as a mechanism for the occurrence of Late Cenozoic volcanism in the northern Tibetan Plateau, *EOS, 76*, F567.

Yin, A., S. Nie, T. M. Harrison, P. Craig, F. J. Ryerson, Y. Geng, and Q. Xianglin, 1997, Late Cenozoic tectonic evolution of the southern Tian Shan, northwestern China, *Tectonics* (in press).

Zeitler, P. K., 1985, Cooling history of the NW Himalaya, *Tectonics, 4*, 127–141.

Zhao W., and W. J. Morgan, 1985, Uplift of the Tibetan Plateau, *Tectonics, 4*, 359–369.

Zhao, W., K. D. Nelson, and Project INDEPTH Team, 1993, Deep seismic reflection evidence for continental underthrusting beneath southern Tibet, *Nature, 366*, 557–559.

Zhou, Z., and P. Chen, 1990, *Paleontology, Stratigraphy, and Geologic Evolution of the Tarim Basin*, Chinese Science Publishing House, Beijing.

Zhou, Q., and J. Zheng, 1990, *Structural Analysis of the Tarim Region*, Science Press, Beijing.

CHAPTER 4

Topographic History of the Western Cordillera of North America and Controls on Climate

Clement G. Chase, Kathryn M. Gregory-Wodzicki,
Judith Totman Parrish, and Peter G. DeCelles

The topography of the earth's surface is crucially important in both climatic and tectonic studies. Numerical models of ancient climate depend with some sensitivity upon the boundary conditions of paleotopography. To interpret paleoclimate indicators in terms of more than local climate requires knowledge of the elevation at which the indicators formed. Despite the relative ease with which present elevations may be measured, quantitative paleoaltimetry is much more difficult to achieve. We review here a number of ways of estimating paleoelevations, and summarize results that bear on the elevation history of the Western Cordillera of North America from Late Cretaceous through Cenozoic time (~80 Ma–Present).

Paleontologic, paleogeographic, tectonic, sedimentologic, and geochemical information can be used to estimate paleoelevations with varying degrees of precision. A majority of the determinations we present here come from paleobotany, in particular from mean annual temperature (MAT) and mean enthalpy estimates derived from the morphology of fossil leaves. These and many of the other paleontological and sedimentological constraints on paleoelevation are themselves climate proxies. One notable exception is tectonic reconstruction of thrust belt topography based on sediment provenance, which provides elevation estimates depending only weakly on climate. Because regional elevation is an essential determinant of local climate, some care is required to avoid circularity in using paleotopography derived from climate proxies as a boundary condition for general circulation climate models. More than just relying on the inferred topography, the models at a minimum should be able to reproduce the climate proxies themselves. In addition, they should be checked against independent observations.

Almost all of our information comes from the central portion of the Cordillera, within the western United States. The topographic evidence is divided into four time slices covering Middle–Early Miocene (14–22 Ma), Oligocene–Late Eocene (22–45 Ma), Middle–Early Eocene (45–53 Ma), and Paleocene–Late Cretaceous (53–80 Ma). There is enough evidence to draw paleotopographic maps for the slices representing Eocene and Oligocene time.

Our primary conclusion from the compiled data is that the Western Cordillera has been at generally high elevations more or less continuously since at least the Late Cretaceous. Paleoelevations of 3–4 km are not unusual throughout the last 80 Ma. The available data agree in general with the hypothesis of Coney and Harms (1984) that high elevations in the western United States are residual from Mesozoic–Early Tertiary compressional thickening of the crust, with subsequent partial collapse of the Basin and Range province attributable to extensional thinning. This implies that general circulation models should not rely upon Late Cenozoic (last 15 Ma) uplift of the western United States to provide impetus towards the Pleistocene ice ages (e.g., Kutzbach et al. (1989), Ruddiman and Kutzbach (1989), Ruddiman et al. (1989)). There is some evidence of Oligocene–Miocene collapse of the Basin and Range province. Scattered estimates of Early Eocene paleoelevations, particularly from Wyoming, tend to be lower than present elevations. Paleobotanical evidence suggests that uplift of the Sierra Nevada of California preceded 15 Ma. Uplift of the Colorado Plateau could have been at any time between initiation of the Grand Canyon at 6 Ma and Cretaceous.

Approaches to Paleoelevations

A number of approaches to paleoaltimetry have been
developed using paleontologic, paleogeographic, tectonic, sedimentologic, and geochemical information.
Paleontologic methods rely largely on the effects of
elevation on local climate and on the relationship between organisms and climate. Thus, for the purposes
of paleoaltimetry, one might suspect these methods of
circularity, but they have been the most extensively
quantified. Estimating paleoelevation from paleogeography relies on the general observation that mountains
in different plate tectonic settings tend to have different average elevations. More detailed approaches
along these lines are the tectonic methods, which deal
with the details of crustal thickening and thermal and
dynamic topography. Sedimentologic information
uses basin analysis to estimate adjacent topography, or
sediment stable isotopic composition to indicate paleotemperatures. Finally, methods using cosmogenic
isotopes and basalt vesicularity show some promise in
providing paleoelevation estimates.

In this section, we discuss each of these methods in
turn, concluding with a discussion of confusion between climatologic signals and tectonic signals in estimating paleoelevation.

Climate-Based Altimeters

The most successful quantitative estimates of paleoelevation to date are provided by analysis of paleoclimatic data. This approach is based on the observation
that some climatic variables vary with elevation. If
one knows the value of such a climatic variable at two
points, the elevation of one of the points, and the rate
at which the variable changes with elevation, one can
estimate the elevation of the second point. When reconstructing paleoelevations, a site at sea level is typically used as the point at which both climate and elevation are known; sea level is identifiable in the geological record because of the distinctive facies associations. Two climate-based paleoaltimeters have been
developed: one which uses variations in mean annual
temperature (MAT), the other which uses variations in
moist enthalpy. Currently, these paleoaltimeters derive
quantitative estimates of paleoelevation with the lowest estimated errors of all the paleoaltimeters.

MAT-based Altimeter

Temperature decreases as elevation increases. The rate
of cooling in a column of free air is the environmental
lapse rate, which is approximately 6°C/km. More
useful to paleoaltimetry is the variable termed gamma
(γ, units: °C/km), defined by Forest et al. (1995) as
"the empirical relationship between mean annual
temperature at the surface and altitude." γ is
equivalent to the "terrestrial lapse rate" of Axelrod
and Bailey (1976) and Wolfe (1994). This factor is
more variable than the environmental lapse rate
because of the variable but generally high heat
capacity of land. Heating of the near-ground atmosphere varies with the albedo, topography, and vegetational cover of the land surface, along with wind
patterns and the nature of the air mass above the surface (Meyer, 1986).

Elevation Z of a terrestrial site can be calculated
using γ after the equation of Axelrod and Bailey
(1976):

$$Z = \frac{MAT_c - MAT_i}{\gamma} + S, \qquad (4.1)$$

where MAT_c = mean annual temperature at sea level
(°C); MAT_i = MAT derived by the same means from a
coeval inland site (°C); and S = ancient sea level
relative to modern sea level (km).

Values for γ have been calculated in two ways: for
local areas (Meyer, 1986, 1992) and for more regional
areas (Axelrod and Bailey, 1976, Wolfe, 1992a).
Meyer (1986) calculated local values for γ by compil-
ing temperature data from climate stations in areas of
high topographic relief from around the world in areas
of 1–2° latitude by 1–5° longitude. However, often
these small areas had no sites at or even near sea level.
His observed local mean γ of 5.89 ± 1.09°C/km is not
tied to sea level and is thus difficult to apply to fossil
studies. A better estimate would start at sea level.

Wolfe (1992a) plotted MAT data for climate stations from the western coast of the United States and
fit a regression line with a gradient of 0.33°C/degree
of latitude. He then compared MAT data from the interior western United States to interpolated coastal
temperature at the same latitude. If the negative values
for γ are deleted from this dataset, the average lapse
rate is 2.84 ± 1.12°C/km. The highest values for γ are
observed on the windward side of mountain ranges,
the lowest in valleys on the leeward side of ranges.

However, the problem with using data from the United States west of the 100th meridian to determine an average value for γ is that today it is an area of elevated base level with elevations generally above 1 km. A large elevated land surface is heated more than a column of air at the same elevation (Parrish and Barron, 1986) and thus terrestrial lapse rates are reduced relative to those in a free air column, because the temperature difference from the coast is reduced for a given elevation. Note that a large area needs to be elevated to cause this effect; an isolated mountain peak will be better equilibrated with atmospheric temperature (Meyer, 1986) (although the potential of preservation of sediments is very low). Axelrod and Bailey (1976) find an average γ of 5.5°C/km for regions closer to sea level, including Brazil and North Carolina.

In order to choose a γ value for equation 4.1, a two-step calculation is used (Wolfe, 1994). First the 5.5°C/km environmental lapse rate is used to obtain a minimum elevation. If the site, or preferably group of sites, appear to have elevations higher than 0.5–1 km above sea level, then the use of 3.0°C/km for γ is more appropriate. Knowledge of the geology is useful in deciding whether an elevation above 1 km for the first calculation for a site or group of sites represents an isolated peak or a large elevated area.

Enthalpy-based altimeters

The paleoaltimeter developed by Forest et al. (1995) is based on energy conservation in the atmosphere. The thermodynamic variable moist static energy, written *h*, is the total specific energy content of air not including kinetic energy. It has two useful qualities: it is nearly constant for a given air parcel, and due to the Earth's rotation, values of mean surface h for North America are fairly constant for a given latitude. Moist static energy can be related to altitude, *Z*, according to:

$$h = H + gZ \qquad (4.2)$$

where *H* is the moist enthalpy, which is a function of temperature and relative and specific humidity, and *g* is the gravitational constant. Given that for a certain latitude, *h* does not vary greatly with longitude in the modern atmosphere, one can estimate the elevation of an inland site by comparing the moist enthalpy at the site to the moist enthalpy for a coastal site at the same latitude according to:

$$Z = \frac{H_{sea\ level} - H_{inland}}{g} \qquad (4.3)$$

Thus, to estimate the paleoelevation of a fossil flora, one calculates the moist enthalpy for the fossil flora and for a coeval paleoflora at a similar latitude on the coast and divides by *g*. Calculated error is less than a kilometer.

This method is based on the laws of thermodynamics, in contrast to the more empirical MAT-based altimeter. The drawback with the enthalpy method is that *h* may be fairly constant for a given airmass, but it varies between airmasses derived from different ocean sources. Thus, one must assume that circulation patterns did not vary significantly in the past to apply this method. This assumption is fairly safe for the western United States, which as of now is the only area to which this method can be applied, but could cause problems if the method is applied to other regions. The original dataset of Forest et al. (1995) omitted some of the higher humidity sites from the dataset of Wolfe (1993) from Japan, Puerto Rico and Panama. This was problematic, because many Tertiary paleofloras from the western United States plot near these sites. However the new dataset of Wolfe et al. (in preparation) eases this problem by adding high humidity sites from Mexico and Japan.

Potential sources of error for climate-based altimeters

The disadvantage of climate-based paleoaltimeters is that the elevation signal is being filtered through climate, a factor which varies in a complex manner through time. For example, if one goes to an area with a cold-temperate climate and sees fossil evidence of a tropical forest, one or more scenarios could apply: (1) global climate could have cooled, either over the long or short term; (2) the latitude could have increased due to plate tectonic movements; (3) the paleogeography could have changed, with perhaps mountain building blocking air masses; or (4) the area could have been uplifted.

In order to address the complication of changing global climate, the fossil interior sites are compared to coastal sites of the same age in both paleoaltimeters. However, even with the accuracy of single crystal $^{40}Ar/^{39}Ar$ dating, one cannot prove that two floras are truly coeval. The error on radiometric dates can be as low as 60 ka, but climate varies on shorter time scales. Until the discovery of a method which can date rocks to within ± 100 yr, the only way to address this problem is to look at as many sites from the coast and interior as possible, and to look at trends, rather than putting too much emphasis on one data point.

A major assumption of the MAT-based altimeter is that the average value for γ does not change significantly with changing global climate. Wolfe (1992a) suggests that Arctic air masses greatly influence modern terrestrial lapse rates in the western United States by decreasing winter temperatures, especially at the higher latitudes. Thus values for γ in the northwestern United States are higher than in the southern United States and/or in protected basins. Frigid Arctic air probably had much less of an influence before the Pliocene, when the Arctic Ocean developed ice cover. Thus, the 3.0°C/km average regional γ might have been even lower, assuming that base level was unchanged.

Another question is whether the value of γ changes when latitudinal temperature gradients change. Data from the Americas for the Last Glacial Maximum suggest not. The ~1 km decrease of snowlines and vegetation zones observed to have occurred in Central America and northern South America suggests a temperature drop of 5°C if modern local values for γ are used (Rind and Peteet, 1985). Stute et al. (1995a) and Stute et al. (1995b) estimate similar amounts of cooling at lower elevations based on concentrations of dissolved noble gas in ancient ground water from aquifers in Brazil, Texas, and New Mexico. The sites, which ranged from 0.2–2.0 km in elevation are interpreted to represent 5.2–5.5°C of cooling. If the value for γ had been higher, then the snowline data should indicate less cooling, perhaps 1.8°C per 1°C/km change in lapse rate. Also, the site from New Mexico should have appeared to cool less than the sites from Texas and Brazil, contrary to observation.

This assumption that the value of γ does not change significantly with changing latitudinal temperature gradients is also supported by values of γ calculated for the Tertiary United States using the enthalpy-based altimeter. If the altitude for a given site is estimated from equation 3 using moist enthalpy values, then one can solve equation 1 for γ. Wolfe et al. (in preparation) show that values of γ for Eocene to Miocene floras from the western United States vary from 1.9°C/km to 3.5°C/km, with the ancient γ value for a given site on average only 0.1°C/km greater than the modern γ value. Thus it appears that using modern γ values for Tertiary sites is reasonable.

Another factor to consider in both climate-based

paleoaltimeters is the change in latitude with time. The paleolatitude of a site is fairly easy to estimate using paleomagnetic data or plate tectonic reconstructions. However, in the fossil record, it is often not possible to find a coastal site for comparison that has the exact same paleolatitude as the interior site. Thus one needs to correct the MAT of the coastal site to the paleolatitude of the interior flora. For some periods there are enough coastal sites that the latitudinal temperature gradient can be estimated. However, temperature does not change smoothly along the coast: there are steps due to variations in width of the continental shelf and other factors (Wolfe, 1992a). Therefore it is important to use many coastal sites to derive coastal MAT. Changes in paleogeography can also change how air masses move, and thus can effect both the value of γ and of enthalpy.

Estimating Climate Variables

Terrestrial MAT and enthalpy, which are the bases for these paleoaltimeters, can be estimated using many proxies, such as floras, faunas, and stable isotopes. Most of the MAT values we use here are based on the leaf morphology of floras.

Foliar physiognomy

The foliar physiognomic method of Wolfe (1993) provides quantitative estimates of climate with the highest estimated precision relative to other techniques. This method is based on the observation that angiosperm leaf physiognomy, that is, morphology, varies with climate. For example, dicotyledonous vegetation in warmer climates tends to have a higher percentage of species with entire-margined, that is, smooth-margined, leaves than vegetation in cooler climates (Bailey and Sinnot, 1916, Wolfe, 1971, Givnish, 1979). Also, leaf size and leaf width tend to be smaller in sunny/dry environments and larger in shady/moist environments (Webb, 1968, Givnish, 1979, 1984). Leaves in very wet environments often have elongated apices, or "drip-tips" (Richards, 1952, Dean and Smith, 1979).

In order to quantify these relationships, Wolfe (1993) collected leaf samples from 106 modern vegetation sites in North America, the Caribbean, and Japan. Each vegetation sample has at least 20 species, which were scored in terms of 31 different leaf physiognomic character states and summed to derive a site

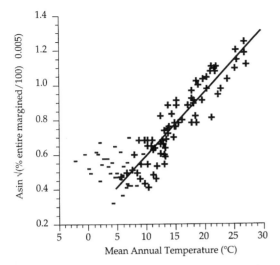

Fig. 4.1. Mean annual temperature versus percent entire margined species for the modern vegetation dataset of Wolfe (1993). Sites with cold month mean temperatures (CMMT) < -2°C (subalpine sites) are plotted with minus signs, sites with CMMTs greater than -2°C are plotted with plus signs. The regression line is for the dataset excluding subalpine sites and has an r^2 of 0.84.

physiognomic score. This dataset shows several trends with climate, the most important being the strong relationship between mean annual temperature and percent entire-margined species (Fig. 4.1).

The paleoclimate of a fossil flora can be estimated by computing the physiognomic score of the fossil forms and comparing it to a physiognomic database based on modern leaf collections and climate stations. Various multivariate statistical methods are used including: correspondence analysis in the Climate-Leaf Analysis Multivariate Program (CLAMP) method of Wolfe (1993); multiple regression analysis (Wing and Greenwood, 1993, Gregory and McIntosh, 1996); and principal components analysis (Forest et al., 1995). Error for estimates of MAT ranges from ± 1°C for CLAMP analysis to ± 1.5°C (Gregory and Chase, 1992) and ± 2.0°C (Wing and Greenwood, 1993) for multiple regression analysis.

Subalpine floras display a different relationship between MAT and percent entire-margined species than warmer climate vegetation. This difference can be seen in Fig. 4.1; the sites with a cold-month mean temperature (CMM) less than -2°C, plotted with minus signs, have higher percentages of entire-margined species than would be predicted given their MAT. Subalpine paleofloras can be identified using CLAMP

analysis or by examining the paleoflora for large (length from 7.6–12.7 cm) evergreen leaves. Such evergreen species are not presently found in areas with a CMM < -2°C (Wolfe, 1979).

If a paleoflora does not appear to be subalpine, then it can be compared to a smaller modern database which excludes subalpine sites; the resulting multivariate analysis yields more accurate results than if the full database was used. The MAT of subalpine samples can be analyzed using CLAMP with a modern database of only subalpine samples (Wolfe, 1993, p. 59).

Greenwood and Wing (1995) suggest that the foliar physiognomic method may tend to underpredict MAT by 2–3°C based on comparisons of paleoclimate derived from foliar physiognomy and that derived from the occurrence of palms. Modern palm roots and apical meristems cannot tolerate sustained freezes, and fossil palms appear to have had similar wood. As long as this bias in MAT estimates is systematic, then it is not a source of error in the elevation calculation because the temperature difference is used in equation (4.1), rather than absolute temperature.

Potential sources of error in the foliar physiognomic method are discussed in Gregory and McIntosh (1996). One important source of error occurs in the sampling process. For example, if a flora for a given area has 100 species and one only samples 40 species, then the resulting leaf physiognomic score for the sample will differ from that of the complete flora. This sampling error varies greatly with sample size, and for samples of less than about 75 species, is greater than the formal error calculated for the multiple regression and CLAMP models. Given the generally small size of fossil floras, the sampling error is a better approximation of the total error of the MAT estimate than the standard deviation of the multivariate models.

In order to calculate the sampling error, Wilf (1996) simplifies the problem by assuming an infinite flora and by only looking at the percent entire-margined species character state. The standard deviation of the MAT estimate can then be calculated using:

$$\sigma_{MAT} = a \sqrt{p \frac{(1 - p)}{k}} \qquad (4.4)$$

where σ_{MAT} = the standard deviation, a = the slope of the MAT versus percent entire margined species linear regression line, p = the observed percent entire-margined species, and k = the number of species collected. This is a reasonable approximation of the sampling error even when all 31 character states are used to estimate MAT, because the percent entire-margined species character state explains the most variation in MAT.

Stable isotope methods

The stable isotopic signature of rainfall and faunas can also be used as paleoclimate and paleoelevation indicators. The δD and δ¹⁸O of modern rainfall is a function of climate and elevation (Drummond et al., 1993) with lighter values indicating cooler temperatures and/or higher elevations. The decrease in stable isotope values with elevation is not purely due to decreasing temperature (Lawrence and White, 1991); decreasing water vapor mass also is a factor (Drummond et al., 1993). Thus, this system is attractive as a possible climate-based paleoaltimeter. Alternatively, if climate is known to have remained constant, elevation can be estimated directly. Paleorainfall level can be recovered directly from fluid inclusions in ore deposits, or can be calculated from minerals, fossils, and paleosols. Very low values for δ¹⁸O isotopic ratios in sediments imply the presence of water derived from snow melt, and therefore either high elevations or cold climate (Dettman and Lohmann, 1993, Norris et al., 1996). However, the interpretation of paleoclimate is complicated by isotopic partitioning and other factors that impact the isotopic composition of rainfall, such as sea surface temperature, airmass trajectories, and amount of rainout (Gregory and Chase, 1994a).

Faunal methods

Paleoclimate can also be estimated from the presence/absence of crocodilians and the carapace size of turtles in fossil faunas. Turtles and crocodilians become sluggish if temperatures become too cold (Huchison, 1982). Today, crocodilians are only found in areas where MAT > 16°C (Markwick, 1994); thus their past distribution is assumed to be controlled by the 16°C MAT isotherm. In turtles, body size is correlated with temperature; non-aquatic turtles with carapaces lengths greater than 30 cm are at present restricted to climates with a mean cold month temperature greater than 13°C (Huchison, 1982).

To some extent, these paleoclimatic inferences are based on the nearest living relative method, in which one infers paleoclimate by looking at the climate distribution of a fossil species' closest relatives. This method has been criticized because, among other things, it does not take into account evolution of lineages or the fact that the observed geographical range of a species is not necessarily equal to its ecological range (Wolfe, 1971, Wolfe, 1979, Wolfe and Schorn, 1990). When this method is combined with some anatomical or physiological-based reasoning (e.g., that certain types of animals are not adapted to the cold) there can be more confidence in projecting the ecological limits of modern species into the past (Wing and Greenwood, 1993).

Paleogeographic Methods

Because the types of information discussed in this section are not available for every ancient mountain range, it is useful to have a method that can be applied consistently over the globe. A "paleogeographic method" was developed by A.M. Ziegler and his colleagues (Ziegler et al., 1985) for use in global paleogeographic reconstructions and is based on elevations of Recent mountain ranges in different plate tectonic settings. Although they (Scotese et al., 1979, Ziegler et al., 1983) presented the reconstructions in very simple terms ("highlands", "lowlands"), the method does allow for greater distinction in paleoelevation estimates. The different types of plate tectonic settings for mountains and the range of elevations for each are listed in Table 4.1, along with some of the diagnostic geologic evidence.

The key to making paleoelevation estimates based on plate tectonic setting is determining the setting, usually involving a combination of paleomagnetic data and basin analytic data. Paleomagnetic data help determine where the continents were relative to each other, whether they were converging, diverging, or sutured together (Ziegler et al., 1985). Basin analysis helps determine the state of a mountain range (see section on sedimentologic methods). Other information, such as abundance and timing of volcanism, also is important in determining the state of a mountain range. Examples and procedures are discussed at length in Ziegler et al. (1985). Reliability of the method depends on Recent mountain ranges being an adequate sample of the possibilities.

Table 4.1. Range of elevations for mountains in different plate tectonic settings

Setting	Elevations	Geologic Evidence
Continent–continent collisions	>4000 m	High-T, high-P metamorphics
Ocean–continent collisions	2000–4000 m	Andesites/granodiorites in continental setting
Island arcs	1000–2000 m	Andesites/granodiorites in marine setting
Rift shoulders		Adjacent fanglomerates
Inland plains	200–1000 m	Extrapolated from adjacent environments
Rift valleys		Basalts, lake deposits in graben
Some forearc ridges		Tectonic mélanges

Source: adapted from Ziegler et al. (1985, Table 2).

Tectonic Constraints on Paleoelevations

The constraints of lithospheric and crustal thickness on elevation of continents are fairly clear, but these constraints can be applied to paleotopography only in a general way. The principle of isostasy almost certainly governs regional elevation. Crustal rocks are less dense than asthenospheric material, so by itself thickening the crust will increase elevation and thinning the crust will lower elevation. The mantle part of the continental lithosphere is in most cases more dense than underlying asthenosphere, so thickening and thinning mantle lithosphere would cause surface lowering and raising respectively. Density contrasts associated with the crust are larger, so the net effect of extending and thinning lithosphere with crust thicker than about 20 km would be a decrease in elevation (McKenzie, 1978). In contrast, we expect compressional orogens to be marked by high elevation: mountain ranges. Lithospheric and asthenospheric temperature and consequent density differences can also be an important component of regional elevations. Also, the addition of magma to the crust can increase crustal thickness independent of horizontal strains.

These principles are simple, and have been applied by Coney and Harms (1984) to estimate paleotopographic evolution of the Western Cordillera. Late Mesozoic and Early Tertiary compression and thrusting would have thickened the crust causing regional uplift. Subsequent extension and thinning of the crust/lithosphere of the Basin and Range province caused a partial collapse of that region. These interpretations can only be general because they require knowledge of the initial thickness of the crust in order to predict the elevation of the final result.

Balanced structural cross sections of thrust belts can be used to reconstruct regional topographic gradients, and provide estimates of paleotopography if tied to ancient sea level by sedimentary evidence. We will use this approach for Paleocene–Cretaceous paleotopography, focusing on the Sevier belt in Utah and several ranges in Wyoming.

Sedimentological Methods

The simplest sedimentological paleoaltitude comes from determining the lateral boundary between coastal marine and terrestrial sediments, and thus paleo-sea level. Beyond this, sedimentologic methods for estimating continental elevations are primarily analyses of transport and depositional gradients as determined from sedimentary structures, clast size, and evolution of clast composition. With the possible exception of intermontane basins, mountains are subject to eventual erosion over their entire surfaces, so wherever sediments are preserved they are at some remove from their source elevations. Therefore, sedimentology can, at best, determine when and roughly where mountains were actively eroding and give a very general sense of relief; any estimates of paleoelevation are exceedingly general. Interpretation of sedimentologic information is hampered by the effects of climate on sedimentology. Some independent understanding of climate is required to differentiate between tectonic and climatic effects on the types of sediments deposited.

Paleoelevations can be estimated from the types of mountains, as discussed in the section on paleogeographic methods. Sandstone composition (Dickinson and Suczek, 1979) can give the general setting of the rocks from which the sediments were derived, although sandstone composition can be severely modified by climate (e.g., Johnsson et al., 1991).

Associated sediment suites also can provide information about tectonic setting and paleoelevation (Table 4.1). For example, island arc mountains are flanked on one side by deep-sea trench sediments and marine forearc basin sediments, whereas ocean–continent convergent boundary (Andean-style) mountains are flanked by deep-sea trench sediments and foreland-basin sediments. Each of these types of sedimentary basins, and the others that occur associated with different types of mountain ranges, has a characteristic suite of sediments and sedimentary environments.

Sandstone composition and basin-type identification methods resemble the paleogeographic method in that both are directed toward determining the type of mountain range and, from that, rough paleoelevations as outlined in Table 4.1. A third sedimentologic method utilizes conglomerates and synsedimentary unconformities to estimate uplift (Heller and Paola, 1989, Heller et al., 1993).

Other Approaches

The quantitative paleoaltimeters, MAT and enthalpy analysis, rely on foliar physiognomy and on stability of certain properties of the atmosphere that might be climatically modulated. Thus it would be desirable to develop independent quantitative techniques that depend on the most certain of elevation-dependent properties, air pressure. In situ cosmogenic isotopes in rocks provide some hope in this direction, because cosmic ray flux in the atmosphere approximately doubles for every km of elevation. We hope to apply this fact in ancient felsic volcanic rocks. $^{40}Ar/^{39}Ar$ determination of the age difference between a flow irradiated at the surface and its capping (and shielding) flow provides the time of exposure. Very careful geologic mapping of the flow contact relationships is necessary to find sample localities that were not seriously eroded during the exposure period. Likely errors are on the order of a kilometer.

Another independent paleoaltimeter that has not yet seen extensive application is based on variation in vesicularity in basalts as a function of atmospheric pressure and therefore altitude (Sahagian and Maus, 1994). In basaltic flows with a simple history, the ratio of bubble sizes between top and bottom of the units should vary with pressure. This technique was applied in Hawaii, and reflected the elevation differences between sea level and Mauna Loa samples from about 4 km with errors less than 1.5 km, usually erring toward low estimates (Sahagian and Maus, 1994).

Tectonic Versus Climatological Signals in Paleoaltimetry

Geomorphic methods for paleoaltimetry have not been listed among the viable methods above. Geomorphology represents a case in which the potential for confusing tectonic and climatic signals is most profound. The idea that the western United States was uplifted in the Late Tertiary was based on geomorphic arguments (Davis, 1911) that the extensive Late Eocene erosion surface in the Southern Rocky Mountains must have been cut near sea level. Regional incision of the surface was then taken as evidence of uplift causing rejuvenation of the landscape. However, geomorphic modeling (Gregory and Chase, 1994b) suggests that the erosion surface could have been developed at high elevations under the influence of a climate characterized by abundant precipitation but absence of large storms. In this interpretation, an erosion surface needn't represent topographic deflation, but temporal and spatial coincidence of relative tectonic quiescence with the right kind of climate. The difficulty of separating climatic and tectonic effects might also apply to sedimentological estimates of paleoelevations, because stream gradients and sediment transport mechanism might well vary with climate (Chase, 1992, Gregory and Chase, 1994b). All of the paleoclimatic paleoaltimeters suffer from the potential for confusion between elevation changes and more global climate changes, because elevation is a powerful determinant of local climate. As we have already indicated, very careful use of coeval sea level sites is necessary to minimize this problem.

Elevation History

We have used the MAT-based paleoaltimeter to derive paleoelevation estimates for three time slices: the Middle–Early Miocene (14–22 Ma), Oligocene–Late Eocene (22–45 Ma), and the Middle–Early Eocene (45–53 Ma). The observations are numerous and consistent enough to plot on nonpalinspastic base maps for the Late Eocene–Oligocene and the Early–Middle Eocene. Data for the Early and Middle Miocene were not plotted on a base map, as paleoelevations appear to change significantly over time within this interval. The Miocene data, along with the Eocene and Oligocene data, were plotted on graphs of elevation versus time. Not enough MAT data were available for the Paleocene or Cretaceous to draw maps, but we used structural evidence to make spot estimates. Considering the spatial resolution now achievable in global circulation models, even with embedded subgrids, these scattered estimates can probably be interpolated with little loss of usefulness.

Methodology for MAT-based Paleoaltitude Maps

Published MAT estimates based on foliar physiognomy were compiled from the literature (Tables 4.2–4.4). These estimates had been calculated using: (1) qualitative physiognomy, in which the general physiognomy of a flora is qualitatively compared with the physiognomy of modern vegetation types (Wolfe, 1979) in order to derive paleoclimate; (2) CLAMP analysis; or (3) multiple regression analysis either using the models of Wing and Greenwood (1993) or Gregory and McIntosh (1996). The error for each MAT estimate was calculated after equation (4.4). If this calculated error was less than the formal error of the multivariate model, then the latter was used.

The floras were then split into coastal floras and interior floras based on longitude and facies associations. The MATs from the interior floras were then subtracted from the coeval coastal MAT. How the coastal MAT was calculated depended on the amount of data available and the observed trends in MAT for the time of interest. For the sake of comparison, coastal MATs were all projected to 45° latitude using the observed temperature gradients of 0.3° C/° latitude and 0.4° C/° latitude for the Eocene and Oligocene–Miocene, respectively (Greenwood and Wing, 1995, Wolfe, 1994) (Fig. 4.2). To derive coastal MAT for an age range during which the coastal MAT remained fairly stable, coastal MATs were averaged. To derive coastal MAT for times without data or with large amounts of variation, the coastal MAT was interpolated based on the neighboring values. For example, to derive the coastal MAT for 34 Ma, the several sites between 33 and 35 Ma with similar MATs were averaged because MAT appears to have been fairly stable during this time. However, for 27 Ma, the value of coastal MAT is interpolated between the average MAT of the 31.5–32.5 Ma group of sites and the MAT for the site at 25 Ma because no data exists specifically for 27 Ma.

The difference in MAT between the coast and the interior site rather than paleoelevation is plotted on the time-slice figures (Fig. 4.3, 4.4). This was done because this variable involves less interpretation and er-

ror than the paleoelevation estimate, which introduces an additional variable, γ, with errors on the order of ± 1°C/km. Total errors for paleoelevation estimates vary depending mostly on the value of γ and the error on the MAT difference term and to a lesser extent on the absolute amount of the MAT difference (Tables 4.2–4.4). Sites with less than 20 leaf forms will have higher errors; these floras are marked with stars in the tables.

Elevation data for the Early and Middle Miocene (Table 4.2) were not plotted on time-slice maps, as MAT difference values appear to change significantly over time within this interval. Instead, the paleoelevations for this interval (Tables 4.2–4.4) were plotted versus time, with data from other regions and time periods added for comparison (Fig. 4.5, 4.6).

Note that paleoelevation estimates have already been published for many of the floras listed in Tables 4.2–4.4. These estimates were not used in this study. Instead, all paleoelevations were recalculated from MAT data using the same method and coastal MAT database for the sake of consistency. As our knowledge of coastal temperatures, floral ages, and γ values improve, elevation estimates will continue to be updated and refined.

Early to Middle Miocene

Most of the data for the Early to Middle Miocene is from Wolfe and Schorn (1994). In this study, paleoelevations were calculated for several floras from western Nevada and the Sierra Nevada using the MAT-based paleoaltimeter. Wolfe and Schorn (1994) interpret the Fingerrock and Pyramid floras, dated as 16.0 and 15.6 Ma, as having MATs of 9.0°C and 7.5°C, respectively, which translate into paleoelevations of 2.9–3.3 ± 1.6 km if the 3.0 °C/km value for γ is used. If taken at face value, these paleoelevations are significantly higher than the modern elevations of 1.3–1.7 km (Table 4.2). The 15.5 Ma Middlegate, Eastgate, and Buffalo Canyon floras and the 15.0 Stewart Valley flora have lower paleoelevations around 2.5 ± 1.5 km, interpretable as indicating an elevational collapse of around a half of a kilometer (Fig. 4.5). The paleoelevations of the 12.5 Ma Aldrich Station and Chalk Hills floras are similar to the present altitudes of 1.7 and 2.0 km, respectively.

The dates of the inland floras from Nevada were calculated by the single crystal $^{40}Ar/^{39}Ar$ method, and thus there is sufficient time resolution to accept that

Table 4.2. MAT data for Early and Middle Miocene floras from North America

Flora	Age	Lat.	Long.	MAT	M	MATc	Pelev5	Pelev3	Melev	Source
			Early & Middle Miocene (11–23.5 Ma)							
Coast										
Cape Blanco, OR	22	42.6	-124.2	16.6	(c)	17.3				44, 45, 48, 50
Eagle Creek, WA	~18–22	45.6	-122.1	12	(c)	13.7				39, 44, 45
Collawash, OR	~18–22	44.6	-122.1	11.7	(c)	13.0				39, 44, 45
Wishkaw, WA	le, em M	47.0	-124.0	11–12	(p)	13.9				39
Weaverville, CA	~14–18	40.6	-123.2	16.2	(c)	15.7				23, 44
Liberal, WA	~13	45.2	-122.8	15.9	(c)	17.1				44, 45
Molalla, OR	~13	45.1	-122.5	15.4	(c)	16.5				39, 44, 45
Weyerhauser, OR	12.9	45.1	-122.8	9.8	(c)	10.9				44, 45
Faraday, OR	~11–12	45.2	-122.4	10.5	(c)	11.6				39, 44, 45
Interior										
Succor Ck., OR	17.1	43.3	-117.1	9.5–11	(p)	4.5	**1.0 ± 0.6**	1.7		16, 39
Latah (Spokane), WA	16.3	47.6	-117.4	9.5–11	(p)	3.3	**0.8 ± 0.6**	1.3	0.6	22, 39, 45
Grand Coulee, WA	15.7–16.8	47.9	-119.0	9.5–11	(p)	3.1	**0.7 ± 0.6**	1.2	0.5	5, 10, 39, 40
Fingerrock, NV	16.0	38.6	-119.0	9.0	(c)	8.3	1.7	**2.9 ± 1.6**	1.7	37, 47, 50
Mascall, OR	15.8	44.3	-119.5	9–10	(p)	5.6	**1.2 ±0.6**	2.0 ± 1.3	0.8	16, 39
Pyramid, NV	15.6	39.9	-119.6	~ 7.5	(c)	9.3	1.8	**3.3 ± 1.7**	1.3	2, 8, 47
Mohawk, CA	15.5	39.7	-120.6					**1.6**	1.6	47
Webber Lk, CA	15.5	39.5	-120.4					**2.1**	2.1	5, 47
Middlegate, NV	15.5	39.3	-117.9					**~ 2.5**	1.7	5, 47
Eastgate, NV	15.5	39.3	-117.4					**~ 2.5**	1.7	5, 47
Buffalo Canyon, NV	15.5	39.2	-117.8					**~ 2.5**	1.8	2, 7, 47
Stewart Valley, NV	15.0	38.6	-119.0	10.6		0.8	1.4	**2.5 ± 1.5**	1.7	37, 47, 50
Chalk Hills, NV	12.5	39.4	-119.6	11.9		0.6	**0.3 ± 0.7**	0.3	1.7	2, 3, 47, 50
Aldrich Station, NV	12.5	38.5	-118.9	13.4		-0.6	**0.0 ± 0.7**	0	1.9	2, 16, 47, 50
Whitebird, ID	Mioc.	45.7	-116.2	9–10	(p)	2.7	**> 0.6**	1.0	0.9	11, 39

Notes for Tables 4.2, 4.3, and 4.4: Flora= name of flora, star after name indicates that the flora contains less than 20 dicot angiosperm species; Lat = latitude, Long= longitude; MAT = mean annual temperature; M = method used to calculate MAT, c= CLAMP, m= multiple regression, p = qualitative physiognomy. MATc =(for coastal floras) MAT projected to 45° latitude, (for interior floras) MAT at coast–MAT of interior flora; Pelev5 = Elevation calculation using equation (4.1), using a γ value of 5.5 °C/km; Pelev3 = Same as Pelev5 except using a γ value of 3°C/km. Preferred value is shown in **bold face**. Error calculated for preferred paleoelevation estimate assuming an error on the lapse rate of ± 1.0°C/km. Errors on MAT estimates explained in text. Melev = modern elevation.

Sources:

1. Axelrod (1944)
2. Axelrod (1956)
3. Axelrod (1962)
4. Axelrod (1966)
5. Axelrod (1985)
6. Axelrod (1990)
7. Axelrod (1991)
8. Axelrod (1992)
9. Becker (1961)
10. Berry (1931)
11. Berry (1934)
12. Bown et al. (1994)
13. Brown (1937)
14. Chaney (1927)
15. Chaney and Sanborn (1933)
16. Evernden (1964)
17. Greenwood and Wing (1995)

18. Gregory (in preparation)
19. Gregory and Chase (1994a)
20. Gregory and McIntosh (1996)
21. Hickey (1977)
22. Knowlton (1926)
23. MacGinitie (1937)
24. MacGinitie (1941)
25. MacGinitie (1953)
26. MacGinitie (1969)
27. MacGinitie (1974)
28. Manchester (1990)
29. Manchester and Meyer (1987)
30. McClammer (1978)
31. Meyer (1986)
32. Potbury (1935)
33. Povey et al. (1994)
34. Sanborn (1935)

35. Wing (1987)
36. Wing and Greenwood (1993)
37. Wolfe (1964)
38. Wolfe (1968)
39. Wolfe (1981a)
40. Wolfe (1981b)
41. Wolfe (1987)
42. Wolfe (1992b)
43. Wolfe (1993)
44. Wolfe (1994)
45. Wolfe and Hopkins (1967)
46. Wolfe and Schorn (1989)
47. Wolfe and Schorn (1994)
48. Wolfe and Tanai (1987)
49. Wolfe and Wehr (1987)
50. J.A. Wolfe, written communication, 1996.

Table 4.3. MAT data for Late Eocene and Oligocene floras from North America

Location	Age	Lat.	Long.	MAT M	MATc	Pelev5	Pelev3	Melev	Source
				Late Eocene (33.5–36.5 Ma)					
Coast									
Sweet Home, OR	~33–34	44.2	-122.3	19.7 (c)	20.9				40, 44
Bilyeu Ck, OR	~33–34	44.1	-122.1	18.2 (c)	19.3				40, 44
Goshen, OR	~33	44.0	-123.0	19.5 (m)	20.6				20, 40
Goshen, OR	~33	44.0	-123.0	19 (c)	20.1				15, 40
Comstock, OR	l Eoc.	43.7	-123.2	19.1 (c)	20.2				34; 44
Interior									
Red Rock Ranch, NM*	36.7	33.4	-107.4	5–10 (p)	13.7	2.7	4.8 ± 2.3	1.83	31
Ruby, MT	l Eoc	44.5	-112.5	11.5 (c)	6.9	1.5	2.5 ± 1.4	1.6	9, 42
Beaver Ck, MT	l Eoc	47.0	-112.8	<13 (p)	>4.7	> 1.1	>1.8 ± 1.3	1.4	41
Drummond, MT	l Eoc	46.7	-113.1	<13 (p)	>4.8	> 1.1	>1.8 ± 1.3	1.4	41
Alvord Ck, OR*	l Eoc	42.2	-118.6	<13 (p)	>5.9	> 1.3	>2.2 ± 1.6	1.37	1, 16, 41
House Range, UT	37–34	39.3	-113.3	13.2 (m)	6.8	1.4	2.5 ± 1.5	1.7	18
Florissant, CO	34.05	38.5	-105.2	12.8 (m)	8.4	1.7	3.0 ± 1.6	2.5	20, 25
Florissant, CO	34.05	38.5	-105.2	10.8 (c)	10.4	2.1	3.7 ± 1.6	2.5	25, 44
Florissant, CO	34.05	38.5	-105.2	13–15 (p)	7.2	1.5	2.6 ± 1.3	2.5	25, 31
Hermosa, NM*	~33.6	33.1	-107.7	0 (p)	23.1	4.4	7.9 ± 3.0	2	31
La Porte, CA	33.2	39.7	-120.0	22.1 (m)	-2.1	0	0	1.5	16, 20, 32, 42
La Porte, CA	33.2	39.7	-120.0	22.3 (c)	-2.3	0	0	1.5	40, 44
Pitch–Pinnacle, CO*	32.9–29	38.4	-106.3	12.7 (m)	8.9	1.8	3.2 ± 1.8	3.1	21
Pitch–Pinnacle, CO*	32.9–29	38.4	-106.3	5–10 (p)	14.1	2.8	4.9 ± 2.5	3.0	31
				Oligocene (23.5–33.5 Ma)					
Coast									
Willamette, OR	32.0	44.0	-123.1	13 (c)	14.5				16, 39, 40, 44
Lyons, OR	31–32	44.8	-123.0	12–13 (p)	12.8				39
Rujada, OR	~31–32	43.8	-122.7	13 (c)	14.3				39, 44
Yaquina, OR	~24–27	44.6	-124.0	15.5 (c)	17.1				30, 44
Sandstone Creek, OR	23–24	45.1	-122.1	12.5–13.5 (p)	14.6				39, 45
Interior									
Bridge Creek, OR	32.2	44.5	-120.2	12 (c)	3.4	0.8 ± 0.6	1.3	~ 1	29, 42
Post, OR	32.2	44.2	-120.5	11.5–12.5 (p)	3.3	0.8 ± 0.6	1.3	~0.9	14, 39
Fossil, OR	32.1	45.0	-120.2	3–10 (p)	5.6	1.2 ± 0.7	2.1	0.8	28, 29
Salmon, ID	l m E	45.2	-113.9	2.6 (c)	11.0	2.2	3.9 ± 1.9	1.6	13, 44
Cow Creek, ID*	l Eoc	44.7	-113.9	<13 (p)	>0.7	>0.3	>0.4 ± 1.3	2	41, 48
Antero, CO*	Orellan	39.1	-105.8	4.5 (p)	11.8	2.3	4.1 ± 2.0	2.7	42
Platoro, CO*	28.9–30.6	37.3	-106.5	5–7.5 (p)	10.8	2.1	3.8 ± 1.9	3.1	31
Hillsboro, NM*	28.1–30.6	32.9	-107.6	0 (p)	19.0	3.6	6.5 ± 2.7	1.7	31
Creede, CO	27.2	37.8	-107.0	4.2 (c)	14.1	2.7	4.9 ± 2.1	3.7	43, 46
Creede, CO	27.2	37.8	-107.0	5–7.5 (p)	12.0	2.3	4.2 ± 1.9	3.7	31

Note: see notes for Table 4.2 for headings key, source list

Table 4.4. MAT data for Early and Middle Eocene floras from North America

Location	Age	Lat,	Long.	MAT M	MATc	Pelev5	Pelev3	Melev	Source
Early Eocene (49–55 Ma)									
Coast									
Kultheith, AK	e Eoc	60*	-132†	19.4 (c)	24.0				44
Interior									
Camels Butte, ND*	53–55	46.8	-102.5	12.9 (m)	8.6	1.8	**3.1 ± 1.5**	0.8	21, 36
Camels Butte, ND*	53–55	46.8	-102.5	11.2 (m)	10.3	2.1	**3.6 ± 1.6**	0.8	36
Upper Willwood, WY	52–53	44.2	-108.2	15.3 (m)	6.9	1.5	**2.5 ± 1.3**	1.5	12, 17, 35
Yellowstone, WY	50–51	44.9	-110.5	10.9 (m)	11.1	2.2	**3.9 ± 1.6**	2.3	36
Wind River, WY	50–51	43.7	-108.8	16.5 (m)	6.0	1.3	**2.2+ 1.1**	1.8	17
Bear Paw, MT	49–51	48.5	-109.8	10.4 (m)	10.3	2.1	**3.6 ± 1.5**	0.9	36
Kisinger Lakes, WY	49–50	43.8	-109.5	15.2 (c)	4.3	1.0	**1.6 ± 1.0**	2.5	27, 44
Kisinger Lakes, WY	49–50	43.8	-109.5	15.7 (m)	3.8	0.9	**1.5 ± 1.0**	2.5	36
Middle Eocene (36.5–49 Ma									
Coast									
Chalk Bluffs, CA	m Eoc	39.2	-121.1	16.5 (c)	17.1				44, 48
Chalk Bluffs, CA	m Eoc	39.2	-121.1	14.4 (m)	15.0				24, 36, 40
Puget 9677, WA	e m Eoc	47.4	-122.0	16.7 (c)	19.8				38, 40, 44
Puget 9841, WA	e m Eoc	47.4	-122.0	18.6 (c)	21.6				38, 40, 44
Puget 9833, WA	m m Eoc	47.4	-122.0	21.5 (c)	24.4				38, 44
Puget 9678, WA	m m Eoc	47.4	-122.0	21.1 (c)	23.8				38, 44
Puget 9694, WA	m m Eoc	47.4	-122.0	22.1 (c)	24.7				38, 44
Puget 9731, WA	m m Eoc	47.4	-122.0	22.6 (c)	25.1				38, 44
Puget 9835, WA	~40	47.4	-122.0	14 (c)	16.3				38, 44
Puget 9835+9681,WA*	~40	47.4	-122.0	14.3 (c)	16.6				33, 38, 39
Puget 9680, WA	~40	47.4	-122.0	15.8 (c)	17.9				38, 39, 44
Puget 9680+11198,WA	~40	47.4	-122.0	15.8 (c)	17.9				33, 38, 39
Interior									
Republic, WA	48–49	48.9	-118.8	11.4 (c)	1.8	**0.5 ± 0.5**	0.8	0.8	44, 49
Chu Chua, BC	e m Eoc	51.4	-117.0	9.2 (c)	4.5	1.0	**1.7 ± 1.2**	0.4	44
Kamloops, BC	e m Eoc	50.7	-120.3	<13 (p)	> 0.8	>0.4	>0.5 ± 1.3	0.4	41
One Mile Ck, BC	e m Eoc	49.5	-119.5	9.3 (c)	4.9	1.1	**1.8 ± 1.2**	0.9	44, 49
Clarno nut beds, OR	48	45.0	-120.4	20+ (p)	0.3	**0.2 ± 0.6**	0.1	0.8	28, 49
Germer Basin, ID*	47–48	44.3	-114.2	9.5 (m)	8.6	1.8	**3.1 ± 1.7**	2.1	19
Green River, CO	45–48	41.0	-109.0	16.7 (c)	6.4	1.4	**2.3 ± 1.3**	2.6	26, 44
Green River, CO	45–48	39.8	-108.5	14.3 (m)	8.8	1.8	**3.1 ± 1.5**	2.6	27, 36
Upper Clarno, OR	~45	45.0	-120.4	~20 (c)	6.1	**1.3 ± 0.6**	2.2	~ 1	16, 42
Thunder Mountain, ID	45	44.9	-115.2	9–11 (p)	12.8	2.5	**4.5 ± 2.1**	2.3	4, 6, 41, 49
John Day Gulch, OR	41.0	44.5	-120.0	16 (c)	1.9	**0.5 ± 0.7**	0.8	0.8	42, 47
Sheep Rock Ck, OR	41.0	43.9	-120.3	<13 (p)	> 4.7	> **1.1 ± 0.7**	1.8	0.8	40, 48
Copper Basin, NV*	40.0	41.8	-115.7	12.1 (c)	4.1	0.9	**1.6 ± 1.6**	2.4	4, 33
Copper Basin, NV*	40.0	41.8	-115.7	8.5 (c)	7.7	1.6	**2.8± 1.8**	2.4	44
Bull Run, NV*	38–42	41.8	-115.7	<13 (p)	> 3.2	>0.8	>**1.3 ± 1.5**		4, 48
South Elko, NV*	40.0	40.5	-115.8	12.3 (c)	4.3	**1.0 ± 0.9**	**1.6 ± 1.6**	1.6	4, 33

Note: see notes for Table 4.2 for headings key, source list. † paleolatitude and paleolongitude given

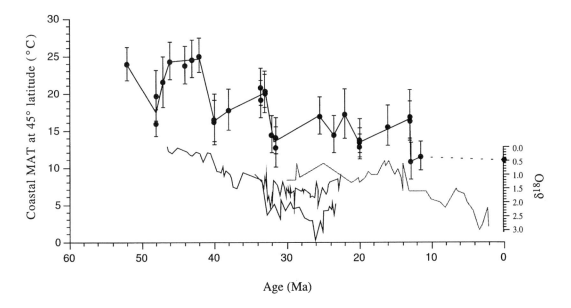

Fig. 4.2. MAT estimates for the coastal floras in Tables 4.2–4.4 in comparison with marine oxygen isotope data. MAT was projected from the paleolatitude of the flora to 45°C using a gradient of 0.3°C/° latitude for the Eocene and 0.4°C/° latitude for the Oligocene-Miocene. Marine isotope data: dark grey line = line 689, southern (Miller, 1992), medium grey line = line 522 Atlantic (Miller et al., 1988), and light grey line = north Pacific (Savin et al., 1985).

MATs in western Nevada increased over a very short period of time. The interpretation that this decline represents an elevational collapse is based on the observation that global climate and thus coastal MAT did not warm a similar amount between 16 and 12.5 Ma. The paleobotanical data suggest that there was a slight warming, and then a sharp drop in temperature after 13 Ma (Fig. 4.2). However, this important period is constrained by only five floras, dated by mollusks or paleobotany. Marine stable isotope data suggest that some global cooling might have occurred around this time (Miller et al., 1991) (Fig. 4.2). More coastal floras and radiometric ages on these floras are needed before the interpretation of elevational collapse, rather than global climate change, is accepted. Nonetheless, Nevada appears to be at or below modern elevations by the Late Miocene.

Miocene floras from Oregon and Washington appear to have had similar or slightly higher elevations than today, and paleoelevations for the Mohawk and Webber Lake floras suggest that the Sierra Nevada has been at modern elevations since about 15.5 Ma (Table 4.2). This conclusion is at odds with the long-held belief, based on paleogradients of Sierran rivers

(Lindgren, 1911), that rapid uplift occurred in the Late Tertiary, since 10 Ma (Huber, 1981). However, numerical models of fluvial erosion and deposition (Chase, 1992, Gregory and Chase, 1994b) imply that stream gradients are sensitive to climatic variables, particularly the magnitude and distribution of precipitation events.

Late Eocene–Oligocene

The majority of data from the Late Eocene–Oligocene time slice is from Colorado. The most important of these sites in Colorado is the 34.05 ± 0.11 Ma Florissant flora from the Front Range of the Southern Rocky Mountains. The flora was originally estimated to have grown at between 0.3–1 km of elevation (MacGinitie, 1953), but the author gave the caveat that there were no known coeval coastal floras which could be used for comparison. Since then, the paleoelevation of the flora has been reinterpreted using the foliar physiognomic method of paleoclimatic analysis and the increased database of coastal floras. Paleoelevation estimates in this study range from 2.6 ± 1.3 km to 3.0 ± 1.6 km to 3.7 ± 1.3 km based on foliar physiognomic

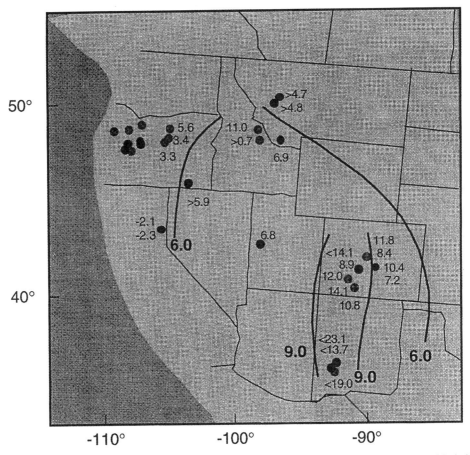

Fig. 4.3. MAT difference between coastal and interior floras (MAT$_{coast}$ - MAT$_{interior}$) for the Late Eocene (black dots) and Oligocene (grey dots). Data listed in Table 3. Coastal floras shown as dots with no accompanying numbers. If the 5.5°C/km value for γ is used, a 3.0°C MAT difference corresponds to a paleoelevation of 0.7 km (adjusting for higher paleo sea level). MAT differences for above this amount should be interpreted using the 3.0°C/km γ value. Using this lower γ, the 6.0 and 9.0°C contours correspond to 2.2 and 3.2 km, respectively. Paleolatitudes calculated by the computer program Paleogeographic Information System (Ross, 1992).

analysis by Meyer (1992), Gregory and McIntosh (1996), and Wolfe (1994), respectively. These estimates vary because each worker uses a different sample of the flora and a different multivariate method of analyzing the foliar physiognomic data.

Forest et al. (1995) estimated a paleoelevation of 2.9 ± 0.7 km for the Florissant flora using the enthalpy-based paleoaltimeter. This estimate is similar to the estimates derived from the MAT-based method using the 3.0°C/km value for γ, suggesting that the use of this γ value is reasonable for Colorado in the Late Eocene. All the paleoclimate estimates are similar to or higher than the modern elevation of 2.5 km.

Besides being well studied, the Florissant flora is important because it was deposited on a regional ero-

sion surface developed on the Front Range and the Denver Basin (Epis and Chapin, 1975, Gregory and Chase, 1994b). Outcrop elevations of the 36.6 Ma Wall Mountain ash-flow tuff suggest that the Denver Basin was only on the order of 200 m lower than the Front Range (Jacob and Albertus, 1985, Leonard and Langford, 1994). Thus the Florissant flora not only constrains the paleoelevation of the Front Range, but also constrains the elevation of the Great Plains to the east.

The other floras from Colorado, including the Pitch-Pinnacle flora from the west flank of the Sawatch Range, the Antero flora from South Park, and the Creede and Platoro floras from the San Juan volcanic field all suggest temperature differences from

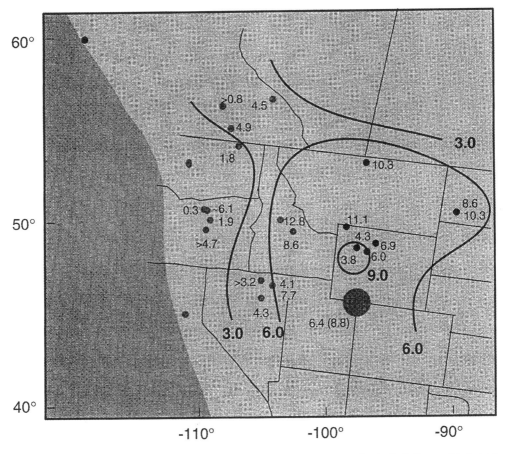

Fig. 4.4. MAT difference between coastal and interior floras (MAT_{coast} - $MAT_{interior}$) for the Early Eocene (black dots) and Middle Eocene (grey dots). Data listed in Table 4.4. Coastal floras shown as dots with no accompanying numbers. Using the 5.5°C/km value for γ, the 3.0°C contour corresponds to an elevation of 0.7 km (correcting for higher Eocene sea level). Above this value, the 3.0°C/km value for γ should be used. Using this lower γ, the 6.0°C contour corresponds to 2.2 km. Paleolatitudes calculated by the computer program Paleogeographic Information System (Ross, 1992).

the coast of around 9°C (Fig. 4.3). If the 3.0 °C/km γ value is used, these sites all have paleoelevations which are similar to their modern elevations (Table 4.3). Thus it appears that some of the Laramide uplifts and basins and volcanic areas of central and southern Colorado were at or above their modern elevations by the Late Eocene. As there is no data for Colorado for the Miocene, it is possible that this area underwent elevational collapse with renewed uplift in the Pliocene, or alternatively underwent uplift and then some elevational collapse during extension. However, the simplest option is that the elevation remained fairly constant through the Miocene and Pliocene (Fig. 4.5).

The MAT estimates for the three New Mexican floras from the present southern Rio Grande rift are all based on small samples and have large errors on the

MAT and paleoelevation estimates. For example, the first estimate of the MAT for the Pitch-Pinnacle flora from Colorado was based on a small sample; qualitative physiognomic analysis suggested that it represented a paleoMAT of 5–10°C (Meyer, 1986) (Table 4.3). However foliar physiognomic analysis of a larger sample suggested that the paleoMAT was 12.7 °C (Gregory and McIntosh, 1996) (Table 4.3). Thus it is likely that the MAT estimates for the New Mexican floras are too cool, and thus the MAT difference is too high. These floras probably grew at elevations similar to or moderately higher than the Colorado floras, but further study is needed. The floras today are at elevations of 2 km or less (Table 4.3), suggesting that Rio Grande extension lowered the regional elevation.

The House Range flora from the Basin and Range

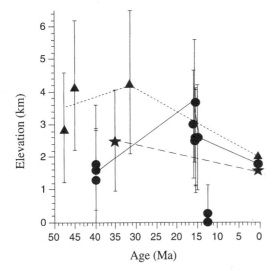

Fig. 4.5. Paleoelevations for sites in the Basin and Range. Sites in Nevada = circles, Utah = stars, and Idaho = triangles. Data given in Tables 4.2–4.4. Paleoelevations were normalized by adding the difference between the modern elevation of a specific site and the average of modern elevations for all the sites from that state (plotted at 0 Ma). The average was plotted if a site had more than one elevation estimate.

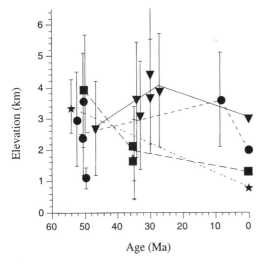

Fig. 4.6. Paleoelevations for sites in the Rocky Mountains. Sites in Colorado = triangles, Wyoming = circles, Montana = squares and North Dakota = stars. Data given in Tables 4.2–4.4, with the addition of the Wyoming paleoelevation estimate for 9.2 Ma, which was based on stable isotope data (Drummond et al., 1993). Paleoelevations were normalized by adding the difference between the modern elevation of a specific site and the average of modern elevations for all the sites from that state (plotted at 0 Ma). The average was plotted if a site had more than one elevation estimate.

province of Utah (Gregory, in preparation) is interpreted to have grown at 2.5 ± 1.5 km, which slightly lower than the elevations estimated for the Colorado floras (Table 4.3, Fig. 4.6). DeCelles (1994) developed a kinematic history for the foreland thrust belt in northeastern Utah based on conglomerate provenance data and cross-cutting relationships. If one assumes a 3–4° regional erosion surface developed on the thrust wedge, then the Paleocene elevation was several kilometers. These estimates are consistent with the estimated elevation. In contrast to Colorado, the paleoelevation of the House Range flora was significantly higher than the modern elevation of 1.7 km (Table 4.3). This downdrop was probably associated with known regional extension.

The Salmon and Cow Creek floras from Idaho are not well dated. They were deposited in grabens developed in the Challis volcanics (Wolfe and Tanai, 1987), and thus the only age constraint is the youngest age on the Challis volcanics, which is 45 Ma (Janecke, 1994). Wolfe has assigned an age of Middle to Late Eocene for these floras based on paleobotanical considerations (Wolfe and Tanai, 1987, Wolfe, 1994).

If these floras grew during 42–46 Ma, the warmest part of the Middle Eocene to Oligocene (Fig. 4.2), then they had elevations from at least 3.5 km to 7 km using the 3.0°C/km γ value. Note that the MAT for the Cow Creek flora was estimated using qualitative physiognomy as less than 13°C. The MAT difference was calculated using 13°C, thus the Cow Creek may have had a larger temperature difference with the coast. If the Salmon and Cow Creek floras grew during the coolest interval, 31–32 Ma (Fig. 4.2), then they had elevations of at least 0.3 to 3.9 km. The latter, more conservative estimates are plotted in the time-slice figure (Fig. 4.3). The high paleoelevation estimated for the Salmon flora as compared to the lower paleoelevation estimated for the Ruby floras of Montana is corroborated by provenance studies which suggest that in the late Middle Eocene to Oligocene, the Idaho rift basin drained to the east to the Renova Basin of southwestern Montana (Janecke, 1994).

There is also some controversy over the age of the Alvord Creek flora, also from the northernmost Basin and Range. Evernden and James (1964) obtain a Early Miocene radiometric age on an overlying basalt, but Wolfe and Tanai (1987) assign the flora to the Late Eocene based on several paleobotanical criteria. This flora, along with the floras from Montana and Idaho, with the exception of the Cow Creek flora, all have paleoelevations significantly greater than their modern elevations.

Early–Middle Eocene

The highest elevations in this time slice are in the Middle Eocene in the fold and thrust belt of Idaho. Paleodrainage analysis suggests that sediment was transported from east central Idaho to the southeast (Janecke, 1995), which is consistent with the paleobotanical data.

Wyoming was also possibly high, but data from the Early Eocene continental interior are difficult to interpret because of the lack of coeval coastal sites. The only known well-dated Early Eocene site, the Kultheith flora from the Gulf of Alaska, is problematic because it is located on the Yakutat block, a suspect terrane which was probably located off the coast of British Columbia in the Early Eocene (Wolfe, 1994). The southernmost paleolatitude for the Yakutat block was used to calculate coastal MAT at 45 ° latitude, as this gives cooler projected MATs and thus more conservative elevation estimates. The projected MAT of 24°C for Kultheith is reasonable because the Early Eocene is considered to be the warmest part of the Eocene, and this estimate is not appreciably warmer than the Middle Eocene temperatures of the Puget group sequence. However, the paleoelevation estimates for the Early Eocene should only be treated as general indications of paleoelevation until more coastal floras are documented.

The Upper Willwood flora from the Bighorn Basin and the Wind River and Kisinger Lakes floras from the Wind River Basin of western Wyoming are estimated to have grown at elevations between 1 and 2 km. The Yellowstone flora from the Absaroka volcanic field appears to have a higher paleoelevation, perhaps as high as 3 km. This paleoelevation datum is consistent with paleocurrent and provenance data of Groll and Steidtmann (1987), which shows that drainage in the Middle Eocene was from the Absaroka province to the south, across what is now the crest of the Wind River Range.

The paleoMATs derived from foliar physiognomy for these floras are corroborated by the occurrence of crocodilians and stable isotope studies. Markwick (1994) shows numerous crocodilian localities from western Wyoming; recall that these are interpreted to indicate paleoMATs of greater than 16°C. Foliar physiognomic estimates of MAT range from 15–17°C

for this area (Table 4.4). Oxygen isotope composition of paleosol carbonates, mammalian tooth enamel, and unionid bivalve shells from the Middle Paleocene–Early Eocene Bighorn Basin suggest paleoMATs of 7–22°C (Koch et al., 1995) which are in agreement with the MAT of 15.3°C derived from the leaf physiognomy of the upper Willwood flora (Greenwood and Wing, 1995).

The Green River flora from the Green River and Washakie basins of Utah appears to have grown at an elevation of 2–3 km, depending on which MAT estimate is used. However, the cooler multiple-regression derived MAT estimate of 14.3 °C (Table 4.4) is inconsistent with data from Markwick (1994) which indicates the presence of crocodilians in this area, though the exact age of these deposits are not given. Also, Dickinson et al. (1988) show that in the Middle Eocene, the Wind River Basin drained to the south into the Washakie and Piceance Basins, so the Green River flora was probably lower than the Wind River and Kisinger Lakes floras. Thus the 2.3 ± 1.3 km estimate is more likely. This paleoelevation is similar to the modern elevation of 2.6 km (Table 4.4). Lacustrine carbonates from the Green River lakes (Norris et al., 1996) have very low oxygen isotope ratios, suggesting that surrounding mountain ranges contributed significant snow melt. This implies that the ranges may have been as much as 3 km above lake level (Norris et al., 1996).

Elevations from the hinterland of the fold and thrust belt in Nevada and British Columbia are moderate, between 1–3 km (Table 4.4). The absence of crocodilians in the faunal dataset of Markwick (1994) supports the suggestion that this area had cool temperatures and thus fairly high elevations in the Middle Eocene.

Note that a majority of the coastal data for the Middle Eocene are from the Puget group floras (Table 4.4). There has been some controversy over the ages of these floras. Wolfe (1968) suggested that this 2 km-thick sequence represents most of Eocene time based on paleobotanical criteria and marine rocks associated with correlative floras. However, radiometric dates of Turner et al. (1983) suggest that the entire sequence was deposited in a few million years. Wolfe (1994) considers such rapid deposition and floral evolution unlikely, and suggests that the radiometric ages from the main body of the Puget were reset by a thermal event.

Paleocene–Cretaceous

Although quantitative paleobotanical estimates of paleoelevations are generally lacking in rocks as old as Paleocene and Cretaceous in the Cordillera, some estimates can be made based on structural cross sections, regional provenance and paleocurrent data from the Cordilleran foreland basin, and from simple isostatic arguments. Crustal shortening in the Sevier thrust belt overlapped in time and space with shortening in the Rocky Mountain foreland region (Laramide style uplifts). In general, deformation swept eastward from central Utah to central Wyoming and Colorado during the period 140–55 Ma.

Regional balanced cross sections in the transport direction across central Utah (Royse, 1993, DeCelles, 1994, Coogan et al., 1995) indicate a minimum of 150 km of horizontal shortening in the Sevier belt, and an additional 50 km of shortening can be estimated for the foreland region (Gries, 1983). Fig. 4.7 depicts a post-Paleocene (pre-regional extension) reconstruction of the Coogan et al. (1995) cross section, with the stratigraphic level of the upper erosional surface determined by regional provenance data (DeCelles et al., 1995, Lawton et al., 1996). At the end of Sevier belt shortening, the region of the present-day Sevier Desert basin in central Utah was occupied by a broad basement-involved antiform, referred to as the Sevier

culmination (DeCelles et al., 1995) or Sevier arch (Harris, 1959). The structural reconstruction indicates that the region west of the Canyon Range was underlain by the Canyon Range thrust sheet, composed of Proterozoic and Lower Paleozoic rocks, with only minor amounts of local relief at the topographic surface (Armstrong, 1968, Gans and Miller, 1983). If we assume that the Late Eocene House Range flora was deposited at an elevation of ~2.9 km (Table 4.3) soon after regional extension had begun (Constenius, 1996, Coogan and DeCelles, 1996), then the topographic crest of the Sevier culmination was probably substantially higher than 3.0 km elevation during Paleocene time. A similar case can be made for a structural transect across northern Utah during Paleocene time (DeCelles, 1994). Restoration of Cenozoic extension rebuilds crustal thickness to ~50 km in central Utah by the end of the Paleocene (Miller et al., 1991). Regional isostatic compensation would have supported ~4 km of regional elevation in the Sevier culmination area.

Local paleotopographic relief adjacent to the Beartooth and Bighorn Mountains uplifts in northwestern Wyoming can be reconstructed by incrementally retrodeforming range-bounding fold pairs within the constraints imposed by syntectonic Paleocene growth strata and provenance data from proximal syntectonic alluvial fan deposits (DeCelles et al., 1991). This type of analysis yields a minimum of 1.5 km of

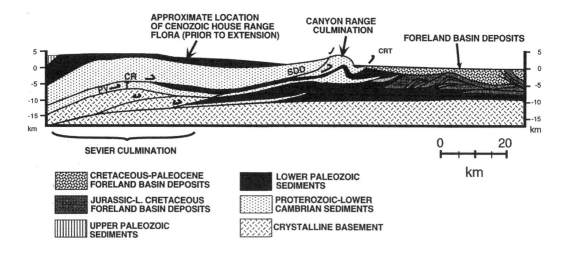

Fig. 4.7. Restoration to post-Sevier thrusting state of regional cross section across central Utah, after Coogan et al. (1995). Approximate location of the future site of the House Range flora is shown. By Late Eocene time (after initial extensional collapse), this location would have been at an elevation of 2.9 ± 1.1 km, presumably somewhat lower than its elevation during Paleocene time. This is used to set sea level for the vertical distance scale.

topographic relief on the eastern flank of the Beartooth uplift, and this value translates into a minimum paleoelevation insofar as the entire region was above sea level. Omar et al. (1994) used fission-track data from the Beartooth Mountains to demonstrate an unroofing event around 60 Ma, well within the Paleocene. They also interpreted fission-track-length data as requiring Mio-Pliocene uplift of about 4 km, but a possible alternative explanation is that the cooling recorded in the track lengths actually represents Late Tertiary incision of the Beartooth Plateau caused by climate change. A similar analysis using Paleocene deposits along the eastern flank of the Bighorn uplift yields nearly identical estimates for minimum paleotopographic relief (Hoy and Ridgway, 1995). Dettman and Lohmann (1993) obtained extremely negative $\delta^{18}O$ values from nonmarine mollusks in Paleocene fluvial channel deposits just north of the Bighorn uplift, suggesting the presence of permanent snowfields in the source drainage basin(s) to the south and west, possibly in the higher elevations of the Beartooth uplift.

Thus, although the data are sparse and several assumptions must be accepted before the cross sections themselves can be constructed, the regional picture seems to suggest that the Sevier belt hinterland was a broad topographic plateau at an elevation of at least 3–4 km. East of the Canyon Range (in central Utah) and the Wasatch Range (in northern Utah) the thrust belt sloped gently eastward into Wyoming and eastern Utah. Individual Laramide uplifts in Wyoming at the end of the Paleocene stood more than 1.5 km above adjacent nonmarine basins. A possible modern analog to the Late Cretaceous–Paleocene paleotopography of the Sevier thrust belt is the Altiplano–Eastern Cordillera–Subandes of southern Bolivia. The Altiplano is a topographic plateau with high elevation but little local relief; the Eastern Cordillera marks a major topographic boundary and a zone of intense thrusting; and the Subandes are similar to the reconstructed frontal part of the Sevier thrust belt.

Persistent Problems

Wyoming Paleoelevations

One of the intriguing aspects of the Early–Middle Eocene MAT data is that the basins in Wyoming are estimated to have stood 1–2.5 km above sea level. This is a reasonable hypothesis; Colorado also appears to be fairly high at this time, and the above tectonic evidence suggests that uplifts in Wyoming were above 1.5 km in the Paleocene. However the paleobotanical basis for this conclusion is not strong: the Wyoming data are speculative because of the poor coastal record of MAT, and only one Early–Middle Eocene site exists in Colorado, though it is very well studied.

The timing of Rocky Mountain uplift has important implications for the topographic history of the western United States. If both Wyoming and Colorado were low in the Early–Middle Eocene, then a major episode of uplift occurred between 45 and 35 Ma, as central and southern Colorado appear to have been at present elevations by the Late Eocene. Higher paleoelevations in the Early to early Middle Eocene would necessitate less late Middle to Late Eocene uplift.

Unfortunately, no floras from the Late Eocene through Miocene have been analyzed from Wyoming, so it is not known when the modern moderate to high elevations were attained in the central Rockies. Late Eocene land snails from east-central Wyoming have been interpreted to indicate paleoMATs of 16.5 °C (Evanoff, 1991), which would translate to a paleoelevation of 1.0–1.6 km depending on which γ value is used. This would suggest that some uplift had occurred by the Late Eocene, however, this MAT estimate is based on the nearest living relative method and thus is not as robust as physiognomic-based estimates.

Steidtmann et al. (1989) see geomorphic, sedimentological, and fission-track evidence for Oligocene uplift in the Wind River Range; perhaps this was a more regional event, and western and central Wyoming underwent uplift at this time. Given the large error bars, the Colorado data do not rule out an episode of minor Oligocene uplift. It seems reasonable to assume that Wyoming and Colorado had similar uplift histories, but it is possible that the Southern Rockies uplifted earlier than the central Rockies. The evidence could also reflect climate change.

Oxygen isotope analysis of 9.2 Ma lacustrine limestone from northwestern Wyoming suggest that the Gros Ventre range just to the west of the Wind River Range was from 0.5–2.3 km higher than the present elevation of 2.8 km (Drummond et al., 1993) (Fig. 4.6). Thus it appears that the high modern elevations were attained by at least the Late Miocene. More data is needed from the Early Eocene of Colorado and the Oligocene and Miocene of Wyoming to better constrain the uplift history of that area.

Colorado Plateau

There are, unfortunately, no MAT data or other quantitative paleoelevation estimates for the Colorado Plateau, which lies at a mean elevation of some 2 km despite its very modest amount of Cretaceous and younger deformation or magmatic activity. One interpretation of Colorado Plateau uplift is that it has occurred within the last 10 million years (Lucchitta, 1979). The Upper Miocene–Pliocene Bouse Formation, classified as marine (Buising, 1990), is now found at elevations above 0.4 km in the Colorado River trough near the Plateau. The apparent uplift of the Bouse is taken as dating Plateau uplift. This presents a strong contrast with evidence for the mountainous areas of western North America reviewed above, which points strongly toward uplift before the Late Tertiary. However, recent Sr isotope measurements on Bouse Formation carbonates by Patchett and Spencer (1995) show lacustrine rather than marine character, impeaching the Bouse as an elevational datum. Though it is reasonable to postulate an origin for uplift of the Plateau related to Laramide subduction (Spencer, 1996), there is no surviving hard evidence of the exact timing. We know only that it must have occurred between the last marine sediments deposited on the Colorado Plateau in Late Cretaceous and initiation of the Grand Canyon at 6 Ma (Lucchitta, 1990). We are not aware of any floras on the Colorado Plateau that might help constrain the uplift history, though there are igneous rocks to which cosmogenic paleoaltimetry or basalt vesicularity might be applied.

Eocene Equability

Sloan and Barron's (1990) results using the National Center for Atmospheric Research Community Climate Model (NCAR CCM) showed highly seasonal temperatures and cold winters for interior North America during the Eocene. The model paleogeographic boundary condition included paleoelevation, although the elevation model was not reported. One of the sensitivity tests performed by Sloan and Barron (1990) reduced the unspecified paleoelevations by half, with little discernible effect. Model results, even with 2X present CO_2, indicated cold-month mean temperatures of <0°C north of 40°N except near the lakes in which the Green River Formation was deposited (Sloan, 1994). With 6X present CO_2, the freezing zone was displaced to about 50°N (Sloan, 1994). In contrast, pa-

leobotanical data, even from lowland regions, are consistent with climates in which the cold-month mean temperature dropped below freezing only in the northernmost sites, in the Canadian Arctic (Greenwood and Wing, 1995). As discussed above, crocodilians are good paleoclimatic indicators. This is for a variety of reasons centered on evolutionary conservatism of physiology (M. Cassiliano, in preparation). Crocodilians are widespread in the Eocene, reaching a paleolatitude of about 55°N in the continental interior (Markwick, 1994). The distribution of crocodilians might provide a rough boundary for the Eocene western interior mountains. On both the western and eastern sides they delineate an area between roughly 110° and 120°W longitude (which was close to the paleolongitude).

Given that some of the paleobotanical estimates of climate in the Eocene are from elevations as high as 2300 m, the discrepancy between the climate-model results and the paleontologic record is striking. Paleoelevation is commonly not included in climate models because many mountain ranges and plateaus are smaller than the grid-cell size for climate-model calculations, which is about 4.5° latitude by 7.5° longitude for the NCAR CCM (Kutzbach et al., 1993, Barron and Moore, 1994). Even where it has been included, as in studies of the effects of the Tibetan and western North American plateaus (Ruddiman and Kutzbach, 1991, Kutzbach et al., 1993), the climate results over the plateaus themselves have not been reported.

Barron and Moore (1994) hypothesized that increased poleward transport of heat by the oceans was responsible for warm temperatures at high latitudes, possibly coupled with changes in deep water generation and in CO_2, depending on change in sub-ice heat flux. They also noted that these changes would not solve the problem of cold continental interiors. A wide range of model experiments consistently failed to warm those regions (Sloan and Barron, 1992). Barron and Moore (1994) offered two explanations. The first was that an interior seaway, similar to that in the Cretaceous, spanned North America and that all remnants have subsequently been destroyed. The other was that the model parameterizations that affect continental interior simulations are wrong. As Eocene nonmarine sediments are widespread in North America, the first explanation is implausible, leaving the conclusion that the model parameterizations are wrong. This represents an interesting departure from previous hypotheses, which questioned the validity of the Eocene climate data (e.g., Sloan and Barron (1991)). The importance of this controversy for this chapter is that if climate was warm even at high altitudes, which have not been simulated, the climate-model results are likely even more suspect.

Conclusions

Paleotopography is an important boundary condition for general circulation models constructed to help understand ancient climates, but we are only just beginning to be able to provide quantitative paleoaltimetry.

Paleontologic evidence depends mostly on the effect of elevation on local climate, and the effects of that climate preserved in organisms. Climate-based paleobotanical methods using leaf morphology provide most of the quantitative estimates of paleoaltitudes, and the smallest estimated errors. Care is needed to avoid circularity in using climate-based paleoelevations as boundary conditions for climate models.

Paleogeographic constraints can give general impressions of regional topography, but they are hard to quantify. Tectonic principles are essential in understanding the origin of topography, but it is difficult to extract precise estimates of elevation from preserved structures. Careful reconstruction of thrust belts can give clues to regional topographic gradients, especially when carefully related to sedimentological data. Sediments themselves can be used to constrain paleoelevational gradients, and they are particularly useful in providing the position of ancient sea level. New, not yet proven methods that depend more directly on air pressure include in-situ cosmogenic isotopes and basalt vesicularity.

We have compiled available paleoelevation data into Middle–Early Miocene (14–22 Ma), Oligocene–Late Eocene (22–45 Ma), Middle–Early Eocene (45–53 Ma), and Paleocene–Late Cretaceous (53–80 Ma) intervals. Data from these intervals show that the Western Cordillera of North America has been ele-

vated since at least the Late Cretaceous. Paleoelevations of 3 to 4 km were not unusual throughout the Tertiary. The Southern Rockies of Colorado have stood at near their present elevation since at least 35 Ma. There is some evidence of Oligocene–Miocene collapse of the Basin and Range province. Sparse estimates of Early Eocene paleoelevations, particularly from Wyoming, tend to be lower than Present elevations. Paleobotanical evidence suggests that uplift of the Sierra Nevada of California preceded 15 Ma. Uplift of the Colorado Plateau could have been at any time between initiation of the Grand Canyon at 6 Ma and Cretaceous. The available data agree in general with the hypothesis of Coney and Harms (1984) that high elevations in the western United States are residual from Mesozoic–Early Tertiary compressional thickening of the crust, with subsequent partial collapse of the Basin and Range province due to extensional thinning.

The persistence of elevation in the Western Cordillera since Mesozoic compression implies that present-day erosion rates are much higher than the average for the Cenozoic, a conclusion that must have climatological implications. In particular, sediment transport efficiencies must have been lower (Gregory and Chase, 1994b), most likely because maximum storm sizes were smaller.

Acknowledgments

We thank Jack Wolfe, Scott Wing, Peter Wilf, Jim Coogan, Gautam Mitra, and Brian Currie for helpful discussions; Jack Wolfe for proofing the data tables; and Jim Coogan for allowing us to modify his regional cross-section. The manuscript was materially improved by suggestions from the two editors, Tom Crowley and Kevin Burke, and three variably anonymous reviewers. We were supported in this work by NASA NAG5-3018 (CGC), NSF EAR93-17078 (KMG), NSF EAR90-23558 (JTP) and NSF EAR93-16700 (PGD).

References

Armstrong, R. L., 1968, Sevier Orogenic Belt in Nevada and Utah, *Geol. Soc. Am. Bull.*, *79*, 429–458.

Axelrod, D. I., 1944, The Alvord Creek Flora: in *Pliocene Floras of California and Oregon* R. W. Chaney (ed.), *Carnegie Inst. Wash. Pub. 553.*, 225–262.

Axelrod, D. I., 1956, Mio-Pliocene floras from west-central Nevada, *Univ. Calif. Publ. Geol. Sci.*, *33*, 321 p.

Axelrod, D. I., 1962, A Pliocene Sequoiadendron forest from western Nevada, *Univ. Calif. Publ. Geol. Sci.*, *39*, 195–268.

Axelrod, D. I., 1966, The Eocene Copper Basin flora of northeastern Nevada, *Univ. Calif. Publ. Geol. Sci.*, *59*, 124 p.

Axelrod, D. I., 1985, Miocene floras from the Middlegate Basin, west-central Nevada, *Univ. Calif. Publ. Geol. Sci.*, *129*, 279 p.

Axelrod, D. I., 1990, Environment of the middle Eocene (45 Ma) Thunder Mountain flora central Idaho, *Nat. Geog. Res.*, *6*, 355–361.

Axelrod, D. I., 1991, The Miocene Buffalo Canyon flora, *Univ. Calif. Publ. Geol. Sci.*, *135*, 128 p.

Axelrod, D. I., 1992, The middle Miocene Pyramid flora of western Nevada, *Univ. Calif. Publ. Geol. Sci.*, *137*, 50 p.

Axelrod, D. I., and Bailey, H. P., 1976, Tertiary vegetation, climate and altitude of the Rio Grande depression, New Mexico-Colorado, *Paleobiology*, *2*, 235–254.

Bailey, I. W., and Sinnot, E. W., 1916, The climatic distribution of certain types of angiosperm leaves, *Am. J. Botany*, *3*, 24–39.

Barron, E. J., and Moore, G. T., 1994, *Climate Model Application in Paleoenvironmental Analysis*, Society of Economic Paleontologists and Mineralogists and Society for Sedimentary Geology, Tulsa, Okla.

Becker, H. F., 1961, Oligocene plants from the upper Ruby River Basin, southwestern Montana, *Geol. Soc. Am. Mem.*, *825*, 127.

Berry, E. W., 1931, A Miocene flora from Grand Coulee, Washington, *U. S. Geol. Surv. Prof. Paper*, *170C*, 31–42.

Berry, E. W., 1934, Miocene plants from Idaho, *U. S. Geol. Surv. Prof. Paper*, *185E*, 97–125.

Bown, T. M., Rose, K. D., Simons, E. L., and Wing, S. L., 1994, Distribution and stratigraphic correlation of upper Paleocene and lower Eocene fossil mammal and plant localities of the Fort Union, Willwood, and Tatman formations, southern Bighorn Basin, Wyoming, *U. S. Geol. Surv. Prof. Paper 1540*, 269 p.

Brown, R. W., 1937, Additions to some fossil floras of the western United States, *U. S. Geol. Surv. Prof. Paper*, *186-J*, 163–206.

Buising, A. V., 1990, The Bouse Formation and bracketing units, southeastern California and western Arizona: Implications for the evolution of the proto-Gulf of California and the lower Colorado River, *J. Geophys. Res.*, *95*, 20111–20132.

Chaney, R. W., 1927, Geology and palaeontology of the Crooked River basin with special reference to the Bridge Creek Flora, *Cont. Palaeont. Carnegie Inst.*, 45–138.

Chaney, R. W., and Sanborn, E. I., 1933, The Goshen Flora of West Central Oregon, *Carnegie Inst. Wash. Pub.*, *439*, 103.

Chase, C. G., 1992, Fluvial landsculpting and the fractal dimension of topography, *Geomorph.*, *5*, 39–57.

Coney, P. J., and Harms, T. A., 1984, Cordilleran metamorphic core complexes: Cenozoic extensional relics of Mesozoic compression, *Geology*, *12*, 550–554.

Constenius, K. N., 1996, Late Paleogene extensional collapse of the Cordilleran foreland fold and thrust belt, *Geol. Soc. Am. Bull.*, *108*, 20–39.

Coogan, J. C., and DeCelles, P. G., 1996, Extensional collapse along the Sevier Desert reflection, northern Sevier Desert basin, western United States, *Geology*, *24*, 933–936.

Coogan, J. C., DeCelles, P. G., Mitra, G., and Sussman, A. J., 1995, New regional balanced cross section across the Sevier Desert region and central Utah thrust belt, *Geol. Soc. Am. Rocky Mtn. Sect., Abst. Prog.*, *27*, 7.

Davis, W. M., 1911, The Colorado Front Range, a study on physiographic presentation, *Assoc. Am. Geog., Annals*, *1*, 21–83.

Dean, J. M., and Smith, A. P., 1979, Behavioral and morphological adaptations of a tropical plant to high rainfall, *Biotropica*, *10*, 152–154.

DeCelles, P. G., 1994, Late Cretaceous-Paleocene synorogenic sedimentation and kinematic history of the Sevier thrust belt, Northeast Utah and Southwest Wyoming, *Geol. Soc. Am. Bull.*, *106*, 32–56.

DeCelles, P. G., Gray, M. B., Ridgway, K. D., Cole, R. B., Srivastava, P., Pequera, N., and Pivnik, D. A., 1991, Kinematic history of foreland uplift from Paleocene synorogenic conglomerate, Beartooth Range, Wyoming and Montana, *Geol. Soc. Am. Bull.*, *103*, 1458–1475.

DeCelles, P. G., Lawton, T. F., and Mitra, G., 1995, Thrust timing, growth of structural culminations, and synorogenic sedimentation in the type area of the Sevier orogenic belt, western United States, *Geology*, *23*, 699–702.

Dettman, D. L., and Lohmann, K. C., 1993, Seasonal change in Paleogene surface water $\delta^{18}O$: Freshwater bivalves of western North America, in *Climate Change in Continental Isotopic Records* P. K. Swart, K. C. Lohmann, J. A. McKenzie, and S. Savin (eds.), pp. 153–163, *Geophysical Monograph 78*.

Dickinson, W. R., and Suczek, C. A., 1979, Plate tectonics and sandstone compositions, *Am. Assoc. Petrol. Geol. Bull.*, *63*, 2164–2182.

Dickinson, W. R., Klute, M. A., Hayes, M. J., Janecke, S. U., Lundin, E. R., McKittrick, M. A., and Olivares, M. D., 1988, Paleogeographic and paleotectonic setting of Laramide sedimentary basins in the central Rocky Mountains region, *Geol. Soc. Am. Bull.*, *100*, 1023–1039.

Drummond, C. N., Wilkinson, B. H., Lohmann, K. C., and Smith, G. R., 1993, Effect of regional topography and hydrology on the lacustrine isotopic record of Miocene paleoclimate in the Rocky Mountains, *Palaeogeogr. Palaeoclimatol. Palaeoecol.*, *101*, 67–79.

Epis, R. E., and Chapin, C. E., 1975, Geomorphic and tectonic implication of the post-Laramide, late Eocene erosion surface in the Southern Rocky Mountains, in *Cenozoic History of the Southern Rocky Mountains,* B. F. Curtis (ed.), pp. 45–74, *Geological Society of America Memoir 144.*

Evanoff, E., 1991, Climatic change across the Eocene–Oligocene boundary: the White River sequence of the Central Rocky Mountains and Great Plains, USA, *Geol. Soc. Am. Abstr. Prog.*, *23*, 301.

Evernden, J. F., and James, G. T., 1964, Potassium-argon dates and the Tertiary floras of North America, *Am. J. Sci.*, *262*, 145–198.

Forest, C. E., Molnar, P., and Emanuel, K. A., 1995, Palaeoaltimetry from energy conservation principles, *Nature*, *374*, 347–350.

Gans, P. B., and Miller, E. L., 1983, Style of mid-Tertiary extension in east-central Nevada: in *Geologic Excursions in the Overthrust Belt and Metamorphic Core Complexes of the Intermountain Region, Nevada,* K. D. Gurgel (ed.), pp. 107–160, *Geological Society of America Field Trip Guidebook, Utah Geological and Mineral Survey Special Studies 59.*

Givnish, T., 1979, On the adaptive significance of leaf form, in *Topics in Plant Population Biology,* O. T. Solbrig, S. Jain, G. B. Johnson, and P. H. Raven (eds.), pp. 375–407, New York, Columbia University Press.

Givnish, T. J., 1984, Leaf and canopy adaptation in tropical forests, in *Physiological Ecology of Plants of the Wet Tropics,* E. Medina, H. A. Mooney, and C. Vazquez-Yanes (eds.), pp. 51–84, The Hague, Dr. W. Junk.

Greenwood, D. R., and Wing, S. L., 1995, Eocene continental climates and latitudinal temperature gradients, *Geology*, *23*, 1044–1048.

Gregory, K. M., in preparation, The late Eocene House Range flora: Paleoclimate and paleoelevation.

Gregory, K. M., and Chase, C. G., 1992, Tectonic significance of paleobotanically estimated climate and altitude of the late Eocene erosion surface, Colorado, *Geology*, *20*, 581–585.

Gregory, K. M., and Chase, C. G., 1994a, Stable isotope study of fluid inclusions in fluorite from Idaho: Implications for continental climates during the Eocene: Comment, *Geology*, *22*, 275–276.

Gregory, K. M., and Chase, C. G., 1994b, Tectonic and climatic significance of a late Eocene low-relief, high-level geomorphic surface, Colorado, *J. Geophys. Res.*, *99*, 20141–20160.

Gregory, K. M., and McIntosh, W. C., 1996, Paleoclimate and paleoelevation of the Oligocene Pitch-Pinnacle flora, Sawatch Range, Colorado, *Geol. Soc. Am. Bull.*, *108*, 545–561.

Gries, R., 1983, North-south compression of Rocky Mountain foreland structures, in *Rocky Mountain Foreland Basins and Uplifts,* J. D. Lowell (ed.), pp. 9–32, Denver, Rocky Mountain Association of Geologists.

Groll, P. E., and Steidtmann, J. R., 1987, Fluvial response to Eocene tectonism, the Bridger Formation, southern Wind River Range, Wyoming, in *Recent Developments in Fluvial Sedimentology,* F. G. Etheridge, R. M. Flores, and M. D. Harvey (eds.), pp. 263–268, *Society of Economic*

Paleontologists and Mineralogists Special Publication 39.

Harris, H. D., 1959, A late Mesozoic positive area in western Utah, *Am. Assoc. Petrol. Geol. Bull.*, *43*, 2636–2652.

Heller, P. L., and Paola, C., 1989, The paradox of Lower Cretaceous gravels and the initiation of thrusting in the Sevier orogenic belt, United States western interior, *Geol. Soc. Am. Bull.*, *101*, 864–875.

Heller, P. L., Beekman, F., Angevine, C., and Cloetingh, S. A. P. L., 1993, Causes of tectonic reactivation and subtle uplifts in the Rocky Mountain region and its effect on the stratigraphic record, *Geology*, *21*, 1003–1006.

Hickey, L. J., 1977, Stratigraphy and paleobotany of the Golden Valley Formation (Early Tertiary) of western North Dakota, *Geol. Soc. Am. Mem.*, *150*, 183 p.

Hoy, R. A., and Ridgway, K. D., 1995, Styles of footwall deformation in Eocene synorogenic conglomerates, east-central flank of the Bighorn Mountains, Wyoming, *Geol. Soc. Am. Rocky Mtn. Sect., Abst. Prog.*, *27*, A14.

Huber, N. K., 1981, Amount and timing of late Cenozoic uplift and tilt of the central Sierra Nevada, California – Evidence from the upper San Joaquin River basin, *U. S. Geological Survey Professional Paper 1197*, 28 p.

Huchison, J. H., 1982, Turtle, crocodilian, and champosaur diversity changes in the Cenozoic of the north-central region of western United States, *Paleobot. Paleoecol. Evol.*, *37*, 149–164.

Jacob, A. F., and Albertus, R. G., 1985, Thrusting, petroleum seeps, and seismic exploration: Front Range south of Denver, Colorado, in *Rocky Mountain Section Field Trip Guide 1985*, D. L. Macke, and E. K. Maughan (eds.), pp. 77–96, Society of Economic Paleontologists and Mineralogists, Tulsa, Okla.

Janecke, S. U., 1994, Sedimentation and paleogeography of an Eocene to Oligocene rift zone, Idaho and Montana, *Geol. Soc. Am. Bull.*, *106*, 1083–1095.

Janecke, S. U., 1995, Possible late Cretaceous to Eocene sediment dispersal along structurally controlled paleovalleys in the MT/ID thrust belt, *Geol. Soc. Am. Abstr. Prog.*, *27*, 16.

Johnsson, M. J., Stallard, R. F., and Lundberg, N., 1991, Controls on the composition of fluvial sands from a tropical weathering environment: Sands of the Orinoco River drainage basin, Venezuela and Colombia, *Geol. Soc. Am. Bull.*, *103*, 1622–1647.

Knowlton, F. H., 1926, Flora of the Latah formation of Spokane, Washington and Coeur d'Alene, Idaho, *U. S. Geol. Surv. Prof. Paper*, *140A*, 17–81.

Koch, P. L., Zachos, J. C., and Dettman, D. L., 1995, Stable isotope stratigraphy and paleoclimatology of the Paleogene Bighorn Basin (Wyoming, USA), *Palæogeogr. Palæoclimat. Palæoecol.*, *115*, 61–89.

Kutzbach, J. E., Guetter, P. J., Ruddiman, W. F., and Prell, W. L., 1989, Sensitivity of climate to late Cenozoic uplift in southern Asia and the American West: Numerical experiments, *J. Geophys. Res.*, *94*, 18393–18407.

Kutzbach, J. E., Prell, W. L., and Ruddiman, W. F., 1993, Sensitivity of Eurasian climate to surface uplift of the Tibetan Plateau, *J. Geol.*, *101*, 177–190.

Lawrence, J. A., and White, J. W. C., 1991, The elusive climate signal in the isotopic composition of precipitation: in *Stable Isotope Geochemistry: A Tribute to Samuel Epstein* H. P. Taylor, Jr., J. R. O'Neil, and J. R. Kaplan (eds.), pp. 169–185, *The Geochemical Society, Special Publication 3*.

Lawton, T. F., Sprinkel, D. A., and Waanders, G. L., 1996, Continental stratigraphy of the Cretaceous thrust belt and proximal foreland, central Utah: in *Recent Developments in the Sevier Thrust Belt, Western USA*, J. G. Schmitt, and D. Lageson (eds.), *Geological Society of America Special Paper* (in review).

Leonard, E. M., and Langford, R. P., 1994, Post-Laramide deformation along the eastern margin of the Colorado Front Range – a case against significant faulting, *Mount. Geol.*, *31*, 45–52.

Lindgren, W., 1911, The Tertiary gravels of the Sierra Nevada of California, *U. S. Geological Survey Professional Paper 73*, 98 p.

Lucchitta, I., 1979, Late Cenozoic uplift of the southwestern Colorado Plateau and adjacent lower Colorado River region: in *Plateau Uplift; Mode and Mechanism* T. R. McGetchin, and R. B. Merrill (eds.), pp. 63–95, *Tectonophysics 61*.

Lucchitta, I., 1990, History of the Grand Canyon and of the Colorado River in Arizona, in *Grand Canyon Geology*, S. S. Beus, and M. Morales (eds.), pp. 311–332, New York, Oxford University Press.

MacGinitie, H. D., 1937, The Flora of the Weaverville Beds of Trinity County California; with description of the plant-bearing beds., in *Eocene flora of Western America,* pp. 83–151, Washington DC, Carnegie Institution of Washington, *Contributions to Palaeontology.*

MacGinitie, H. D., 1941, *A Middle Eocene Flora from the Central Sierra Nevada,* Carnegie Institution of Washington Publication *534,* 178 pp.

MacGinitie, H. D., 1953, *Fossil Plants of the Florissant Beds, Colorado,* Carnegie Institution of Washington Publication 599.

MacGinitie, H. D., 1969, The Eocene Green River flora of northwestern Colorado and Northeastern Utah, *Univ. Calif. Publ. Geol. Sci., 83,* 140 p.

MacGinitie, H. D., 1974, An early middle Eocene flora from the Yellowstone-Absaroka volcanic province, northwestern Wind River basin, Wyoming, *Univ. Calif. Publ. Geol. Sci., 108,* 103 pp.

Manchester, S. R., 1990, Eocene to Oligocene floristic changes recorded in the Clarno and John Day Formations, Oregon, USA, in *Proceedings of the Symposium on Paleofloristic and Paleoclimatic Changes in the Cretaceous and Tertiary,* E. Knobloch, and Z. Kvacek (eds.), pp. 183–187, Prague, *International Geological Correlation Programme Project no. 216.*

Manchester, S. R., and Meyer, H. W., 1987, Oligocene fossil plants of the John Day Formation, Fossil Oregon, *Oregon Geol., 49,* 115–127.

Markwick, P. J., 1994, "Equability," continentality and Tertiary "climate': The crocodilian perspective, *Geology, 22,* 613–616.

McClammer, J. U., 1978, *Paleobotany and stratigraphy of the Yaquina flora (latest Oligocene-earliest Miocene) of western Oregon,* M.A. thesis, University of Maryland, College Park.

McKenzie, D., 1978, Some remarks on the development of sedimentary basins, *Earth Planet. Sci. Lett., 40,* 25–32.

Meyer, H. W., 1986, *An evaluation of the methods for estimating paleoaltitudes using Tertiary floras from the Rio Grande Rift vicinity, New Mexico and Colorado,* Ph.D. dissertation, University of California at Berkeley.

Meyer, H. W., 1992, Lapse rates and other variables applied to estimating paleoaltitudes from fossil floras, *Palæogeogr. Palæoclimat. Palæoecol., 99,* 71–99.

Miller, K. G., 1992, Middle Eocene to Oligocene stable isotopes, climate, and deep-water history:

The terminal Eocene event?, in *Eocene-Oligocene climate and biotic evolution,* D. R. Prothero, and W. A. Berggren (eds.), pp. 160–177, Princeton, New Jersey, Princeton University Press.

Miller, K. G., Feigenson, M. D., Kent, D. V., and Olsson, R. K., 1988, Oligocene stable isotope ($^{87}Sr/^{86}Sr$, $\partial^{18}O$, $\partial^{13}C$) standard section, Deep Sea Drilling Project Site 522, *Paleoceanography, 3,* 223–233.

Miller, K. G., Wright, J. D., and Fairbanks, R. G., 1991, Unlocking the ice house: Oligocene-Miocene oxygen isotopes, eustasy, and margin erosion, *J. Geophys. Res., 96,* 6829–6848.

Norris, R. D., Jones, L. S., Corfield, R. M., and Cartlidge, J. E., 1996, Skiing in the Eocene Uinta Mountains? Isotopic evidence in the Green River Formation for snow melt and large mountains., *Geology, 24,* 403–406.

Omar, G. I., Lutz, T. M., and Giegengack, R., 1994, Apatite fission-track evidence for Laramide and post-Laramide uplift and anomalous thermal regime at the Beartooth overthrust, Montana-Wyoming, *Geol. Soc. Am. Bull., 106,* 74–85.

Parrish, J. T., and Barron, E. J., 1986, *Paleoclimates and Economic Geology,* Society of Economic Paleontologists and Mineralogists Short Course 18.

Patchett, P. J., and Spencer, J. E., 1995, Sr isotopic evidence for a lacustrine origin for the Upper Miocene to Pliocene Bouse formation, Lower Colorado River trough: Implications for timing of Colorado Plateau uplift, *EOS, Trans. Am. Geophys. Union, 76,* F604–F605.

Potbury, S. S., 1935, The La Porte Flora of Plumas County, California, *Carnegie Institution of Washington Publication 465 pt. 2,* 29–81

Povey, D. A. R., Spicer, R. A., and England, P. C., 1994, Palaeobotanical investigation of early Tertiary palaeoelevations in northeastern Nevada: Initial results, *Rev. Paleobot. Palynol., 81,* 1–10.

Richards, P. W., 1952, *The Tropical Rain Forest,* Cambridge, University Press.

Rind, D., and Peteet, D., 1985, Terrestrial conditions at the Last Glacial Maximum and CLIMAP sea-surface temperature estimates: are they consistent?, *Quatern. Res., 24,* 1–22.

Ross, M. I., 1992, Paleogeographic Information Systems, Austin, Texas.

Royse, F., Jr., 1993, An overview of the geologic structure of the thrust belt in Wyoming, northern Utah, and eastern Idaho: in *Geology of Wyoming* A.

W. Snoke and others (eds.), pp. 272–311, *Geological Survey of Wyoming Memoir 5.*

Ruddiman, W. F., and Kutzbach, J. E., 1989, Forcing of late Cenozoic Northern Hemisphere climate by plateau uplift in southern Asia and the American West, *J. Geophys. Res.*, *94*, 18409–18427.

Ruddiman, W. F., and Kutzbach, J. E., 1991, Plateau uplift and climatic change, *Sci. Am.*, *264 (#3)*, 66–75.

Ruddiman, W. F., Prell, W. L., and Raymo, M. E., 1989, Late Cenozoic uplift in southern Asia and the American West: Rationale for general circulation modeling experiments, *J. Geophys. Res.*, *94*, 18379–18391.

Sahagian, D. L., and Maus, J. E., 1994, Basalt vesicularity as a measure of atmospheric pressure and paleoelevation, *Nature*, *372*, 449–451.

Sanborn, E. I., 1935, The Comstock flora of west central Oregon: in *Eocene flora of Western America* E. I. Sanborn, S. S. Potbury, and H. D. MacGinitie (eds.), pp. 1–28, *Carnegie Institution of Washington Contributions to Palaeontology 465.*

Savin, S. M., Abel, L., Barrera, E., Hodell, D. A., Keller, G., Kennett, J. P., Killingley, J. S., Murphy, M., and Vincent, E., 1985, The evolution of Miocene surface and near-surface marine temperatures: Oxygen isotopic evidence: in *The Miocene ocean: Paleoceanography and biogeography,* J. P. Kennett (ed.), pp. 49–82, *Geological Society of America Memoir 163.*

Scotese, C. R., Bambach, R. K., Barton, C., van der Voo, R., and Ziegler, A. M., 1979, Paleozoic base maps, *J. Geol.*, *87*, 217–277.

Sloan, L. C., 1994, Equable climates during the early Eocene: Significance of regional paleogeography for North American climate, *Geology*, *22*, 881–884.

Sloan, L. C., and Barron, E. J., 1990, "Equable" climates during Earth history?, *Geology*, *18*, 489–492.

Sloan, L. C., and Barron, E. J., 1991, Reply to comments on "'Equable' climates during Earth history?", *Geology*, *19*, 540–542.

Sloan, L. C., and Barron, E. J., 1992, A comparison of Eocene climate model results to quantified interpretations, *Palæogeogr. Palæoclimat. Palæoecol.*, *93*, 183–202.

Spencer, J. E., 1996, Uplift of the Colorado Plateau due to lithospheric attenuation during Laramide low-angle subduction, *J. Geophys. Res.*, *101*, 13595–13609.

Steidtmann, J. R., Middleton, L. T., and Shuster, M. W., 1989, Post-Laramide (Oligocene) uplift in the Wind River Range, Wyoming, *Geology*, *17*, 38–41.

Stute, M., Clark, J. F., Schlosser, P., Broecker, W. S., and Bonani, G., 1995a, A 30,000 yr continental paleotemperature record derived from noble gases dissolved in groundwater from the San Juan Basin, New Mexico, *Quatern. Res.*, *43*, 209–220.

Stute, M., Forster, M., Frischkorn, H., Serejo, A., Clark, J. F., Schlosser, P., Broecker, W. S., and Bonani, G., 1995b, Cooling of tropical Brazil (5°C) during the last glacial maximum, *Science*, *269*, 379–383.

Turner, D. L., Frizell, V. A., Triplehorn, D. M., and Naeser, C. W., 1983, Radiometric dating of ash partings in coal of the Eocene Puget Group, Washington: implications for paleobotanical stages, *Geology*, *11*, 527–531.

Webb, L. J., 1968, Environmental relationships of the structural types of Australian rain forest vegetation, *Ecology*, *49*, 296–311.

Wilf, P., 1996, How good are dicot leaves as thermometers?, *Abstracts Volume, Fifth Int. Org. of Palobot Conference*, *5*, 112.

Wing, S. L., 1987, Eocene and Oligocene floras and vegetation of the Rocky Mountains, *Ann. Missouri Botan. Gard.*, *74*, 748–784.

Wing, S. L., and Greenwood, D. R., 1993, Fossils and fossil climate: the case for equable continental interiors in the Eocene, in *Palaeoclimates and Their Modelling with Special Reference to the Mesozoic Era*, J. R. L. Allen, B. J. Hoskins, B. W. Sellwood, and R. A. Spicer (eds.), pp. 243–252, London, *Philosophical Transactions of the Royal Society, London B. Biological Sciences 341.*

Wolfe, J. A., 1964, Miocene floras from Fingerrock Wash southwestern Nevada, *US Geological Survey Professional Paper 454-N,* p.

Wolfe, J. A., 1968, Paleogene biostratigraphy of nonmarine rocks in King Country, Washington, *U. S. Geological Survey Professional Paper 571,* 33 p.

Wolfe, J. A., 1971, Tertiary climatic fluctuations and methods of analysis of Tertiary floras, *Palaeogeography Palaeoclimatology Palaeoecology*, *9*, 27–57.

Wolfe, J. A., 1979, Temperature parameters of humid to mesic forests of eastern Asia and relation to forests of other regions of the Northern Hemisphere and Australasia, *U. S. Geological Survey Professional Paper 1106*, 37 p.

Wolfe, J. A., 1981a, A chronologic framework for

Cenozoic megafossil flora of northwestern North America and its relation to marine geochronology, *Geol. Soc. Am. Spec. Pap.*, *184*, 39–47.

Wolfe, J. A., 1981b, Paleoclimatic significance of the Oligocene and Neogene floras of the northwestern United States, in *Paleobotany, Paleoecology, and Evolution*, K. J. Niklas (ed.), pp. 79–101, New York, Praeger Press.

Wolfe, J. A., 1987, An overview of the origins of the modern vegetation and flora of the northern Rocky Mountains, *Ann Missouri Bot. Gard.*, *74*, 785–803.

Wolfe, J. A., 1992a, An analysis of present-day terrestrial lapse rates in the western conterminous United States and their significance to paleoaltitudinal estimates, *U. S. Geol. Surv. Bull.*, *1964*, 35 p.

Wolfe, J. A., 1992b, Climatic, floristic, and vegetational changes near the Eocene–Oligocene boundary in North America, in *Eocene-Oligocene climate and biotic evolution*, D. R. Prothero, and W. A. Berggren (eds.), pp. 421–436, Princeton, Princeton University Press

Wolfe, J. A., 1993, A method of obtaining climatic parameters from leaf assemblages, *U. S. Geol. Surv. Bull.*, *2040*, 71 p.

Wolfe, J. A., 1994, Tertiary climate changes at middle latitudes of western North America, *Palaeogeogr, Palaeoclim., Palaeoec.*, *108*, 195–205.

Wolfe, J. A., and Hopkins, D. M., 1967, Climatic changes recorded by Tertiary land floras in northwestern North America, in *Tertiary Correlation and Climatic Changes in the Pacific*, K. Hatai (ed.), pp. 67–76, Tokyo, Pacific Science Congress.

Wolfe, J. A., and Schorn, H. E., 1989, Paleoecologic, paleoclimatic, and evolutionary significance of the Oligocene Creede flora, Colorado, *Paleobiology*, *15*, 180–198.

Wolfe, J. A., and Schorn, H. E., 1990, Taxonomic revision of the spermatopsida of the Oligocene Creede Flora, Southern Colorado, *U. S. Geol. Surv. Bull.*, *1923*, 40 p.

Wolfe, J. A., and Schorn, H. E., 1994, Fossil floras indicate high altitude for west-central Nevada at 16 Ma & collapse to about present altitudes by 12 Ma, *Geol. Soc. Am. Abstr. Prog.*, 521.

Wolfe, J. A., and Tanai, T., 1987, Systematics, phylogeny, and distribution of Acer (maples) in the Cenozoic of western North America, *Hokkaido Univ. J. Fac. Sci.*, *22*, 1–246.

Wolfe, J. A., and Wehr, W., 1987, Middle Eocene dicotyledonous plants from Republic, northeastern Washington, *U. S. Geol. Surv. Bull.*, *1597*, 25 p.

Ziegler, A. M., Rowley, D. B., Lottes, A. L., Sahagian, D. L., Hulver, M. L., and Gierlowski, T. C., 1985, Paleogeographic interpretation: with an example from the mid-Cretaceous of North America, *Ann. Rev. Earth Planet. Sci.*, *13*, 385–425.

Ziegler, A. M., Scotese, C. R., and Barrett, S. F., 1983, Mesozoic and Cenozoic paleogeographic maps, in *Tidal Friction and the Earth's Rotation, II*, P. Brosche, and J. Sündermann (eds.), pp. 240–252, Berlin, Springer-Verlag.

CHAPTER 5

Effects of Tropical Mountain Elevations on the Climate of the Late Carboniferous: Climate Model Simulations

Bette L. Otto-Bliesner

The Westphalian (~305 million years ago [Ma]), during the Late Carboniferous, is interesting from both a geologic and a climatic standpoint. The southern supercontinent of Gondwana and the tropical continent of Laurussia had collided closing the seaway that had separated them during the Devonian and the Earlier Carboniferous (Ziegler et al., 1979; Scotese, 1994). As a result of this collision, the Central Pangean Mountains, consisting of the Appalachian-Mauretanide-Hercynian orogenic belts, formed at tropical latitudes. Based on high-temperature/high-pressure metamorphics found for the late

Paleozoic Appalachian mountains, these mountains probably reached over 3 km in altitude (Ziegler et al., 1985).

The Westphalian was also the time when the extensive coal deposits of eastern North America, western Europe, and the Donetz Basin of the Ukraine formed in the (then) tropical zone between 12°S and 10°N (Fig. 5.1) (Ronov, 1976; Parrish et al., 1986). The presence of these coal deposits has been cited as evidence of a broad latitudinal band of tropical wetness (Raymond et al., 1989). The prerequisite for tropical coal deposits: luxuriant

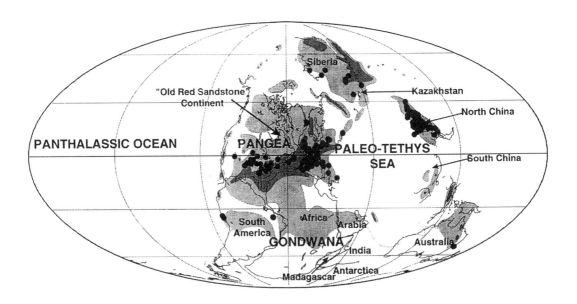

Fig. 5.1. Evidence for extensive Carboniferous coal deposits (Kraus et al., 1993) in the tropical zone between 12°S and 10°N along the northern flank of the Central Pangean Mountains. Figure illustrates Westphalian (Late Carboniferous; ~305 Ma) paleogeography and present-day continental outlines (Scotese, 1994). Shading and symbols: dark gray = mountains, medium gray = lowlands, light gray = land ice, white = oceanic areas, •'s = coals.

plant growth, swamp formation, and peat preservation, are at present generally met in only a narrow latitudinal band about the equator with monthly continuity of rainfall (Ziegler et al., 1987). Studies of well-preserved wood in Westphalian deposits reveal either faint or no growth rings indicating that the growth of the coal floras was not interrupted by seasonal dry spells (Creber and Chaloner, 1984; Chaloner and Creber, 1990).

Topography is known to be an important climatic control (Meehl, 1992). Careful attention needs to be given to prescription of paleotopography in modeling climate on tectonic time scales. Hay et al. (1990), using a pole-to-pole idealized continent with a north-south mountain range as the lower boundary condition in a Global Climate Model, found zonal patterns of continental precipitation and soil moisture and concluded that the present-day nonzonal climatic patterns may be related to large asymmetric topographic features such as the Himalayan-Tibetan Plateau complex. This mountain complex in southern Asia is known to enhance the summer Asian monsoonal circulation and precipitation. With a much lower Tibetan Plateau (Ruddiman and Kutzbach, 1990), a condition which existed before the Neogene, GCM simulations give a much weaker summer monsoon system (e.g., half-mountain (HM) results, Kutzbach et al., 1989; Ruddiman and Kutzbach, 1989). The uplift history of this plateau remains fragmentary although uncertainties are being reduced through application of a variety of methodologies (Harrison et al., this volume, chapter 3).

The extensive Central Pangean Mountain Belt has been proposed as a possible focus of year-round precipitation over the broad tropical zone of Westphalian coal formation, in particular, on the windward side of the mountain belt (Rowley et al., 1985). No present-day analogs exist for this tropical mountain belt. Present-day mountains at tropical latitudes occupy a much smaller areal extent and are significantly lower. The Himalayan-Tibetan Plateau complex is at subtropical to mid-latitudes. The simulations in this study are undertaken to provide an assessment of the thermodynamic and dynamic response of the atmosphere to an extensive tropical mountain belt. In particular, the sensitivity of the response to the elevation of the mountain belt is investigated. These simulations illustrate the importance of accurate reconstructions of paleotopography in applying global climate models to paleoclimate problems.

Details of Numerical Model

A set of sensitivity simulations for the Late Carboniferous was conducted using the National Center for Atmospheric Research (NCAR) GENESIS Global Climate Model, version 1.02 (Thompson and Pollard, 1995). The model consists of components for the atmosphere, ocean, and land-surface. The atmospheric component is based on the NCAR Community Climate Model version 1 (CCM1). The atmosphere is described by the equations of fluid motion, thermodynamics, and mass continuity along with representations of radiative and convective processes, cloudiness, precipitation, and the influence of mountain belts. The CCM1 code has been modified to incorporate new model physics for solar radiation, water vapor transport, convection, boundary layer mixing, and clouds. The horizontal resolution is R15 which corresponds to a 4.5° latitude by 7.5° longitude spectral transform grid, on average. A sigma-coordinate system is used in the vertical with 12 levels.

The ocean component of GENESIS contains a 50-meter deep mixed-layer ocean coupled to a thermodynamic sea-ice model. The ocean representation crudely captures the seasonal heat capacity of the ocean mixed layer but ignores salinity, upwelling, and energy exchange with deeper layers. Poleward oceanic heat transport is included in present-day simulations using the prescribed latitudinal variation from the "0.5 X OCNFLX" case of Covey and Thompson (1989). A six-layer sea-ice model, patterned after Semtner (1976), predicts local changes in sea ice thickness through melting of the upper layer and freezing or melting on the bottom surface. Fractional sea ice coverage is included, as is a reduction in the albedo as the surface temperature of the sea ice approaches 0°C.

The land-surface component (Pollard and Thompson, 1995) incorporates a land-surface transfer model (LSX) which accounts for the physical effects of vegetation, a three-layer thermodynamic snow model, and a six-layer soil model. The soil model layers are chosen so that the top layer is thin enough (5 cm) to capture diurnal temperature variations, while the total thickness (4.25 m) is capable of resolving seasonal variations. Heat is diffused linearly and moisture nonlinearly, with a provision for soil ice. Water can infiltrate into the soil at a rate capped by the rate of downward soil drainage and diffusion in the upper layer. Any excess rainfall minus evaporation and any excess snowmelt become surface runoff.

The GENESIS model adequately reproduces observed present-day patterns of precipitation over tropical land with the Intertropical Convergence Zone (ITCZ) precipitation maximum centered at 10°S latitude in January shifting northward to 8°N in July (Thompson and Pollard, 1995). Soil moisture patterns and their seasonal variations are strongly controlled by the model precipitation. The model's global and annual mean precipitation (4.5 mm day^{-1}) is considerably greater than observed (3.1 mm day^{-1}). The zonal mean values are approximately 60% greater than observed because of high precipitation associated with ITCZ convection. This overestimate occurs over both land and ocean but is a bias in only the magnitude not the location of the precipitation maxima. It has been corrected in a newer version of GENESIS with decreased surface drag over oceans.

Climate Sensitivity Simulations and Boundary Conditions

The sensitivity of the Late Carboniferous tropical climate was examined in a series of sensitivity simulations exploring the effects of four different tropical mountain elevations (Table 5.1). The baseline Carboniferous simulation (hereinafter referred to as TMNONE) was performed with tropical land elevations uniform at 0.5 km. The sensitivity simulations involved progressively elevating the Central Pangean Mountains to test the sensitivity of the tropical climate to uncertainties in the reconstructed height of these mountains. All other boundary conditions are kept the same in the simulations (Table 5.2).

Paleogeography and Topography

Paleogeography for the model simulations adopted the ~305 Ma reconstruction (Fig. 5.1) of C.R. Scotese and J. Golonka (Scotese, 1994). The plate tectonic reconstruction is based primarily on the paleomagnetic data compiled by Van der Voo

Table 5.1. Topography Specification

Experiment	Prescribed Values
Tropical No Mountain TMNONE	500 m lowlands 500 m tropical mountains
Tropical Low Mountain TMLOW	500 m lowlands 1000 m tropical mountains
Tropical Mountain TMREAL	500 m lowlands 1000–3000 m tropical mountains
Tropical High Mountain TMHIGH	500 m lowlands 5000 m tropical mountains

(1993) with modifications over Siberia (Khramov and Rodionov, 1980). The Late Carboniferous marks the beginning of the formation of Pangea. The southern supercontinent of Gondwana, consisting of South America, Africa, Australia, Antarctica, and parts of Asia, and the tropical continent of Laurussia, made up of North America and parts of Eu-

Table 5.2. Boundary Conditions for GENESIS Model Simulations (see text and figures for details)

Boundary Conditions	Prescribed Values
Solar luminosity	1328.9 W m^{-2} (3% less than present)
Orbital parameters eccentricity/obliquity/ precession	0 / 23.4° / 0
Atmospheric CO_2	Present-day
Ozone mixing ratio	Present-day
Land-sea distribution	Scotese (1994) (Fig. 5.1)
Topography	Table 5.1, Fig. 5.2
Land ice	South Pole to ~45°S 500–3000 m elevation (Figs. 5.1 and 5.2)
Soil characteristics texture/color	Intermediate between sand and clay / Intermediate
Vegetation type	No vegetation—bare soil
Oceanic poleward heat flux	Present-day averaged to be symmetric about equator

rope, had collided closing the seaway that had separated them earlier. As a result of this Himalayan-type collision, a large tropical mountain range was formed. Data also indicate mountainous belts over the Northern Hemisphere land masses and at high southern latitudes. In these simulations, nonmountainous land is set to an elevation of 0.5 km, while mountain ranges outside the tropics are assigned elevations ranging up to 3 km using topographic and plate tectonic analogs (Ziegler et al., 1985).

With a Himalayan-type collision between Gondwana and Laurussia, considerable uplift is conceivable but large uncertainties exist. Tropical mountain heights were thus varied among the simulations. In the TMNONE simulation, tropical mountain heights were set to 0.5 km, the same as the surrounding lowlands. The TMLOW, TMREAL, and TMHIGH simulations prescribe successively higher elevations to the tropical mountain range (Table 5.1). The TMREAL simulation allows variation in the height of these mountains and is considered the most "realistic." The TMHIGH simulation is an extreme orography run. The model representation (spectrally smoothed to R15 resolution) of the land elevations is given in Figure 5.2.

These initial sensitivity simulations all assume no vegetation and soil with texture intermediate between sand and clay and with intermediate color. At ocean points, poleward oceanic heat transport was modified for the Late Carboniferous simulations by

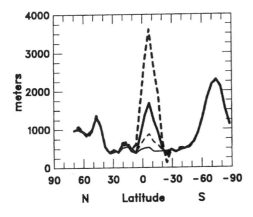

Fig. 5.2. Comparison of model representation of land elevations (in meters). Figure illustrates latitudinal distribution, averaged for continental areas only, for TMNONE (thin solid), TMLOW (thin dashed), TMREAL (heavy solid), and TMHIGH (heavy dashed) simulations. The TMREAL simulation is considered the most realistic based on topographic and plate tectonic analogs.

averaging Covey and Thompson's (1989) data to make it symmetric about the equator.

Ice Sheet and Snowcover

Glacial deposits (tillites) dated to the Late Carboniferous have been found on the continents of Australia, Africa, South America, and Antarctica as well as India (Crowell, 1983; Caputo and Crowell, 1985; Veevers and Powell, 1987) and suggest the possibility of four ice sheets with total areal extent of 18×10^6 km^2 (Parrish et al., 1986; Veevers and Powell, 1987; Crowley and Baum, 1991). A more extensive single ice sheet covering nearly 70% of the continent of Gondwana from the South Pole to roughly 45°S latitude and with areal extent of 42×10^6 km^2 has been proposed by Scotese and Golonka (personal communication, 1993). The rationale for a single massive ice sheet assumes extensive destruction of glacial evidence, especially during highly erosional post-glacial environments. This proposed glacial ice reconstruction gives an estimated glacio-eustatic sea level fluctuation of 190 m which is on the high end of the range of sea level estimates (70–200 m) from cyclothem fluctuations (Crowley and Baum, 1991). For comparison, the area of the Laurentide ice sheet at 21,000 BP is estimated to be about 12×10^6 km^2 (Crowley and Baum, 1991). This extensive ice sheet was prescribed in all simulations (Fig. 5.1) and ice-sheet heights ranging up to 3 km were assigned using the area-volume relationship employed in the CLIMAP studies (Paterson, 1972) (Fig. 5.2).

Atmospheric Carbon Dioxide

Models of the geochemical cycle (Walker et al., 1981; Garrels and Lerman, 1984; Budyko et al., 1987; Berner, 1994) and independent isotopic measurements of paleosols (ancient soils) (Mora et al., 1996) and marine sediments (Freeman and Hayes, 1992; Yapp and Poths, 1992) suggest that atmospheric CO_2 has varied considerably over the last 600 m.y. because of imbalances in the effects of weathering, organic burial, and metamorphism. A geochemical model of Berner estimated atmospheric CO_2 levels as high as 17 times present concentrations for the Phanerozoic (Berner, 1994). For 305 Ma atmospheric CO_2 concentrations are predicted to be similar to the present (Berner, this volume, chapter 12). Input of CO_2 into the atmosphere by global degassing might have been low if seafloor spreading rates diminished with the formation of Pangea, while a significant draw down of atmospheric CO_2 is expected to have occurred between 380 and 350 Ma in response to enhanced burial of organic carbon as a consequence of the rise and spread of vascular land plants (Berner, this volume, chapter 12). Atmospheric CO_2 concentrations equal to those in the present-day control run of GENESIS (340 ppm) were adopted for these simulations.

Solar Luminosity

The solar constant can vary because of changes in the sun's output and the Earth's orbital dynamics. Astrophysicists calculate that the sun has brightened over the age of the Earth ($\sim$ 4.6 billion years), as hydrogen has been converted to helium in the sun's core (Endal and Sofia, 1981), having been 30–40% dimmer when the Earth formed. Over the Phanerozoic, this translates to an approximate 6% increase in solar luminosity (Crowley et al., 1991). The solar luminosity at 305 Ma has been set to 3% less than present (present-day value: 1370 W m^{-2}) in all simulations. Milankovitch cycles of the Earth's orbital dynamics have been documented in records for the Pleistocene, but the periods and magnitude of this forcing before the Pleistocene are uncertain (Berger, 1978; Berger et al., 1989). A circular solar orbit (eccentricity = 0) with present-day obliquity (23.4°) is therefore assumed. This orbital configuration omits any differences in solar forcing between the Northern and Southern Hemispheres in their respective seasons.

Climatology of the Tropics at Present

The present-day tropical climate between 30°S and 30°N is influenced by the equatorial low pressure belt, the subtropical highs, and the trade winds. Near the equator, the climate is characterized by high humidity and cloud cover and daily showers of heavy, localized rainfall. Warm temperatures and abundant rainfall year-round combine to favor dense, broadleaf, evergreen forest vegetation. Rainfall in these regions is the result of the converging air flow of the trade winds from the North- and Southern Hemispheres into the equatorial low pressure belt. This convergence forces warm, moist,

unstable air upward forming clouds and generating showers. This narrow zone of converging air is termed the ITCZ. Subtropical high pressure belts of subsiding air, clear skies, and sparse precipitation dominate at 25–30° latitude.

Because of the tilt of the Earth's axis with respect to the Earth's orbital plane, the sun is more directly overhead in the Northern Hemisphere in July and in the Southern Hemisphere in January. The zone of maximum surface heating in the tropics shifts seasonally with it. In response, the major pressure systems and wind regimes affecting tropical climate shift north in July and south in January. Thus, the ITCZ and associated abundant rainfall is concentrated north of the equator at $\sim$5–10°N in July and south of the equator at $\sim$5–10°S in January (Fig. 5.3). The region that receives abundant rainfall year-round is thus confined to a band 8–10° wide near the equator (Fig. 5.4). On either side of this rainforest zone, a wet summer–dry winter climate prevails because of the seasonal shifts of the ITCZ and subtropical highs which alternate in controlling the precipitation regime (Fig. 5.3).

Results of Sensitivity Experiments

The TMREAL simulation was started from a zonally-symmetric wind and temperature initial state and integrated through 18 seasonal cycles. Approach to equilibrium is discussed in Otto-Bliesner (1996). All other simulations were started from year 12 of the TMREAL simulation and integrated for 6 years. Mean climate statistics represent 5-year averages calculated from years 14–18 of the simulations. The standard deviation of the monthly average statistics for the individual years from the 5-year average is also computed. This is a measure of the interannual variability inherent in the model statistics. The standard deviations can be used in conjunction with the Student t-test (Panofsky and Brier, 1965) to separate signal from noise (Chervin and Schneider, 1976) when comparing simulations. For this set of simulations, the difference between two simulations is statistically significant at the 95% level if the change is greater than 1.75 times the standard deviation. That is, there is 5% chance that this difference could occur by chance. Those differences less than this limiting value are not considered statistically different and fall within the year-to-year variability of the model results.

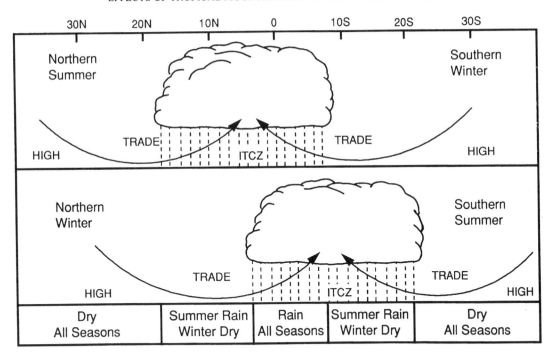

Fig. 5.3. Schematic of pressure and air flow and resultant precipitation during the course of the year. Migration of zone of tropical convergence and rising air into summer hemisphere results in zones of winter dryness banding the tropical everwet zone (Adapted from Trewartha and Horn, 1980).

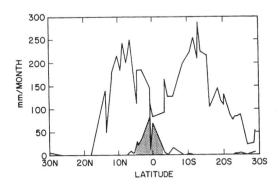

Fig. 5.4. Present-day observed January and July rainfall versus latitude for Africa. Note that the position of the ITCZ and thus maximum rainfall is located at 13°S in January and 6°N in July resulting in only a narrow band at the equator (3°S–5°N) where precipitation is greater than 20 mm in both months. (After Ziegler et al., 1987).

The tropical surface patterns and hydrologic response are strongly influenced by the latitudinal migration of the sun with seasons. Small changes in surface heating lead to substantial excursions of surface convergence and precipitation. The location of the Central Pangean Mountains just south of the equator result in these mountains having their largest effect during the boreal summer (June–July–August) when maximum solar heating occurs north of the equator. Model comparisons for July will thus be emphasized.

Temperature

Increasing the elevation of the Central Pangean Mountains significantly cools July surface temperatures over the mountainous areas (Fig. 5.5). In the TMNONE simulation (Fig. 5.5a), July surface temperatures at 7°S over land average 18°C, a cooling of 5°C compared to the present because of the lowered solar luminosity at 305 Ma. Raising the mountain heights from 0.5 to 1 km further lowers July temperatures over elevated tropical land by ~2°C in the TMLOW simulation, a statistically significant cooling compared to the small interannual temperature variability at these latitudes (Figs. 5.5b). The "realistic" mountain heights in the TMREAL simulation lead to a pronounced cooling of up to 14°C (Fig. 5.5f), although July temperatures remain above freezing even over the highest elevations (Fig. 5.5c). Temperature gradients on the northern side of the mountains are greatly enhanced. July temperatures are below freezing over the higher elevations in the TMHIGH simulation with surface temperatures as low as -20°C and average land temperatures at 7°S of -4°C (Figs. 5.5d, 5.5g). Cooling over the higher elevations is significant in all simulations, but these changes are local. Significant remote changes in temperature do not develop. These results are in accord with an effective length scale of temperature perturbations of ~1500 km found in previous modeling studies (North et al., 1992; Crowley et al., 1994).

Sea Level Pressure and Surface Winds

Temperature changes induced by the Central Pangean Mountains induce changes in the July sea level pressure patterns over the tropical land areas and force changes in the surface wind patterns (Fig. 5.6). In the TMNONE simulation, the equatorial low pressure belt follows the sun shifting north of the equator in July (Fig. 5.6a). Surface flow is predominantly from the east with appreciable cross-equatorial surface flow from the Southern to the Northern Hemisphere (Fig. 5.6e). Convergence of the surface winds place the ITCZ at approximately 20°N latitude. Onshore flow from the eastern Panthalassa Ocean at 5°N leads to weak surface convergence into the east-west oriented trough of low pressure.

The addition of mountains over the tropical land areas affects the seasonal excursion of the equatorial low pressure belt and thus the regions of surface convergence. In the TMLOW simulation the effect is small (Figs. 5.6b, 5.6f). Much larger changes occur in the TMREAL simulation (Figs. 5.6c, 5.6g). A band of low pressure and resulting surface convergence over the tropical continental area is still located north of the equator, but a pronounced trough

of low pressure also extends southwestward along the northern edge of the Central Pangean Mountain Range, which serves as an elevated heat source in July. As a result, strong low-level convergence into this low pressure system from both the eastern and western Panthalassa Ocean occurs in the vicinity of the mountains. Large gradients in the prescribed topography in the TMHIGH simulation result in less organized (and noisier) patterns in sea level pressure in their vicinity (Fig. 5.6d), although the general pattern is similar to the TMREAL simulation. In addition, surface flow is disrupted with cold, downslope flow along the southern edge of these high mountains (Fig. 5.6h).

Vertical Motion

Surface convergence patterns can be better delineated by examining vertical motion fields. In Figure 5.7, latitudinal cross sections of vertical motion along 0° longitude give the predicted vertical motion from the surface to the top of the atmosphere. This longitude transects the Central Pangean Mountains and the region of tropical coal deposits. The shift of the ITCZ north of the equator in July in the TMNONE and TMLOW simulations leads to maximum upward motion at 12°N (Figs. 5.7a, 5.7b). Sinking motion occurs south of the equator. Upward vertical motion covers a much broader latitudinal band in the TMREAL and TMHIGH simulations extending from 20°N to 10°S as the mountains force converging surface flow upward (Figs. 5.7c, 5.7d).

Precipitation

Maximum precipitation (Fig. 5.8) over the tropical land areas coincides with regions of maximum surface convergence and upward vertical motion. In the TMNONE simulation, a band of precipitation extends from 5°N over western Laurussia to 15°N over eastern Laurussia (Fig. 5.8a). Little precipitation falls from 5–25°S. Although precipitation increases over tropical land areas in July in the TMLOW simulation, the changes are not significant (Figs. 5.8b, 5.8f). Raising the mountain peaks to 2–3 km elevation significantly enhances upslope precipitation on the northern flanks of the Central Pangean Mountains (Figs. 5.8c, 5.8e, 5.8g). The precipitation pattern has a bimodal distribution with a band of precipitation (continental average of

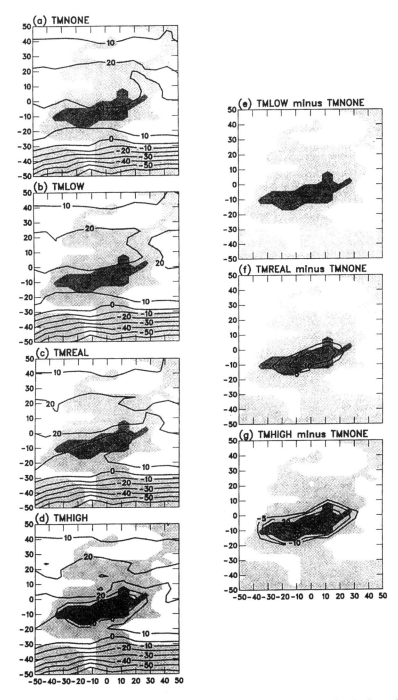

Fig. 5.5. Comparison of July surface air temperatures (°C) (temperature at 2 m above the surface in the model simulations). The elevations of the tropical mountains in the TMREAL and TMHIGH simulations result in substantial cooling in their vicinity. Surface temperatures remain below freezing year-round over the higher elevations in the TMHIGH simulations. The location of the Central Pangean Mountains is shaded.

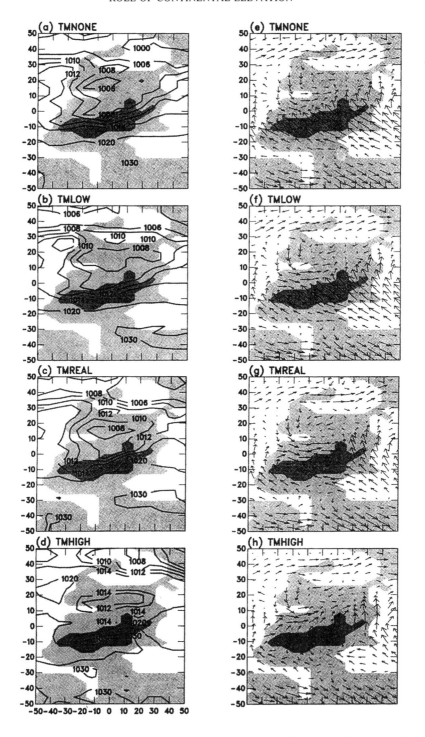

Fig. 5.6. Changes in the July sea level pressure (mb) patterns force changes in the surface wind patterns (vector length: 1°/1 m s⁻¹). Strong cross-equatorial flow from the Southern to Northern Hemisphere dominates in the TMNONE and TMLOW simulations. Greater mountain heights in the TMREAL simulation induce converging surface winds on the northern edge of the Central Pangean Mountains but leads to unrealistic downslope winds in the TMHIGH simulation.

(a) TMNONE

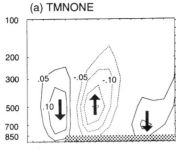

(b) TMLOW

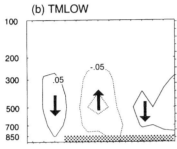

(c) TMREAL

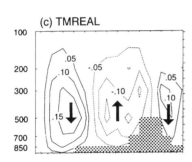

(d) TMHIGH

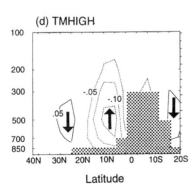

Latitude

Fig. 5.7. Converging surface winds in the tropics result in rising vertical motions. Figure shows latitudinal cross-sections of vertical motion (Pa s^{-1}, negative indicates upward motion) in July at 0° longitude. Rising air occurs at 10–15°N in July in all the simulations. The higher mountains in TMREAL and TMHIGH induce rising motion at 0–5°S (the northern flank of the Central Pangean Mountains), a region of sinking air in July in the TMNONE and TMLOW simulations.

9 mm/day at 10°N) associated with the main portion of the ITCZ north of the equator and a localized maximum of precipitation (continental average of 6 mm/day) over the northeastern flanks of the mountains. Precipitation decreases north of the latter maximum although not significantly. These precipitation decreases are more pronounced in the TMHIGH simulation and significant along the coastal regions at 10°N (Figs. 5.8d, 5.8h). Because of the disruption of low-level flow patterns by the unrealistically steep mountains prescribed in the TMHIGH simulation, the precipitation increases along the northern edge of the mountains in a less well-organized but still significant way.

Soil Moisture

Soil moisture predictions by the model give the time-integrated effects of precipitation and evaporation on the surface moisture (Fig. 5.9). July soil moisture is moderate to high along the eastern tropical coastlines of the Westphalian continent in the TMNONE simulation but is very low in the interior regions south of the equator (Fig. 5.9a). Soil moisture increases in the western regions, in the TM-LOW simulation, but is still low (Fig. 5.9b). Increased precipitation along the northern flanks of the mountains in the TMREAL simulation results in significantly increased soil moisture in those regions (Fig. 5.9c). The soils in western Laurussia are still quite dry south of the equator. Although precipitation does not increase as greatly in the TMHIGH simulation as in the TMREAL simulation, soil moisture is significantly enhanced with a broad region > 80% as a result of significantly decreased evaporation over these mountainous regions with their cold surface temperatures (Fig. 5.9d). Soil moisture in these regions consists of ice crystals in the soil pores. The TMHIGH simulation gives unrealistically low soil moisture over eastern Laurussia north of the equator.

Seasonal Variation

The coals of the Westphalian are thought to have been formed under conditions of year-round wetness (Ziegler et al., 1987; Chaloner and Creber, 1990). In figure 5.10, the seasonal variation of precipitation, evaporation, and soil moisture on the north-

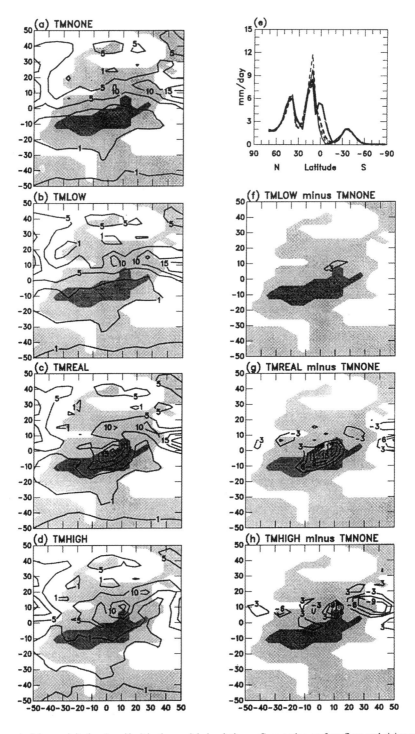

Fig. 5.8. Changes in July precipitation (mm/day) in the model simulations. Converging surface flow and rising motion at 0–5°S in the TMREAL simulation significantly enhances precipitation. The actual precipitation amounts are given in (a) – (d) with precipitation differences from the TMNONE simulation given in (f) – (h). The average continental precipitation as a function of latitude is given in (e). See Fig. 5.2 for legend.

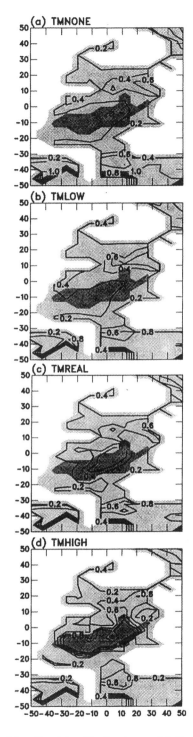

Fig. 5.9. July soil moisture (fractional amount) indicates dry conditions in the tropical band just south of the equator in the TMNONE and TMLOW simulations. Enhanced precipitation in the TMREAL simulation and much cooler temperatures in the TMHIGH simulation result in wetter soil conditions over the Central Pangean Mountains.

ern edge of the mountains (latitude 2.3°S, longitude 0°) delineates how well each simulation meets this requirement. In the TMNONE simulation, precipitation, evaporation, and soil moisture all exhibit large seasonal cycles with maximum amounts from November–March and minimum amounts from June–September. Soil moisture varies considerably from a high of 0.5–0.6 in February–March to a low of 0.1–0.2 from June–September. Precipitation is negligible from June–September. The TMLOW simulation also exhibits large seasonal variations. Precipitation increases in this area in September in the TMLOW simulation, but much of this increase is removed from the soils by increased evaporation. Soil moisture exhibits very little seasonal variation

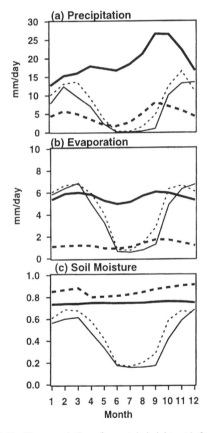

Fig. 5.10. The prescription of mountain heights at 1–3 km in the TMREAL simulations results in flow patterns such that the minima in precipitation, evaporation, and soil moisture in June–July–August in the TMNONE and TMLOW simulations are eliminated. This is consistent with interpretations of year-round wetness in coal-forming regions. This figure shows monthly averages for a locale on the northern edge of the Central Pangean Mountains (2.3°S latitude, 0° longitude). See Fig. 5.2 for legend.

at this location in the TMREAL and TMHIGH simulations averaging 0.75 in the TMREAL simulation and 0.85 in the TMHIGH simulation. Precipitation is greatest in the TMREAL simulation with amounts in excess of 12 mm/day in all months and maximum values during the boreal fall. Precipitation is significantly less in the TMHIGH simulation, but due to greatly reduced evaporation, it is adequate to maintain moist soils.

Discussion and Conclusions

Although accurate determination of land-sea distribution is important for estimating past climates (Fawcett and Barron, this volume, chapter 2), the role of topography also needs to be considered. Because of uncertainties in finding adequate tectonic analogs, precisely determining the height of mountain ranges million of years ago is difficult. These simulations suggest that not only the presence of mountains but the height of the mountains can have a significant effect on the surface flow patterns and hydrologic cycle.

The TMREAL simulation with tropical mountain heights ranging from 1–3 km gives the best fit to the tropical coal data. The Central Pangean Mountains act to impede the seasonal excursions of the Intertropical Convergence Zone. In July, a trough of low pressure remains anchored along the northern edge of the mountains and draws in moisture-laden air from both the western and eastern Panthalassa oceans. This air converges and is forced upward with the uplift enhanced by the mountains. The local increase in July precipitation compared to that predicted in the TMNONE simulation is 680%. Soil moisture remains high year-round in all regions of coal deposits except far western Laurussia. Discrepancies in this region may be due to an underestimation of the mountain peaks in the region. (The TMHIGH simulation with its higher peaks in western Laurussia gives much moister soils in this region.)

Raising the tropical mountain heights to only 1 km in the TMLOW simulation is not adequate to significantly alter the July flow patterns from the TMNONE simulation. The ITCZ shifts north of the equator. Tropical land areas south of the equator are dominated by sinking air and negligible precipitation. Prescribing mountains at too great an elevation and steepness can mask the climatic significance of the mountains. In addition, numeric problems with handling a very steep feature may come into play (Crowley et al., 1994). Sea level pressure patterns become less organized in the TMHIGH simulation disrupting low-level convergence. Precipitation, although enhanced, is less well organized. In addition, downstream effects are more pronounced. Eastern Laurussia at 10°N, a region of considerable coal deposits, receives significantly less precipitation in July and has lower soil moisture in the TMHIGH simulations than in any of the other three simulations.

Similar results are evident for climate model simulations for other time periods during the evolution of the supercontinent of Pangea. By the Permian, the Central Pangean Mountains had shifted to ~10°N with a predominantly east-west orientation (Ziegler, 1990). In climate model simulations, these mountains anchor a lobe of high precipitation year-round increasing precipitation by 25% compared to an idealized Pangea with no topography (Kutzbach and Ziegler, 1993; Kutzbach, 1994). Similarly for the Triassic, Wilson et al. (1994) found that most of the seasonal precipitation in the model falls on the major tropical highlands of Pangea.

The amplification effect of topography on precipitation is dependent on season and the latitudinal location of the mountain range. The Central Pangean Mountains located at tropical latitudes in the Late Carboniferous act to increase local precipitation by 20% in January and 680% in July. The amplification factor in July is tied to the influence of these tropical mountains on the seasonal excursions of the ITCZ. In contrast, climate model simulations with present-day geography suggest that the Tibetan Plateau, located at subtropical to mid-latitudes, enhances summer precipitation in its vicinity by only a factor of 3–4 (Kutzbach et al., 1993). In this case, the subtropical mountains act to enhance the monsoonal circulation associated with the ITCZ rather than its latitudinal excursion. The simulated present-day summer Indian monsoon is sensitive to the height of the Himalayas (Thompson and Pollard, 1995). Increasing the peak heights by 1–1.5 km to be in better agreement with actual heights produces a more realistic monsoon season with the correct timing of June through September.

The simulations in this study include an extensive ice sheet over southern Gondwana as a lower

boundary condition. The presence of permanent polar ice has been proposed as another mechanism for
increased year-round equatorial rainfall (Ziegler et
al., 1987). They propose that permanent polar
high-pressure over this ice sheet would have sup-
pressed seasonal excursions of the ITCZ and confined it to a narrow latitudinal belt creating a broad,
year-round warm and wet equatorial climatic zone.
The sensitivity of the seasonal excursions of the
ITCZ to the presence and nature of the Gondwanan
glaciation has been explored in an additional simulation (Otto-Bliesner, 1996). With no mountains
(all land elevations set to sea level) and solar luminosity set to its present-day value (to keep the
southern supercontinent of Gondwana free of permanent snowcover persisting over summer), there
is no significant difference in either the seasonal
excursions or latitudinal width of the belt of equatorial precipitation than in the simulation with a prescribed single ice sheet. Crowley et al. (1994) had
similar results in their simulations with and without
the Greenland ice sheet. These results counter the
hypothesis that polar ice cover confines the Hadley
cells and channels the equatorial rainy belt (Ziegler
et al., 1987).

The general agreement between the geological
record of extensive coals over a broad tropical latitudinal band in the Westphalian and our results
when the Central Pangean Mountain Belt is included
in a global climate model simulation as a lower
boundary condition establishes the crucial role of
this tropical mountain range in altering the tropical
circulation patterns. Simulations exploring other
mechanisms suggest that they can have an additive
effect to that of the mountains. Crowley et al.
(1995) found that equatorial precipitation during the
late Carboniferous varied with Milankovitch orbital
configuration. Coal deposits coincided with model-
simulated temperate everwet conditions over Laurussia with better correlation when cold summer orbit conditions occurred in the Southern Hemisphere.

The robustness of the climatic response to tropical mountains will need to be tested further to understand its sensitivity to model prescriptions and parameterizations. Considerable uncertainty exists in
the paleogeographic reconstructions (Heckel,
1995), representation of topographic roughness in
climate models (Valdes, 1993), and atmospheric
CO_2 (Berner, 1994). In addition, the application of
high-resolution (~50 km or less) regional climate
models (Giorgi, 1995), which can resolve patterns
of precipitation and storm tracks that depend upon
the detailed topography and coastlines, and proxy
formation models (Wold and DeConto, 1998),
which simulate the physical, chemical, and/or biological conditions necessary for the formation of
climate-sensitive sediments, to paleoclimate problems can better define our understanding of the importance of tectonic controls on paleoclimate.

References

Berger, A.L., 1978, Long-term variations of caloric
insolation resulting from the Earth's orbital elements, *Quat. Res., 9*, 139–167.

Berger, A.L., M.F. Loutre, and V. Dehant, 1989, Influence of the changing lunar orbit on the astro-
nomical frequencies of pre-Quaternary insolation
patterns, *Paleoceanography, 4*, 555–564.

Berner, R.A., 1994, GEOCARB II: A revised model
of atmospheric CO_2 over Phanerozoic time, *Am.
J. Sci., 294*, 56–91.

Budyko, M.I., A.B. Ronov, and Y.L. Yanshin,
1987, *History of the Earth's Atmosphere*,
Springer-Verlag, New York.

Caputo, M.V., and J.C. Crowell, 1985, Migration
of glacial centers across Gondwana during the Paleozoic Era, *Geol. Soc. Am. Bull., 96*, 1020–
1036.

Chaloner, W.G., and G.T. Creber, 1990, Do fossil
plants give a climatic signal?, *J. Geol. Soc.,
147*, 343–350.

Chervin, R.M., and S.H. Schneider, 1976, On determining the statistical significance of climate
model experiments with general circulation models, *J. Atmos. Sci., 33*, 405–412.

Covey, C., and S.L. Thompson, 1989, Testing the
effects of ocean heat transport on climate,
Palaeogeogr., Palaeoclimatol., Palaeoecol., 75,
331–341.

Creber, G.T., and W.G. Chaloner, 1984, Influence
of environmental factors on the wood structure of
living and fossil trees, *Bot. Rev., 50*, 357–448.

Crowell, J.C., 1983, Ice ages recorded on Gondwanan continents, *Trans. Geol. Soc. S. Afr., 86*,
237–262.

Crowley, T.J., and S.K. Baum, 1991, Estimating
Carboniferous sea-level fluctuations from
Gondwanan ice extent, *Geology, 19*, 975–977.

Crowley, T.J., S.K. Baum, and W.T. Hyde, 1991,

Climate model comparison of Gondwana and Laurentide glaciations, *J. Geophys. Res.*, *96*, 9217–9226.

Crowley, T.J., K.-J. Yip, and S.K. Baum, 1994, Effect of altered Arctic sea ice and Greenland ice sheet cover on the climate of the GENESIS general circulation model, *Glob. Planet. Change, 9*, 275–288.

Crowley, T.J., K.-J. Yip, S.K. Baum, and S.B. Moore, 1995, Modeling Carboniferous coal formation, *Palaeoclimates: Data and Modelling, 2,* 159–177.

Endal, A.S., and S. Sofia, 1981, Rotation in solar-type stars, I, Evolutionary models for the spin-down of the sun, *Astrophys. J., 243*, 625–640.

Freeman, K.H., and J.M. Hayes, 1992, Fractionation of carbon isotopes by phytoplankton and estimates of ancient CO_2 levels, *Global Biogeochem. Cycles*, *6*, 185–198.

Garrels, R.M., and A. Lerman, 1984, Coupling of the sedimentary sulfur and carbon cycles — an improved model, *Am. J. Sci.*, *284*, 989–1007.

Giorgi, F., 1995, Perspectives for regional earth system modeling, *Glob. Planet. Change, 10*, 23–42.

Hay, W.W., E.J. Barron, and S.L. Thompson, 1990, Results of global atmospheric circulation experiments on an Earth with a meridional pole-to-pole continent, *J. Geol. Soc. (London)*, *147*, 385–392.

Heckel, P.H., 1995, Glacial-eustatic base-level — Climatic model for Late Middle to Late Pennsylvanian coal-bed formation in the Appalachian basin, *J. Sedimen. Res.*, *B65*, 348–356.

Khramov, A.N., and V.P. Rodionov, 1980, Paleomagnetism and reconstruction of paleogeographic positions of the Siberian and Russian plates during the Late Proterozoic and Palaeozoic, *J. Geomagn. Geoelectr., 32*(suppl. III), SIII23–SIII37.

Kraus, J.U., C.R. Scotese, A.J. Boucot, and C. Xu, 1993, Lithologic indicators of climate: Preliminary report. Paleomap Project, Progress Report #33, Department of Geology, University of Texas, Arlington, Tex.

Kutzbach, J.E., 1994, Idealized Pangean climates: Sensitivity to orbital change, in *Pangea: Paleoclimate, Tectonics, and Sedimentation during Accretion, Zenith and Breakup of a Supercontinent*, G. D. Klein (ed.), pp. 41–56,

Geological Society of America Special Paper 288, Boulder, Colo.

Kutzbach, J.E., P.J. Guetter, W.F. Ruddiman, and W.L. Prell, 1989, The sensitivity of climate to Late Cenozoic uplift in southeast Asia and the American southwest: Numerical experiments, *J. Geophys. Res.*, *94*, 18,393–18,407.

Kutzbach, J.E., and A.M. Ziegler, 1993, Simulation of Late Permian climate and biomes with an atmosphere-ocean model: Comparisons with observations, *Phil. Trans. R. Soc., Series B, 341*, 327–340.

Kutzbach, J.E., W.L. Prell, and W.F. Ruddiman, 1993, Sensitivity of Eurasian climate to surface uplift of the Tibetan plateau, *J. Geol., 101*, 177–190.

Meehl, G.A., 1992, Effect of tropical topography on global climate, *Ann. Rev. Earth Planet. Sci., 20*, 85–112.

Mora, C.I., S.G. Driese, and L.A. Colarusso, 1996, Middle to late Paleozoic atmospheric CO_2 levels from soil carbonate and organic matter, *Science, 271*, 1105–1107.

North, G.R., K.-J. Yip, L.-Y. Leung, and R.M. Chervin, 1992, Forced and free variations of the surface temperature field, *J. Clim., 5*, 227–239.

Otto-Bliesner, B.L., 1996, The initiation of a continental ice sheet in a global climate model (GENESIS), *J. Geophys. Res., 101*, 16,909–16,920.

Panofsky, H.A., and G.W. Brier, 1965., *Some Applications of Statistics to Meteorology*, Pennsylvania State University, University Park, Pa.

Parrish, J.M., J.T. Parrish, and A.M. Ziegler, 1986, Triassic paleogeography and paleoclimatology and implications for Therapsid distribution, in *The Ecology and Biology of Mammal-like Reptiles*, N. Hotton II, P. D. MacLean, J. J. Roth, and E. C. Roth (eds.), pp. 109–131, Smithsonian Institution Press, Washington, D.C.

Paterson, W.S.B., 1972, Laurentide ice sheets: Estimated volumes during late Wisconsin, *Rev. Geophys., 10*, 885–917.

Pollard, D., and S. L. Thompson, 1995, Use of a land-surface-transfer scheme (LSX) in a global climate model (GENESIS): The response to doubling stomatal resistance, *Glob. Planet. Change, 10*, 129–161.

Raymond, A., P. H. Kelley, and C. B. Lutkin, 1989, Polar glaciers and life at the equator: The history of Dinantian and Namurian (Carboniferous) climate, *Geology*, *17*, 408–411.

Ronov, A.B., 1976, Global carbon geochemistry, volcanism, carbonate accumulation, and life, *Geoch. Int.*, *13*, 172–195.

Rowley, D. B., A. Raymond, J. T. Parrish, A. L. Lottes, C. R. Scotese, and A. M. Ziegler, 1985, Carboniferous paleogeographic, phytogeographic and paleoclimatic reconstructions, *Int. J. Coal Geol.*, *5*, 7–42.

Ruddiman, W. F., and J. E. Kutzbach, 1989, Forcing of late Cenozoic Northern Hemisphere climate by plateau uplift in southern Asia and the America west, *J. Geophys. Res.*, *94*, 18,409–18,427.

Ruddiman, W. F., and J. E. Kutzbach, 1990, Late Cenozoic plateau uplift and climate change, *Trans. R. Soc. Edinburgh: Earth Sci.*, *81*, 301–314.

Scotese, C. R., 1994, Late Carboniferous paleogeographic map, Figure 3, in *Pangea: Paleoclimate, Tectonics, and Sedimentation During Accretion, Zenith and Breakup of a Supercontinent*, G. D. Klein (ed.), p. 6, Geological Society of America Special Paper 288, Boulder, Colo.

Semtner, A. J., Jr., 1976, A model for the thermodynamic growth of sea ice in numerical investigations of climate, *J. Phys. Oceanogr.*, *6*, 379–389.

Thompson, S. L., and D. Pollard, 1995, A global climate model (GENESIS) with a land-surface transfer scheme (LSX). Part 1: Present climate simulation, *J. Clim.*, *8*, 732–761.

Trewartha, G. T., and L. H. Horn, 1980, *An Introduction to Climate,* McGraw-Hill, New York.

Valdes, P., 1993, Atmospheric general circulation models of the Jurassic, *Phil. Trans. R. Soc. Lond., Series B,341,* 317–326.

Van der Voo, R., 1993, *Paleomagnetism of Atlantis, Tethys, and Iapetus*, Cambridge University Press, Cambridge.

Veevers, J. M., and C. M. Powell, 1987, Late Paleozoic glacial episodes in Gondwanaland reflected in transgressive-regressive depositional sequences in Euramerica, *Geol. Soc. Am. Bull.*, *98*, 475–487.

Walker, J. C. G., P. B. Hays, and J. F. Kasting, 1981, A negative feedback mechanism for the long-term stabilization of Earth's surface temperature, *J. Geophys. Res.*, *86*, 9776–9782.

Wilson, K. M., D. Pollard, W. W. Hay, S. L. Thompson, and C. N. Wold, 1994, General circulation model simulations of Triassic climates: Preliminary results, in *Pangea: Paleoclimate, Tectonics, and Sedimentation During Accretion, Zenith and Breakup of a Supercontinent*, G. D. Klein (ed.), pp. 91–116, Geological Society of America Special Paper 288, Boulder, Colo.

Wold, C. N., and R. M. DeConto, 1998, Proxy formation model used to predict the locations of Late Cretaceous evaporites, in *The Evolution of Cretaceous Ocean/Climate Systems*, E. Barrera and C. Johnson (eds.), Geological Society of America Special Publication, (in press).

Yapp, C. J., and H. Poths, 1992, Ancient atmospheric CO_2 pressures inferred from natural goethites, *Nature*, *355*, 342–344.

Ziegler, A. M., 1990, Phytogeographic patterns and continental configurations during the Permian Period, in *Palaeozoic Palaeogeography and Biogeography*, W. S. McKerrow and C. R. Scotese (eds.), pp. 363–379, *Geol. Soc. London Mem.*

Ziegler, A.M., C.R. Scotese, W.S. McKerrow, M.E. Johnson, and R.K. Bambach, 1979, Paleozoic paleogeography, *Annu. Rev. Earth Planet. Sci.*, *7*, 473–502.

Ziegler, A. M., D. B. Rowley, A. L. Lottes, D. L. Sahagian, M. L. Hulver, and T. C. Gierlowski, 1985, Paleogeographic interpretation: With an example from the Mid-Cretaceous, *Annu. Rev. Earth Planet. Sci.*, *13*, 385–425.

Ziegler, A. M., A. Raymond, T. C. Gierlowski, M. A. Horrell, D. B. Rowley, and A. L. Lottes, 1987, Coal, climate, and terrestrial productivity: The present and early Cretaceous compared, in *Coal and Coal-Bearing Strata: Recent Advances*, A. C. Scott (ed.), pp. 25–49, Geol. Soc. London Spec. Pub., London.

CHAPTER 6

The Role of Mountains and Plateaus in a Triassic Climate Model

William W. Hay and Christopher N. Wold

The Triassic is a unique episode in Earth history (Trümpy, 1982), characterized by the most arid climates of the Phanerozoic. Although the entire Triassic was arid, the Early Triassic (Scythian) was unique in leaving no record of equatorial forests or soils that would reflect the high rainfall that must occur in the Intertropical Convergence Zone (ITCZ). The amalgamation of Laurussia and Gondwana to form Pangea had occurred at the end of the Pennsylvanian Period (290 Ma), and the breakup of Pangea that led to formation of the modern ocean basins occurred during the Jurassic, starting about 180 Ma. The Triassic represents the middle of the interval during which Pangea existed. The general paleogeography of the Triassic was not greatly different from that of the Permian and Early Jurassic. Pangea extended almost from pole to pole, having a nearly meridional western margin and a large embayment, the Tethys, on its eastern side. The entire continental mass drifted northward from the Pennsylvanian through the Jurassic. At the beginning of the Permian it was asymmetric with respect to the equator. The South Pole lay near the center of Gondwana, and Siberia did not reach the North Pole. By the Early Jurassic, the distribution of land had become more symmetric about the equator. Hay and Southam (1977) proposed that Pangea had a much higher average elevation (about 1500 m) than the Present Day average continental elevation (740 m). This was based on the very large mass of sediment offloaded from the continents onto their continental margins during the Mesozoic and Cenozoic, equivalent to a layer of sedimentary rock several kilometers thick spread over the entire continental blocks. Hay et al. (1987) estimated that during the Triassic the mean elevation of Pangea might have been about 1200 m.

What was unique about the conditions on Earth during the Triassic that produced such extreme aridity? Factors that are generally considered important to promoting aridity are: (1) a large contiguous land area; (2) high plateau elevations; (3) cool global temperatures; (4) a small area of warm tropical ocean; and (5) a lack of vegetation. The continuous land area was not appreciably larger than in the Permian or Jurassic. The average elevation of the continent during the Triassic was probably intermediate between that of the Permian and Jurassic. The global temperature was warmer than today, but not as warm as during the Cretaceous or Eocene. The area of the ocean in the tropics was large, which should have increased rainfall. Land plants had undergone a catastrophic reduction in diversity and numbers at the end of the Permian, but the Triassic aridity was so pervasive there were only a few areas where land plants could thrive and influence the climate. None of these factors reaches a maximum in the Triassic compared to earlier or later times, suggesting that there is something else responsible for Triassic aridity.

Previous Investigations of Triassic Paleoclimate with Atmospheric General Circulation Models

Two sets of atmospheric general circulation model sensitivity experiments had been conducted on idealized geographies approximating Triassic Pangean conditions, Kutzbach and Gallimore (1989), and Hay et al. (1990a). Kutzbach and Gallimore (1989) used the National Center for Atmospheric Research's Community Climate Model 1 (NCAR CCM1) with a seasonal cycle to explore conditions on an Earth with a single broad continent extending from 75°N to 75°S latitude with a meridional western margin and an indented eastern margin. The supercontinent extended through 180° of longitude in high latitudes but the eastern embayment reduced its width to 90° of longitude in the equatorial region. They experimented with both a low and high plateau continent, and found that in both cases there was a very strong monsoonal circulation; the high plateau model resulted in intensification of the monsoon and greater aridity. For the second set of experiments Hay et al. (1990a) used an older version of the NCAR Community Climate Model, CCM0, to simulate the climate on a meridional continent with a land area of 148 x 10⁶ km², equal to the total land area of the Earth today. The sides of the supercontinent were meridians 104° of longitude apart, everywhere.

Two experiments were conducted comparing two flat continents, one with a "low" average elevation of 750 m and the other with an average elevation of 1500 m. The Northern Hemisphere was specified to be ice-free and the Southern Hemisphere was specified to be ice-covered poleward of 70°S, so that each configuration of the meridional continent consisted of two sensitivity experiments. The simulations used a "swamp ocean," that is an ocean that did not circulate and had no heat transport. The models were run with mean annual conditions (no seasonal cycle). As had been found by Kutzbach and Gallimore (1989), the higher continent resulted in greater aridity, but there was still an equatorial belt of precipitation across the continent. Another important effect of the plateau continent was a change in sea level pressure of about 20 hPa reflecting the displacement of air by the high plateau continent and resulting in an increase of the sea level pressure contrast between highs and lows by a factor of two. Two other experiments used the low meridional continent with 3 km high mountain ranges along the western and eastern margins. The primary effect of the meridional mountain ranges was to intercept the zonal transport of moisture into the continental interior. For the model with mountains along the western margin of the continent, the mountains served as a focus for precipitation from the trade winds and from the ITCZ. The runoff calculated in the simulation indicated that the high precipitation localized on the eastern side of the mountain range could serve as a source for a large river, like the modern Amazon.

At mid-latitudes, the precipitation from the westerlies was trapped on the western side of the mountain range and an extensive rain shadow developed to the east. For the model with mountains along the eastern margin of the continent, the precipitation from the trade winds and ITCZ was trapped on the eastern slope of the range where it would flow back into the ocean. Without this moisture, the interior of the continent was very dry, but an intense low pressure system which developed over the equatorial interior of the continent drew in moisture from the ocean to the west and caused precipitation on the low western margin of the continent. The mid-latitude westerlies carried some moisture into the continental interior, but the overall condition was one of extreme aridity. This led to the conclusion that if mountains were present along the eastern margin of a continent, the result would be extremely arid conditions over almost all of the land area, as exists in Australia today.

Selecting Paleogeographies to Test the Effects of Topography

As an initial test of the GENESIS (version 1.02a; Pollard and Thompson, 1995; Thompson and Pollard, 1995a, b) Atmospheric General Circulation Model (AGCM) in the early 1990s, we chose to simulate paleoclimates using two more realistic and detailed paleogeographic reconstructions, one for the Early Triassic (Scythian, 245 Ma) and another for the Late Triassic (Carnian, 225 Ma). The topography of many areas of the Earth changed as the rifting leading to the breakup of Pangea was initiated. The results of the simulations have been described briefly by Hay et al. (1994) and in detail by Wilson et al. (1994). In this paper we concentrate on discussion of the effects of topographic relief on those climate model results.

The paleogeographies of the Scythian and Carnian were similar, but their paleoclimates were different. The Scythian is regarded as the most arid stage of the Triassic, and probably the most arid time during the Phanerozoic (Habicht, 1979; Crowley, 1983; Ronov et al., 1989). There was extensive evaporite deposition (Gordon, 1975; Habicht, 1979; Busson, 1982, Ronov et al. 1989). Eolian sands are widespread, indicating extensive desert conditions (Habicht, 1979; Ronov et al. 1989). Red terrestrial sediments are characteristic; they imply a low groundwater table permitting oxidation of the sediments. There is no evidence for equatorial coals, although some coals formed at high latitudes. The Carnian was also arid, but less extreme than the Scythian, and contains evidence for more precipitation such as extensive fluvial deposits, kaolinite, and karst development.

Regional Uplifts, Mountains, Aridity, and Climate Models

Regional uplifts and mountains have long been suspected as a major factor in promoting and localizing aridity. They affect the climate at different spatial scales by changing the radiation balance, interrupting the flow of winds, and causing the temperature and humidity of the air to change as it rises and descends. Until recently, only the largest of these features could be represented in the coarse resolution grid used by most climate models. The NCAR Community Climate Models (CCM0 and CCM1) used from the 1970s until the mid 1990s used a resolution of 4.5° latitude x 7.5° longitude. This translates into grid cells 500 km of latitude x 833 km of longitude at the equator or 500 km of latitude x 589 km of longitude at 45° latitude. This coarse grid recognized only the Tibetan-Himalaya uplift, western North and South American uplifts, and East African-Kalahari uplifts. The Alps, Atlas, and most other modern mountain ranges are too narrow to be represented at this resolution. At first it would appear that this would introduce a major flaw in any attempt to model the climate, but in examining observational climate data it becomes evident that the climatic effects of topographic features diminishes sharply as their size decreases. Climate model experiments with and without the Earth's major uplifts are often termed experiments with and without mountains. This is misleading in that the uplifts may be broad elevated areas or long narrow mountain ranges, and the effect of the two kinds of topography on the climate are very different.

Effects of Regional Uplifts

Broad uplift results in a plateau over which there is a shorter path for both incoming and outgoing radiation. This produces seasonal climate extremes, enhanced seasonal reversals of atmospheric pressure, and strong monsoonal circulation.

In the middle of this century it was suggested that the large mid-latitude arid regions of Asia and North America might be due to large-scale perturbations of the atmospheric flow by large uplifts rather than the direct expression of the rain shadow effect of individual mountain ranges (Charney and Eliassen, 1949; Bolin, 1950). Simulations of an Earth with and without topography by Kasahara and Washington (1969), Kasahara et al. (1973), Manabe and Terpstra (1974), Barron and Washington (1984), Barron (1985), Hay et al. (1990a), Manabe and Broccoli (1990), Ruddiman and Kutzbach (1990), and Prell and Kutzbach (1992) all showed that the major effect of regional uplifts is to disrupt the zonal circulation, increase vorticity, and create climatic contrasts. There are three effects: (1) to form a barrier to global winds; (2) to alter the radiation balance; and (3) to displace air, changing the sea level pressure and the pressure contrast between highs and lows. These impacts have been discussed at length in a series of papers on the relation of Late Cenozoic climate change and plateau uplift by Kutzbach et al. (1989, 1993), Ruddiman and Kutzbach (1989, 1990, 1991a, b), and Hay (1996).

Monsoonal circulation is a reversal of the winds and ocean currents with the seasons. It is largely a response to the presence of a large landmass at mid-latitudes. Summer insolation warms air over the land developing a low pressure system that draws in air. Winter conditions result in a high-pressure system developing over the plateau and winds that flow outward. The idea that the high topography of southern Asia increases the strength of the monsoons was introduced by Flohn (1950). Prell and Kutzbach (1992) and Kutzbach et al. (1993), using NCAR's CCM0 and CCM1 models, explored the effect of changing the elevation of the Tibetan Plateau on the Indian monsoon. They conducted simulations using a low elevation (0 km), half of the Present Day elevation (2.1 km), and Present Day average elevation (4.2 km). Strong monsoons appeared in experiments with the elevation half that of the Present Day Tibetan Plateau, suggesting that an elevation of 2.1 km will induce strong topographic forcing. The monsoons that develop today over the Arabian Sea are a special case because of the existence of another topographic feature. They are very strong not only because the Tibetan Plateau develops extremes of high and low pressure in winter and summer, but also because they are steered by the uplift associated with the East African Rift system. During the Northern Hemisphere summer, the Tibetan low pressure system pulls the southeasterly trade winds of the Southern Hemisphere across the equator. The equator-crossing winds are turned to the northeast by the combined effect of the Coriolis Force and the barrier presented by the East African Rift ranges (Findlater, 1966). North of the equator they turn to the northeast and flow toward the Tibetan low. As a result of topographic steering by the African uplift, the winds form a low-level jet stream, now termed the Somali or Findlater Jet, that flows to the northeast across the Arabian Sea (Findlater, 1974, 1977). The jet forces upwelling in the northwestern Arabian Sea and downwelling in the southeastern Arabian Sea (Brock et al., 1992).

Effects of Mountain Ranges

Narrow mountain ranges act as obstructions to the flow of air, creating orographically induced climatic effects on a smaller scale than that of regional uplifts. The regions affected are much more restricted and hence the local climate is much more readily related to its cause by human observers.

The most familiar orographic effects are the precipitation and rain shadow effects due to cooling and moisture loss of rising air and heating of the descending air after it has passed over the crest of a mountain range (Barry, 1981; Barry and Chorley 1982). In investigating why fresh water lakes formed during the Triassic on the North American margin downwind of the Proto- Central Atlantic Rift where salt was being deposited, Hay et al. (1982) described the orographic effects in quantitative terms. A more detailed general discussion of the rain shadow effect in terms of adiabatic changes in air temperatures, water content and precipitation has been presented by Hay (1996). A mountain range, especially if perpendicular to the prevailing winds, acts as a barrier to atmospheric circulation. It also acts as a trap for water vapor converting latent to sensible heat, resulting in precipitation on the windward slopes. As the drier air passes the crest of the range and descends, its evaporative potential increases. The "rain shadow" on the lee side of a mountain range is not only the result of a deficiency of moisture, but is enhanced because the dry air has a dehydrating effect on soils and vegetation.

The GENESIS (Version 1.02a) Paleoclimate Model

GENESIS (Pollard and Thompson, 1995; Thompson and Pollard 1995a, b) is a climate system model specifically designed to facilitate paleoclimatologic research by accommodating boundary conditions very different from those of the Present Day. It was also designed to simulate processes at the Earth's surface at a resolution that can be related to geologic deposits. GENESIS (1.02a) used a Land-Surface Transfer Model (LSX) to describe the interaction of the atmosphere with the land surface on a 2° x 2° grid (222 x 222 km at the equator and 222 x 157 km at 45° latitude), while still calculating the processes in the atmosphere using a coarser 4.5° x 7.5° grid.

GENESIS (1.02a) was developed with support from the Earth Sciences Section of the National Science Foundation. It began with modification of NCAR's CCM1 by S. Thompson and D. Pollard at NCAR with input from K. Wilson and W. Hay at the University of Colorado. GENESIS (1.02a) differed from CCM1 in several respects important to the simulation of Triassic climate. The atmospheric part of the model was a heavily modified version of CCM1. It used semi-Lagrangian transport for water vapor, rather than advecting it in spectral transform space as in CCM1. At each step the semi-Lagrangian method interpolates the water vapor field back to the departure points of the model grid, which preserved sharp gradients better and avoided problems with calculation of negative water vapor inherent in the earlier spectral method (Williamson and Rasch, 1989; Williamson, 1990). GENESIS (1.02a) had an explicit plume model for atmospheric convection and the planetary boundary layer which produced more realistic vertical penetration than the earlier bulk convective adjustment used in CCM1. It had the Thompson et al. (1987) solar radiation scheme for aerosols, used a more sophisticated cloud parameterization following Slingo and Slingo (1991), allowing for stratus, anvil cirrus, and convective clouds. GENESIS (1.02a) was coupled to a slab ocean and included more realistic treatment of vegetation, soil, snow, and sea ice. The ocean was represented by a thermodynamic slab 50 m thick that crudely approximated the seasonal heat capacity of the real ocean mixed layer. Ocean heat transport was prescribed as a function symmetric about the equator, based on Present Day observations (Covey and Thompson, 1989). Up to two vegetation layers (trees and grass) could be specified at each grid point. The radiative and turbulent fluxes through these layers to the ground surface were calculated. Precipitation could be intercepted by the vegetation to subsequently drip to the ground or evaporate. The soil had up to six layers, extending to 4.25 m depth. Each layer could have different porosity and permeability, and infiltration was governed by nonlinear equations. Soil moisture and ice were computed as proportions of the pore space in each layer. Finally, GENESIS (1.02a) had independent AGCM and surface grids, transferring values between the fields by interpolation or averaging. Both the surface and atmospheric grids had a diurnal cycle, with solar radiation calculations performed every 1.5 hours. Full AGCM calculations were performed every 0.5 hours except for absorptivities and emmissivities of H_2O, CO_2, and O_3, which were calculated once every 24 hours.

Boundary Conditions

Insolation, Atmospheric Composition, Soils, Ocean Heat Transport, Heat Convergence Under Sea Ice

The general boundary conditions used for the simulations described here were discussed in greater detail in Wilson et al. (1994). The solar constant was specified to be 1,343 Wm^{-2} (2% less than Present Day), the eccentricity of the Earth's orbit to be 0°, and the obliquity of the axis of rotation to be 23.4°. The CO_2 concentration of the atmosphere was specified to be 1,360 ppm (about four times Present Day) based on the geochemical model of Berner (1980). The soil was prescribed to be everywhere as a light colored loam (#2 in the color table, #6 in the texture table of BATS, the Biosphere-Atmosphere Transfer Scheme of Dickinson et al., 1986). The oceanic heat flux was defined as -30 Wm^{-2} at the equator, increasing to 30 Wm^{-2} at 50° and decreasing to 16 Wm^{-2} poleward of 60°. The heat convergence under sea ice was specified to be 8 Wm^{-2} for 100% ice cover.

Paleogeography

The paleogeographic boundary conditions used for the climate model simulations were established in the following steps: (1) reconstruction of the positions of the continental blocks and terranes relative to each other; (2) placement of the relative reconstruction into a latitudinal reference frame; (3) establishing of paleotopographic contours for the land areas; (4) conversion of the paleogeographic maps to the 2° x 2° grid used by the GENESIS (1.02a) Land Surface Transfer Model (LSX).

Paleocontinental Reconstructions

The paleocontinental reconstructions used as base maps (Fig. 6.1) were based largely on Wilson (1989) and Wilson et al. (1989), with modifications as described in Wilson et al. (1994). They used 110 terranes and blocks digitized from the UNESCO Geologic World Atlas (Choubert and Faure-Muret, 1976). We made a tighter fit of the continents using outlines that represented the continental margins prior to stretching and rifting. A detailed discussion of the terranes, blocks, digitization methods, and corrections for stretching and bending of the continental margins during rifting is included in Hay et al. (in press). In their reconstructions, Wilson et al. (1994) tried to account for every block or terrane larger than 2° x 2° (222 x 222 km), that is to define the paleogeography at the same resolution as used by GENESIS (1.02a) for calculations of land surface processes.

Paleolatitudinal Reference Frame

The paleolatitudes were established using the paleomagnetic reference frame of Van der Voo (1990).

Paleotopographic Reconstruction Methods

Paleotopographic reconstruction is the most uncertain aspect of the paleogeography. Topography is a result of the following factors: (1) thickness of crust (continental, arc or oceanic) having a density lower than that of the upper mantle; (2) temperature of the lithosphere; (3) stresses that prevent isostatic equilibrium from being achieved; and (4) the time delay involved in isostatic adjustment.

The differences in thickness of crust produce not only the largest-scale topographic features, the continental blocks and ocean basins, but also uplifted areas of intermediate size, such as the Tibetan

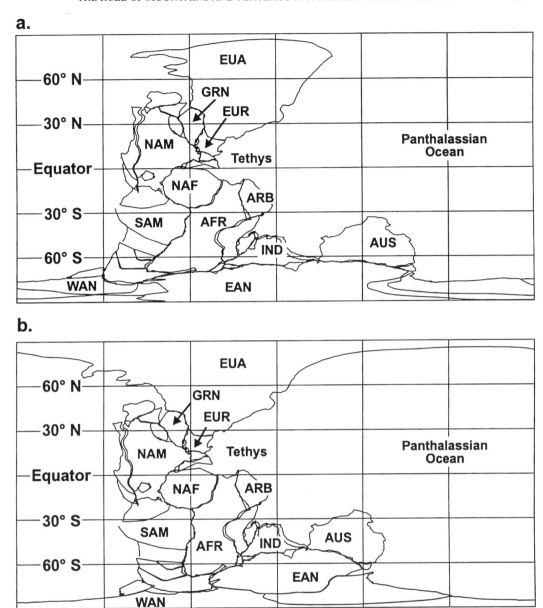

Fig. 6.1. Plate tectonic reconstructions for the Triassic shown on an Equidistant Cylindrical map projection. The names of the largest continental blocks have been abbreviated as follows: AFR = northeast, central, and southern Africa, ARB = Arabia, AUS = Australia, EAN = East Antarctica, EUA = Eurasia, EUR = Europe, GRN = Greenland, IND = India, NAF = Northwest Africa, NAM = North America, SAM = South America, WAN = West Antarctica. (a) The Early Triassic (Scythian, 245 Ma) global reconstruction, and (b) the Late Triassic (Carnian, 225 Ma) global reconstruction. The smaller, island terranes in the Panthalassian Ocean shown in Figures 6.2 and 6.3 were added later.

Plateau, and smaller-scale features such as mountain ranges. The continental crust has a density of about 2750 kg/m^3, and old, cool upper mantle has a density of about 3300 kg/m^3. Cratonic continental crust is typically about 35 km thick and has an elevation of about 450 m; isostatically balanced 50 km thick continental crust would have an elevation of about 3 km. Mountain ranges are supported by a "root" of rock less dense than the adjacent upper mantle. In continental regions thickening of the crust results from compression which may produce folding, overthrusting or underthrusting of sheets of continental rock. The surficial area of a volume of crust is reduced and its thickness increased, resulting in formation of a mountain range. In exceptional cases, one block of crust may be shoved beneath another; this has been suggested as an explanation for the great crustal thickness beneath Tibet. In island arcs the thickness of the crust is increased by both compression of material scraped off subducting ocean floor and by underplating as the ocean crust and sediments are subducted (Von Huene and Scholl, 1991). The ocean crust thickens with age as igneous material is added from below.

Compression in the continents and island arcs leaves a geologic record in the form of folded and faulted rocks. Overthrusting is also easily detected from the surface geology, but underthrusting and overriding of one block by another may be difficult to demonstrate. It may be suspected by the topographic anomaly, and this interpretation is supported by deep seismic studies, but proof remains elusive.

High mountain ranges are rapidly eroded. The loss of elevation due to erosion (denudation) is partially compensated by isostatic adjustment, so that metamorphic rocks formed at considerable depths are brought to the surface. The original thickness of the overlying rock can be determined from the degree of metamorphism permitting an estimate of the initial height of the mountain range.

Stretching thins continental crust and results in a depression. During continental breakup stretching first through plastic deformation and then through brittle failure, reduces the thickness of the continental crust producing first a sag and then a rift valley. When the continental crust thickness is reduced to zero, the ocean crust appears, seafloor spreading is initiated, and the continental blocks begin to drift apart.

The second factor producing topography is the temperature of the lithosphere. This is responsible for the 3 km elevation of the mid-ocean ridge over old ocean crust, as shown in Stein and Stein (1992). From the elevation difference, it can be calculated that the warmer lithosphere beneath a mid-ocean ridge has a density of about 3200 kg/m^3; it is about 3% lighter than lithosphere beneath 180 m.y. old ocean crust. The elevation difference produced by the warmer lithosphere is the same as that of many mountain ranges, but there is a major difference in their width. The mid-ocean ridge in the slow-spreading Atlantic is 2000 km wide and that in the fast-spreading Pacific is more than twice as wide. Warm lithosphere can affect elevations in continental areas as well as in the ocean. The high elevation of western North America is not a result of thickened continental crust; the crust there is only about 20 km thick, and the elevation must be due to a lithosphere that is warmer and less dense than that beneath the craton to the east. Temperature differences in the lithosphere do not necessarily leave a direct geologic record, but they produce an indirect record because they drive the erosion-sedimentation system. Broad uplifts caused by warm lithosphere provide source areas for sediment, and are the locus of the headwaters of rivers. Areas where the lithosphere has cooled and become more dense are basins. Paradoxically, areas underlain by warm lithosphere are often sites of continental rifting, as in the western United States and in East Africa. The subsidence due to rifting is restricted to a narrow axis, whereas the uplift takes place over a much broader area on the adjacent margins.

The third factor producing topography is stress that holds the rocks out of isostatic equilibrium. This occurs at subduction zones, where the subducting oceanic plate is first arched gently upward as it approaches the ocean trench, and then bent sharply downward to descend beneath the overriding plate. It also occurs in regions where intraplate stress is prominent, causing large scale arches and troughs, such as those on the Indian Ocean floor west of Ninetyeast Ridge (Cloetingh and Wortel, 1985). The only geologic record left by topographic anomalies due to stress will be in the erosion and deposition of sediment.

The fourth factor affecting topography is the time delay associated with isostatic adjustment. The response time of the Earth's surface to a suddenly emplaced load is on the order of 20,000 to 40,000 years. Ice sheets form topographic uplifts that are of the same order of scale as the Tibetan Plateau. However, unlike lithospheric features, they can grow and decay in tens of thousands of years; they change more rapidly than the Earth can adjust to their load. In the region of the ice sheet the isostatic response may produce topographic changes of the order of hundreds of meters. This is a very important consideration in trying to reconstruct topography during the last glacial maximum and during the deglaciation.

Of the four factors influencing topography, only compression and stretching of crust leave an obvious geologic record at the Earth's surface. Changes in the density (warmth) of the lithosphere may even produce apparently contradictory geologic evidence (stretching and rifting of elevated areas, compression of subsiding areas). All affect the erosion-sedimentation system, which should be the best recorder of topographic change. Unfortunately, topography also affects climate, and climate affects the erosion-sedimentation system.

Several different methods for paleotopographic reconstruction have been proposed, based on plate tectonics and presumed changes in mantle density (Hay, 1981, 1983; Hay et al. 1987), nature of the rocks (Ziegler et al., 1985), and sedimentary mass balance (Shaw and Hay, 1990; Hay et al. 1989, 1990b, Hay and Wold, 1993, Wold et al., 1993).

Hay (1981, 1983) noted that the NCAR map of Present Day global topography at 4.5° latitude x 7.5° longitude horizontal resolution and 1 km vertical resolution showed only the Tibetan-Himalaya uplift, western North and South American uplifts, and East African-Kalahari uplifts. He proposed that the major uplifts shown on the NCAR map were related to three tectonic processes: (1) subduction of warm young lithosphere, which is responsible for the uplift of western North and South America; (2) collisions of continental fragments with continental blocks, which is responsible for the uplift of Tibet and the surrounding areas; and (3) thermal uplift prior to and following continental rifting, which is responsible for the uplift of East Africa and also may be responsible for the high elevation of the Kalahari Plateau. These ideas were further developed and quantified by Southam and Hay (1981), who discussed the process and effects of rifting in detail. They assumed: (1) that the subduction of oceanic

lithosphere less than 20 m.y. old produces an uplift of 1.5 km in the adjacent continental margin; (2) that a colliding block will cause uplift proportional to its size, for example, a block having an area of 10^6 km^2 would cause a 1 km uplift, and a 4 x 10^6 km^2 block (e.g., the Indian subcontinent) would cause 4 km of uplift; and (3) that the thermal uplift associated with a continental-scale rift is 2000 km across and mimics the shape of a mid-ocean ridge. Uplift associated with subduction of young ocean lithosphere is a result of replacement of higher density older cold lithosphere by lower density warm lithosphere. The elevation differences produced by collision of blocks of different sizes was suggested by inspection of block sizes, collision times, and elevations in southern Asia. The uplift history associated with continental-scale rifts is based on the North America-Africa separation and the East African Rift. At the time of separation, the density of the lithosphere at the site of breakup is the same as that of lithosphere at a mid-ocean ridge. The entire process of separation of continental blocks requires 100 m.y. from its initiation until drift occurs; involving first arching, then starting about 30 m.y. before separation, development of a longitudinal sag, and finally, about 15 m.y. before separation, development of a fault-bounded rift valley. Following Kinsman (1975), they assumed that the replacement of cold continental lithosphere by warmer lithosphere produced marginal uplifts having an incremental elevation of 1.75 km over the original surface at the time of separation. The rate of uplift of the rift margins was assumed to be the inverse of postrift subsidence. Because the uplift accelerates as the time of separation approaches, it would not become a major source of sediment until about 40 m.y. prior to separation. They assumed that only long rifts between major continental blocks have a history of uplift. Many smaller features, like the Bay of Biscay and Gulf of Suez never experience uplift because they formed passively as a result of changes in motions of the larger adjacent plates. To illustrate the use of this method of paleotopographic reconstruction, Hay et al. (1987), presented global paleotopographic reconstructions at 5° latitude x 5° longitude resolution for 40, 80, 120, and 160 Ma.

The use of rock type and metamorphic grade may assist in estimating the heights of ancient mountain ranges, because it gives an idea of the amount of erosion that has taken place. Ziegler et al. (1985) used rock types exposed today to make quantitative estimates of elevations in the past. Their elevation ranges 9–5 represent environments above sea level. Their range 9 environments are characterized by high temperature–high pressure metamorphics indicating these rocks are the remains of collisional mountains with elevations between 4 and 10 km. Range 8 environments are characterized by andesites/granodiorites in a continental setting indicating Andean-type peaks with elevations between 2 and 4 km. Range 7 environments are characterized by andesites/granodiorites in a marine setting with adjacent fanglomerates; they indicate island arc peaks and rift shoulders with elevations between 1 and 2 km. Range 6 consists of deposits that lie between these previously listed range 7–9 "mountain indicators" and their range 5 alluvial and coastal plain complexes, but also include basalts, lake deposits in graben settings, and tectonic melanges. The environments corresponding to range 6 are inland plains, rift valleys, and some forearc ridges; their elevations lie between 0.2 and 1 km. Range 5 is characterized by alluvial complexes, major floodplain complexes, swamps and channel sands. They indicate coastal plains, lower river systems, and delta tops lying at elevations between 0 and 0.2 km above sea level. They noted that some of these indicators, such as the andesites in a continental setting, would be indicative of the height of individual peaks, and that the regional elevation might be one elevation category lower. However, applied to the Present Day, the Ziegler et al. (1985) method for paleotopographic reconstruction fails to recognize three of the broadest and climatologically most significant uplifts on Earth today; the Tibetan Plateau, the general uplift of western North America, and the Altiplano of South America. The elevation of these areas would not be detected because they are largely covered by young unconsolidated sediment even though they are high plateaus.

We believe that the best method for regional reconstructions is that based on sedimentary mass balance (Shaw and Hay, 1990; Hay et al. 1989, 1990b, Hay and Wold, 1993, Wold et al., 1993). Unfortunately, sedimentary mass balance requires detailed knowledge of the stratigraphy and construction of isopach maps. It also requires that material lost through subduction be accounted for. Because

of these limitations it has not yet been applied globally.

Triassic Paleotopographic Reconstructions

Because of the problems of interpretation of the geologic evidence in terms of elevation we applied the method of Hay et al. (1987), based on generalizations from plate tectonics and changes in mantle density, to produce the topographic reconstructions used as boundary conditions for the Triassic paleoclimate simulations. The reconstructions can be regarded as theoretical and speculative, but they are an attempt to take into account the surface expression of mantle convective and plume processes that are responsible for the large scale features of the topography observed today.

Two steps were involved in determining the paleotopographies specified as boundary conditions for the simulations. First, establishment of a relative topography based on plate tectonic and sedimentologic considerations, and second, adjustment to an estimate of the global mean elevation of land based on subsequent offloading of sediment from the continents.

The initial paleotopographic maps, assuming the elevations are based on tectonic factors alone, were sketched on the plate tectonic base maps using a 1 km contour interval to indicate uplift above the global average elevation. As this work was nearing completion we received a copy of the Lithologic Paleogeographic Atlas for the Mesozoic and Cenozoic by Ronov et al. (1989). Their paleogeographic maps include qualitative indications of the elevations of source areas of sediment based on the masses of sediment that accumulated and their grain sizes. We compared the elevated areas indicated on their maps for the Early and Late Triassic to areas we had designated as tectonic uplifts and found that generally, they corresponded well. However, they indicated some additional areas of uplift, most notably in the northeastern Africa-Arabian region and the Altai, that were added. In some regions, such as the margin of South America and Africa, they showed a greater uplift than we had assumed. We added and corrected these areas, assigning them an arbitrary average elevation of 1.5 km.

The second step in paleotopographic reconstruction involved adjusting the elevations estimated from tectonic considerations to conform to a general scheme for decline in continental elevations since the end of the Paleozoic as proposed by Hay and Southam (1977). They noted that very large masses of sediment had accumulated in the passive margins formed during continental breakup in the Mesozoic and Cenozoic and on the deep seafloor, particularly in the abyssal plains off trailing passive margins. They estimated that this was equal to a solid layer 1.76 km thick added to the continental blocks. Taking isostatic adjustment and pore space into account they came to the conclusion that the mass of sediment on the continental blocks at the end of the Paleozoic would have added 820 m to the Present Day average elevation of the ice-free continents of 740 m. This is equivalent to an average elevation of land of about 1560 m if the pore space were air filled, or about 1400 m if the pore space were water filled. The large Mesozoic-Cenozoic sediment mass in continental margins has since been documented by three independent estimates using different methods (Southam and Hay, 1981; Gregor, 1985; Ronov et al., 1986; also reported in Budyko et al., 1987).

Based on tectonic and sedimentologic considerations alone, the average elevation of the Scythian land was about 900 m and the Carnian land about 800 m. Assuming a linear thinning and loss of elevation of the continental blocks from 255 Ma to Present Day, the average elevation of land during the Scythian would have been about 1375 m, decreasing to about 1250 m during the Carnian. Incremental elevations were added to the initial paleotopographic reconstructions to produce these average elevations.

It can be argued that the average elevations of the continents at the end of the Paleozoic and during the early Mesozoic were not as high as was suggested by Southam and Hay (1981), but then the source of the sediment that has accumulated in passive margins and on the deep seafloor needs to be explained. Two possibilities are apparent: (1) that the continents have become thicker through time by some sort of "underplating"; and (2) that the mountains formed adjacent to subduction zones are the main source of sediment. There is little evidence to support or reject the hypothesis of continental growth by underplating. The volumes of sediment shed from orogenic belts into the trenches associated with subduction zones and the amount of basement abraded have been estimated by Von Huene and Scholl (1991, 1993). Assuming that the other side of the mountain range shed a similar amount of sediment which then accumulated on the passive margins on the far side of the continent, the masses are an order of magnitude too small to account for the passive margin and ocean floor accumulations.

Conversion of the Paleogeographic Maps to a 2° x 2° Grid

Detailed outlines of the paleogeographic reconstructions were converted to the 2° x 2° grid required by GENESIS (1.02a) for land surface process calculations. To understand the changes that occur in representation of the Earth's surface at different grid scales, we compared average elevation maps of North America, South America, and Africa at 1° x 1°, 2° x 2°, 3° x 3°, 4° x 4°, and 5° x 5° resolution. It is important to realize that even at 1° x 1° resolution only a few individual mountain ranges can be recognized, but regional uplifts and marginal seas, for example, the Red Sea, are well defined. At 2° x 2° resolution there is some further loss of detail, but all of the major features can still be discerned. Major degradation takes place at lower resolution. At resolutions of 3° x 3° and coarser, only the broad features of the paleogeography can be detected. The Red Sea is represented by only two grid cells at 3° x 3° resolution, a single grid cell at 4° x 4° resolution and cannot be detected at all at 5° x 5° resolution. On the other hand, all of the topographic "noise" of small features disappears at these coarser resolutions, while large-scale features become more obvious. The change in width of the North American Cordillera at about 50°N, the Altiplano, and the East Africa Rift – Kalahari uplifts are enhanced.

At the 4.5° x 7.5° resolution used by the NCAR topographic boundary condition map, Texas and France were each represented by a single elevation. Coarse as it may seem, the 4.5° x 7.5° resolution of the topography has been used by many climate models and allows the modern global climate to be simulated well. The reason such a coarse topography can be used is that large-scale atmospheric features such as high and low pressure systems, are adequately resolved at the 4.5° x 7.5° scale.

Geologists seek to describe the surface of the

Earth with as much detail as the observations will allow, whereas atmospheric scientists prefer to generalize the topography to a scale appropriate to the general circulation of the atmosphere. Averaging the topography over large areas obscures all of the features with which humans are familiar but reveals the large-scale topographic trends that affect the Earth's radiation balance and large-scale circulation of the atmosphere. Selection of a 2° × 2° resolution for the GENESIS (1.02a) land-surface transfer model (LSX) was made with the knowledge that it is a scale relevant to the interpretation of geologic deposits.

Scythian Paleotopography

The Scythian (245 Ma) paleogeographic reconstruction used for the climate simulation has been discussed in detail in Wilson et al. (1994). The paleotopographic reconstruction at the 2° × 2° resolution used by GENESIS (1.02a) for land surface processes is shown in Figure 6.2a and at the 4.5° × 7.5° resolution used for the internal atmospheric processes in Figure 6.2b. Only the broadest of the mountain ranges that formed in the Late Paleozoic were still evident: the Urals and Altaids in Asia, the Variscan mountains in Europe, and the Appalachians in North America. The Cape Fold Belt of South Africa blends into the general uplift of the southern margin of the continent. In reconstructing the paleotopography we assumed that there was an uplift with a median rift valley between North America and Northwest Africa and between North and South America prior to separation of the continental blocks at 180 Ma to form the Atlantic Ocean and Gulf of Mexico. The rift valleys were too narrow to show in the 2° × 2° resolution. We assumed that there was a smaller uplift between South America and Africa preceding the separation of these continental blocks at 140–120 Ma, and a higher uplift between Africa and the Madagascar-India-Australia-Antarctica block, assumed to separate at about 120 Ma. We also assumed that there was still uplift along the entire northern margin of Australia, India and East Africa as a remnant of the rifting of the Cimmerian blocks, and the southeast Asian terranes from the Gondwanan margin in the Late Permian. We included uplift along the eastern (now southern) margin of Asia as a result of the earlier collision of the Kopet Dagh

and southern Chinese blocks. The result was a configuration of Pangea with mountains along its eastern margin.

The reconstruction for the Induan (237 Ma) of Golonka et al. (1994) is not greatly different from ours. The major difference in the paleocontinental reconstruction is that they show the southeast Asian blocks forming a continuous arc across the eastern Tethys. Their paleotopography distinguishes only between lowlands and mountains, but the uplifted areas are essentially the same as we show with the exception of the margins of northwestern and western Australia, and northeastern and northern Africa, where they show only lowlands.

Carnian Paleotopography

The Carnian (225 Ma) tectonic reconstruction used for the climate simulation has also been presented in Wilson et al. (1994). The paleotopographic reconstruction at the 2° × 2° resolution is shown in Figure 6.3a and the 4.5° × 7.5° resolution in Figure 6.3b. The mountain ranges that formed in the Late Paleozoic are still evident, but lower. We assumed that their heights declined following the loss of elevation scheme of Wold et al. (1993). They proposed that, once formed, mountain ranges continuously lose elevation to erosion, with the loss partially compensated by isostatic adjustment. Because the denudation rate is related to elevation, the topography decays following the exponential curve that results from applying a constant relation between denudation (erosion) rate and elevation through time. As the relation between denudation rate and elevation, we used their value of 0.113×10^{-6} m per meter of elevation per year.

The uplifts between North America and northwest Africa, between North and South America, between South America and Africa, and between Africa and the Madagascar-India-Australia-Antarctica block, are all higher. A deep rift valley existed between North America and Africa, but is not resolved at the 2° × 2° resolution. The uplift along the northern margin of Gondwana had largely subsided. Uplift along the eastern (now southern) margin of Asia is higher as a result of collisions of blocks and continental fragments. The general configuration of Pangea is that of a continent with broad interior arch uplifts.

The Norian (216 Ma) reconstruction published

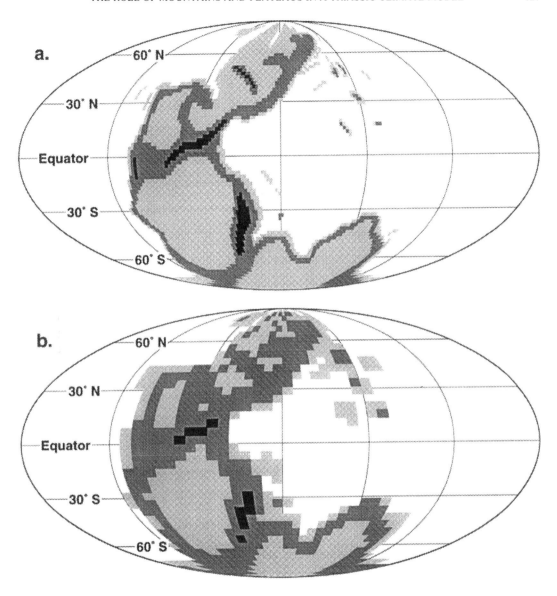

Fig. 6.2. Early Triassic (Scythian, 245 Ma) paleogeography, with shorelines and elevations interpreted from plate tectonics, theoretical considerations of the relation between plate tectonics and elevations, and from geologic data on the early Triassic map of *Ronov et al.* (1989). The maps are shown on a Mollweide projection. The land areas and their elevations are indicated by the shaded areas on the maps: light gray = 0–1000 m elevation above sea level, dark gray = 1000–3000 m, and black = 3000 m and higher. (a) The 2° × 2° grid resolution used by the GENESIS Land Surface Transfer (LSX) scheme. (b) The 4.5° × 7.5° grid resolution used for calculations within the atmosphere.

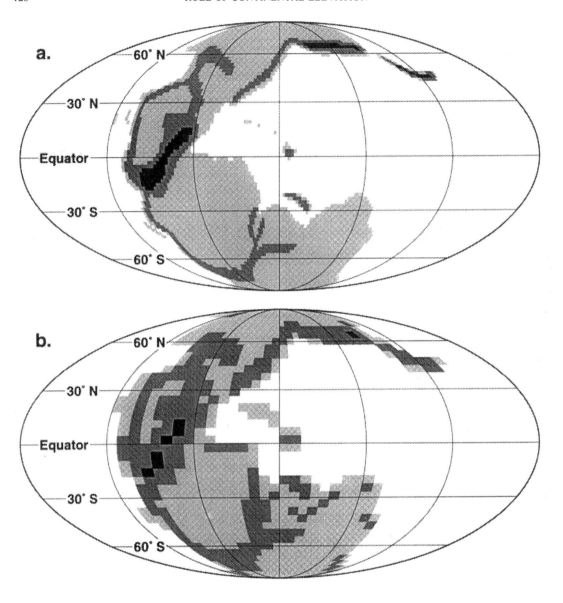

Fig. 6.3. Late Triassic (Carnian, 225 Ma) paleogeography, with shorelines and elevations interpreted from plate tectonics, theoretical considerations of the relation between plate tectonics and elevations, and from geologic data on the Late Triassic map of *Ronov et al.* (1989). Land area elevations are indicated with the same shading as in Figure 6.2. (a) The 2° × 2° grid resolution used by the GENESIS Land-Surface Transfer model (LSX). (b) The 4.5° × 7.5° grid resolution used for calculations within the atmosphere.

by Golonka et al. (1994) is similar to ours, differing only in that we suggest some uplift between India and Antarctica, and a larger area of uplift in the North Sea region.

Results

Global maps showing the distribution of physical and paleontologic climate indicators for the Triassic have been presented by Habicht (1979), Parrish et al. (1986), and by Ronov et al. (1989). Maps showing these and other physical and paleontologic climate indicators used to validate our GENESIS (1.02a) Triassic climate simulations were presented in Wilson et al. (1994). The general results of the GENESIS (1.02a) simulations have been presented by Wilson et al. (1994) and Hay et al. (1994). The discussion here is limited to the effects of the topography on the simulated climate. The two simulations can be regarded as sensitivity experiments for two configurations of Pangea differing chiefly in the extent of marginal uplift. As shown in Figure 6.2a the Scythian paleogeography has a continuous uplift along its eastern margin and along its western margin south of 45°N. The Carnian paleogeography has uplift on the eastern margin of the continent only north of 30°N, but uplift along its western margin to 65°N.

In both cases, the effects of topography on the paleoclimate are very important in producing widespread aridity. The uplifts along the eastern margin of the continent are probably the most important factor in producing the extreme aridity of the Scythian.

Scythian

The surface pressure field for the Northern Hemisphere winter is shown in Figure 6.4a. A strong low is developed over the northern end of the uplift in eastern Africa where separation of Madagascar will occur. In the central part of the Panthalassian Ocean, the low of the ITCZ remains in the Northern Hemisphere.

Figure 6.4b shows the surface pressure field for the Northern Hemisphere summer. There is a well-developed low pressure system north of the site of the future North American-African margins. The low

pressure system serves as a target for the southeastern monsoons. There are well developed highs over the northern and southern tropics in the Panthalassian Ocean.

Figure 6.5a shows the near surface winds for the Northern Hemisphere winter. The winds over the Panthalassian Ocean are quite zonal, with the core of the low latitude easterlies at about 15°, and that of the Northern Hemisphere westerlies at about 40°. The southern part of the Panthalassian Ocean is only about 70° wide, and the winds over it are perturbed by the adjacent land into two subtropical highs. Along the east-west trending East African, Indian, and western Australian segment of the Gondwana margin, south of 30°S, the winds blow from west to east parallel to the coast. The winds are much more irregular over the continents. Wind speeds are generally slow in the Northern Hemisphere, and they are clearly guided by the uplifts in the continental interiors. Except for a region just north of the equator (Europe), the mountains along the eastern margin of the continent prevent penetration of the moisture-bearing easterlies into the interior of the continent, just as they had in the simulations with idealized geography (Hay et al., 1990a). South of the equator, the winds are drawn in from the west, flowing in a band across the continent to a low situated over the north end of the uplift on the future site of separation of Madagascar from Africa. The winds flow across the mountains on the northern margin of Gondwana between India and Australia and into the continental interior.

The winds for the Northern Hemisphere summer are shown in Figure 6.5b. Winds over the Panthalassian Ocean are similar to those for the Northern Hemisphere winter, but the winds over the continents have reversed almost everywhere, corresponding to the "megamonsoons" simulated by Kutzbach and Gallimore (1989) with an idealized Pangean geography. Again, the mountains along the eastern margin of Pangea block the easterlies from entering the continental interior. A band of westerlies flows from the equatorial region across North America and Europe, guided by the aging Appalachians and the uplift on the site of the future separation of North America and Africa.

Precipitation for the Northern Hemisphere winter is shown in Figure 6.6a. It is very low everywhere

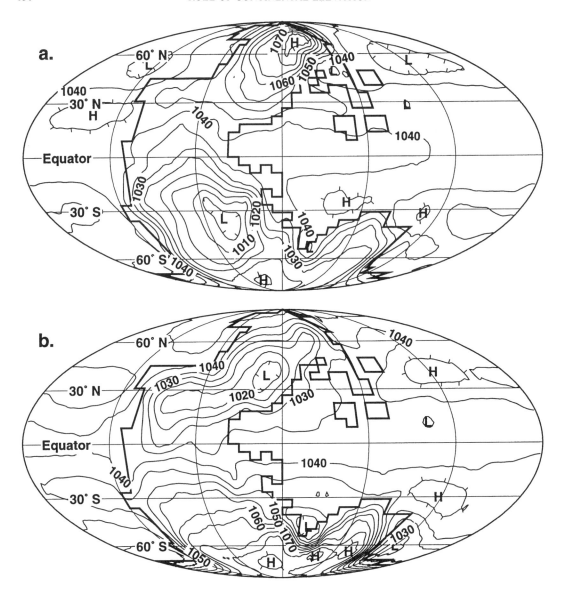

Fig. 6.4. Atmospheric pressure at sea level (mb) simulated by GENESIS (1.02a) for the Early Triassic (Scythian, 245 Ma). Low pressure centers have short segmented lines on the inside of the lowest pressure contour, and centers of high pressure have short lines on the outside of the highest pressure contour. (a) the Scythian Northern Hemisphere winter (DJF; monthly mean values averaged for December, January, and February), and (b) Scythian Northern Hemisphere summer (JJA; monthly mean values averaged for June, July, and August). The land areas are outlined at the 4.5° x 7.5° resolution used for calculations of the sea level pressure field.

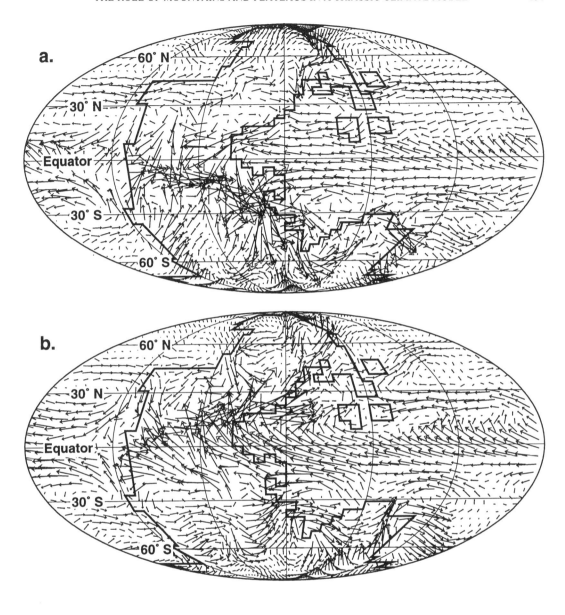

Fig. 6.5. Near-surface wind vectors at the 926 hP level (~600 m) simulated by GENESIS (1.02a) for (a) the Scythian Northern Hemisphere winter (DJF), and (b) Scythian Northern Hemisphere summer (JJA). The land areas are outlined at the 4.5° x 7.5° resolution used for calculations of the wind at this level.

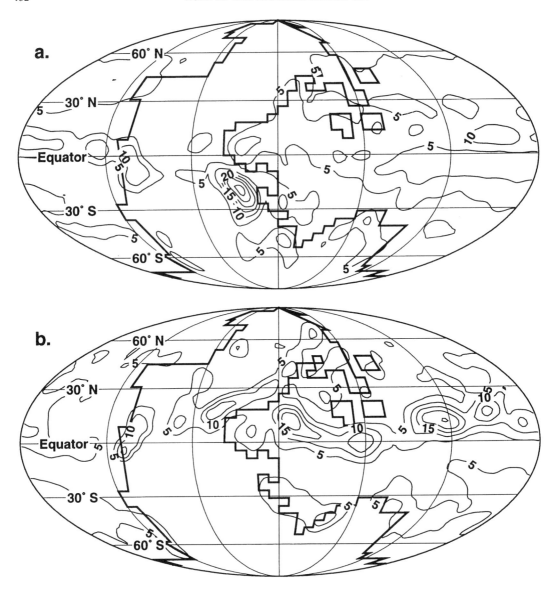

Fig. 6.6. Total atmospheric precipitation (mm/day) simulated by GENESIS (1.02a) for (a) the Scythian Northern Hemisphere winter (DJF), and (b) Scythian Northern Hemisphere summer (JJA). The outlines of the land areas are shown at the 4.5° x 7.5° resolution used for calculations within the atmosphere.

over the continents except on the western margin of the continent about 10° south of the equator and in northeast Africa and Arabia, where the highest precipitation at any season of the year occurs. The high precipitation is centered over Somalia, at the northern end of the uplifted region where Madagascar will separate from Africa. The precipitation for the Northern Hemisphere summer is shown in Figure 6.6b. The Intertropical Convergence is represented by a zone with sites of higher rainfall at about 15°N. The areas with highest precipitation are again on the western margin of the continent near the equator and in Europe, adjacent to the northwestern part of the Tethyan embayment. Hay et al. (1994) noted that the rainfall in Europe was essentially limited to the summer season. The extreme seasonality of rainfall may be a major factor in allowing the oxidation of the terrigenous Buntsandstein sediments to produce their red color.

Carnian

The surface pressure field for the Carnian Northern Hemisphere winter is shown in Figure 6.7a. The low developed over the uplift in eastern Africa, where separation of Madagascar will occur, is less intensively developed than in the Scythian simulation because of the lower elevation prescribed in the paleogeography. In the central part of the Panthalassian Ocean the low of the ITCZ remains in the Northern Hemisphere.

Figure 6.7b shows the surface pressure field for the Northern Hemisphere summer. As in the Scythian simulation a strong low is developed north of the uplift on the site of separation of North America and Africa. The low pressure system again serves as a target for the southeastern monsoons. There are well developed lows over the northern tropics in the Panthalassian Ocean and the ITCZ.

Figure 6.8a shows that the winds in the Carnian Northern Hemisphere winter simulation are somewhat more zonal than in the Scythian simulation. The winds over the Panthalassian Ocean indicate a series of subtropical highs in both hemispheres; this is related to the effect of Australia and of southeast Asia in partially enclosing the Tethys. The more zonal winds across Pangea are largely the result of prescribing lower mountains along the eastern margin of the continent. A band of strong westerlies extends from western equatorial Pangea across

Africa and along the Indian-Australian margin. These are joined and reinforced by cross-equatorial flow that is initially from the Tethyan embayment to the northeast. Flow of the winds around Gondwana is cyclonic, inducing upwelling all along the margin.

As in the Scythian simulation, the winds over the Pangea largely reverse during the Northern Hemisphere summer. In the Southern Hemisphere the strong westerlies flow in an undulating pattern across Gondwana, influenced by the topography (Fig. 6.8b). They move equatorward over uplifts and poleward over depressions, being steered by the topography. The column of air shortens as it goes over uplift. In order for its potential vorticity to remain constant, its planetary vorticity must decrease. This is accomplished by moving equatorward. Flow around the Gondwanan margin is anticyclonic, suppressing upwelling. The Appalachians and uplift on the site of the separation of North America and Africa guide winds flowing in from the western margin of the continent to Europe to the northeast. Winds over Asia are also clearly guided by the topography.

Precipitation in the Northern Hemisphere winter, shown in Figure 6.9a, is concentrated on the western margin of the continent just south of the equator and over Somalia. In the Carnian simulation, the precipitation over Somalia is less intense than it was in the Scythian simulation.

Precipitation in the Carnian Northern Hemisphere summer is shown in Figure 6.9b. A very well developed wet ITCZ extends across the Tethys and Panthalassian Ocean and across the continent at 15°N. This coincides with the areal distribution of geologic evidence for the Carnian "pluvial" episode (see map in Wilson et al., 1994). The Petrified Forest of Arizona (USA) was located on the equator just inland from the western shore. The high precipitation shown on the uplift where North America and Africa are about to separate masks the fact that conditions in the developing rift valley there were quite arid (see Hay et al., 1982).

Validation of the Climate Models with Geologic Evidence

Wilson et al. (1994) made a detailed comparison of the results of the climate simulations with physical, geologic and paleontologic evidence. The most

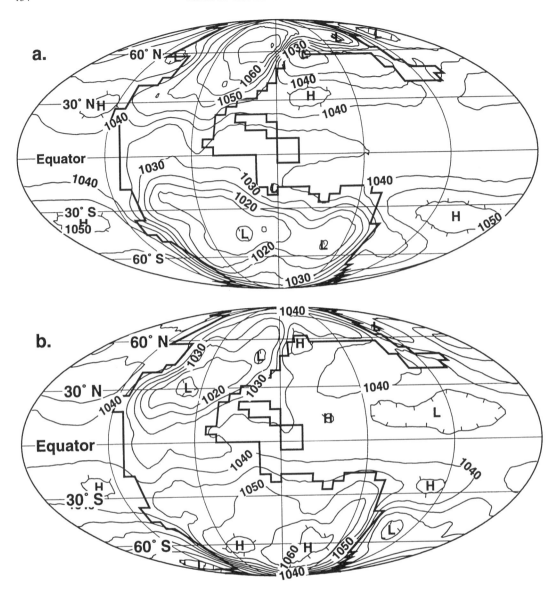

Fig. 6.7. Surface pressures at sea level (mb) simulated by GENESIS (1.02a) for (a) the Carnian Northern Hemisphere winter (DJF), and (b) Scythian Northern Hemisphere summer (JJA). The land areas are outlined at the 4.5° × 7.5° resolution used for calculations of the near surface pressure field.

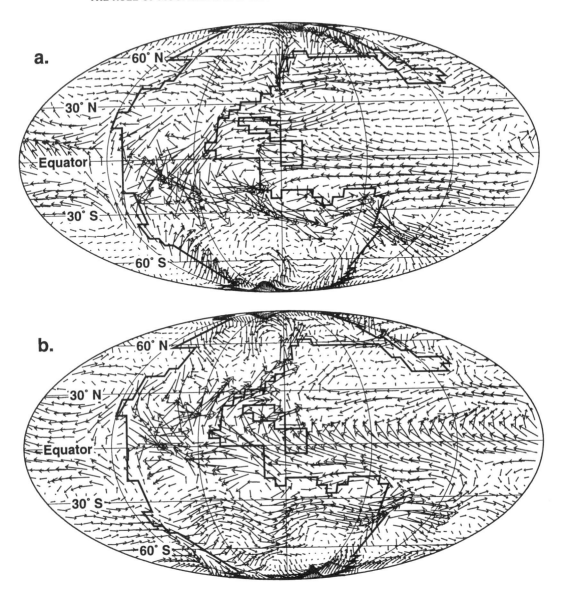

Fig. 6.8. Near surface wind vectors at the 926 hP level (~600 m) simulated by GENESIS (1.02a) for (a) the Carnian Northern Hemisphere winter (DJF), and (b) Carnian Northern Hemisphere summer (JJA). The land areas are outlined at the 4.5° x 7.5° resolution used for calculations of the wind at this level.

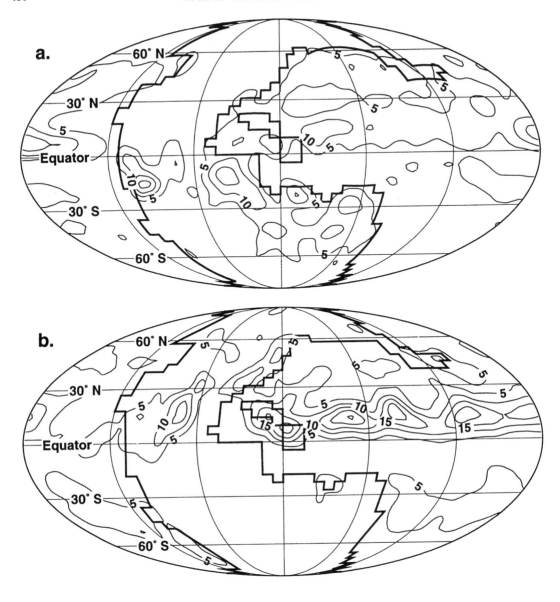

Fig. 6.9. Total atmospheric precipitation (mm/day) simulated by GENESIS (1.02a) for (a) the Carnian Northern Hemisphere winter (DJF), and (b) Carnian Northern Hemisphere summer (JJA). The outlines of the land areas are shown at the 4.5° x 7.5° resolution used for calculations within the atmosphere.

easily evaluated paleoclimate proxies are glacial deposits and evaporites. The results indicate that no ice cap would develop over the land, and there is no permanent sea ice. The simulations suggest that permafrost existed at depths below 1.75 m in the soils poleward of 60° latitude in both hemispheres. There is no geologic record of Triassic glacial deposits (till and diamictite). The occurrence of evaporites implies a significant excess of evaporation over precipitation, and is the most solid evidence for aridity.

Scythian

According to our Scythian paleogeographic reconstruction, evaporites occur on the eastern margin of Pangea from 5° to 20°N (central Europe) and from 15° to 25°S (eastern Arabia), and on the western margin from 5° to 15°N (western United States). The evaporite deposits are mostly gypsum and anhydrite, halite occurs only rarely.

The high precipitation simulated by GENESIS (1.02a) for western Arabia and northeastern Africa would seem to preclude the formation of evaporites in this region, and the presence of extensive Scythian evaporites was initially taken as a falsification of the simulation. However, Pollard and Schulz (1994) explored conditions in this area in more detail using a proxy formation model for the deposition of evaporites. Instead of interpreting the climate conditions at the site of evaporite deposition to be those indicated for the 2° x 2° grid cell within which the deposit lies, the proxy formation model allows the model-simulated climate to act on a specific environment within the grid cell and too small in areal extent to influence the regional climate. For the evaporite proxy formation model they considered the effects of the climate on small, 50-m deep bodies of saline water isolated from the ocean and thus unable to exchange water or heat with it. Because evaporation is a function of both the temperature and salinity of the water, the evaporation precipitation-balance for such isolated bodies of water can deviate sharply from that calculated by the model for the ocean. They found that the evaporation-precipitation balance for isolated bodies of water in Arabia, North Africa, and Europe differed significantly from that calculated for the water grid cells adjacent to the shore, which assumed heat exchange with the ocean. The area of excessive evapo-

ration outlined the known region of evaporite deposition startlingly well. They found that the high rainfall on the uplift to the southwest and an excess of evaporation over precipitation at the site of evaporite deposition were compatible and could be taken as a validation of the model.

Paleontologic evidence also supports the paleoclimate simulation, although little is known about the adaptations and requirements of animals and plants in the Triassic. The labyrinthodonts were large amphibians that probably required standing water; their occurrence correlates with high soil moisture. Lungfishes (dipnoans) and conchostrachans are thought to be indicators of seasonal drying of the river courses, and their occurrences are in areas predicted by the simulation to have strong seasonal contrasts in precipitation and runoff. There are possible problems in relating the occurrence of theraspid (mammal-like) reptiles and the Dicroidium (seed-fern) flora of the Southern Hemisphere to the simulated paleoclimate because the seasonal temperature changes seem extreme.

Carnian

The Carnian geologic record contains extensive evaporites, including halite, and the widespread occurrence of another paleoclimate proxy, coals. The evaporites are located on the eastern margin of Pangea, from 30°N to 10°S (from Central Europe through Iberia to western North Africa, and in eastern Arabia). Evaporites were deposited in the deep rift between North America and West Africa at paleolatitudes from 10°N to 10°S. Halite also occurs on the East African margin at a paleolatitude of 45°S (in Tanzania), in a rift basin associated with the separation of Madagascar from Africa. Some evaporites are located on the western margin of Pangea at 15° to 25°N and at 35°S. The coal deposits in both the Northern and Southern Hemisphere tend to lie along and immediately equatorward of the edge of the soil permafrost boundaries at 60°N and S.

The Pollard and Schulz (1994) proxy formation model for evaporites predicts all of these occurrences except those on the sites of the rift valleys between North America and Africa and Africa and Madagascar. The Carnian paleotopographic maps show these as highland areas. Neither the 4.5° x 7.5° nor the 2° x 2° map allows the rift valleys to be resolved. The salts that accumulated in the rift valleys were halite and anhydrite, so that they must have been at or below sea level. The walls of the rifts would have been in the order of 2–3 km (similar to the relief of the modern Dead Sea rift). Wold et al. (1994) modified the Pollard and Schulz (1994) proxy formation model for evaporites to allow sub-grid cell investigation of the local climate within mountainous regions. The air temperatures over the highland areas are adjusted using a lapse rate of 6.5°C/km to estimate the temperature and humidity at sea level in the rift valley. The 4.5° × 7.5° elevation profiles, estimated rift topography and simulated evaporation minus precipitation (E minus P) balances are shown in Figure 6.10. Using the estimated sea level air temperature, they calculated the evaporation rate for a brine with an initial salinity of 175, the salinity at which gypsum begins to precipitate from a brine concentrated from sea water. The precipitation is the annual mean from the GENESIS (1.02a) simulation for the 4.5° × 7.5° grid cell. A positive E minus P indicates the potential for gypsum formation, a negative E minus P indicates freshening of the brine. The modified proxy formation model predicts both the presence of evaporites in the rift and the potential presence of fresh-water lakes (such as those in the Newark and Connecticut Valley Basins) on the western margin of the rift.

As was the case with the Scythian simulation, the paleontologic evidence offers support for much of the climatology produced by the simulation, but there are some areas, particularly in the southern polar region, where fossil floras may be incompatible with the temperature extremes indicated by the model. An extensive discussion is given in Wilson et al. (1994).

Discussion

Almost all of the uplifts included in the paleogeographic reconstructions are more than 2° wide, so that they appear on both maps at 2° × 2° and 4.5° × 7.5° resolution. No long narrow mountain ranges are resolved by the 2° × 2° resolution, much less by the 4.5° × 7.5° resolution.

The effects of the paleotopography on the atmospheric circulation simulated for the Triassic are pervasive. Uplifts enhance the high and low pressure systems that develop in response to heating and cooling of the land surface. These pressure effects in turn guide the winds and focus precipitation. The effect is most well developed at low latitudes, and is also expressed as an intensification of the monsoonal circulation.

In all of the simulations the larger uplifts located in the tropics and subtropics develop low pressure systems near 20–30° latitude during the summer and these serve as a target for the trade winds from the east and westerly winds over the continent. The broader uplifts exert a greater attraction for the winds in summer than narrower uplifts. There may be a threshold size for an uplift above which it becomes significant for the global atmospheric circulation, and below which it exhibits only local effects.

All of the simulations also show a flow of air into the continental interior just north or south of the equator during the Northern and Southern Hemisphere summers, respectively. This air is the source of moisture that falls on uplifts near the equator on the western margin of the continents.

In all of the simulations the precipitation occurs during the summer and is focused on the mountains that lie between 10–20° latitude. Lesser amounts of precipitation occur during summer on uplifts on the eastern margins of the continents in latitudes to 60°, and on the western margin of the continent near the equator. The precipitation over the land is much reduced in winter.

In both the Scythian and Carnian simulations the ITCZ is well developed during the Northern Hemisphere summers, when it lies about 15°N. It is more strongly developed in the Carnian than in the Scythian simulation. We believe that this is mostly a result of the paleogeographic distribution of the landmasses, but the paleotopography may also play a role. The generally greater elevations in the Northern Hemisphere may create a global pressure differential that would intensify the ITCZ when it is in the Northern Hemisphere.

One of the paleogeographic boundary conditions specified for the simulation of Scythian (Early Triassic, 245 Ma) climate using the GENESIS (1.02a) model (Wilson et al., 1994) was uplift along most of the eastern as well as along the western margin of Pangea. Even though the uplift was low, it trapped most of the precipitation coming in from the ocean. This contributed to the extreme aridity throughout

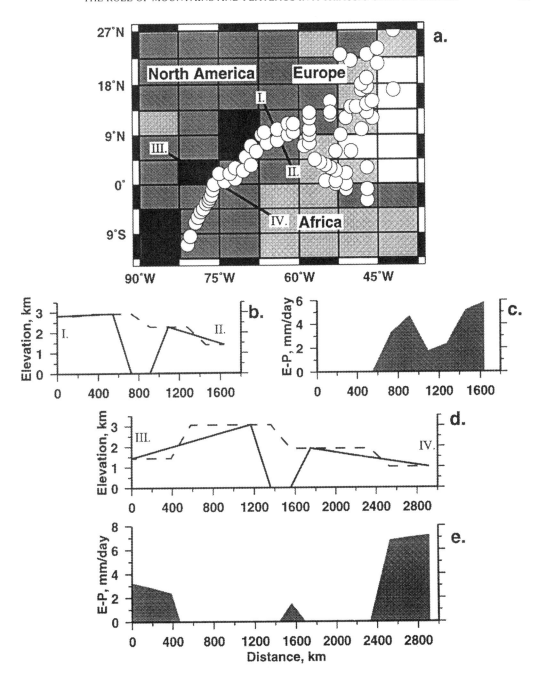

Fig. 6.10. Proxy formation model for evaporites run using monthly mean meteorological fields from the Carnian paleoclimate simulation. (a) 4.5° x 7.5° elevation (from Figure 6.3b) for the region around the proto-Atlantic rift. The land area elevations are shaded as in Figure 6.3. The evaporite locations are indicated by the white circles and two vertical profiles perpendicular to the rift are indicated by Roman numerals. (b) On the profile I–II, the original 4.5° x 7.5° elevation is indicated by the dashed line and our estimation of the rift topography is shown by the solid line. (c) The shaded regions in the diagram show where E minus P was positive (E calculated by the PFM and P was the mean annual precipitation from the GENESIS 1.02a simulation) indicating the potential for precipitation of gypsum. (d) On the profile III–IV, the original 4.5° x 7.5° elevation is indicated by the dashed line and our estimation of the rift topography is shown by the solid line. (e) The shaded regions in the diagram show where E minus P was positive indicating the potential for precipitation of gypsum.

much of the continent, and may explain the greatest peculiarity of Triassic paleoclimatology, the absence of an equatorial rain belt over the land area.

Summary and Conclusion

Topography has a strong influence on the atmospheric circulation simulated for these two stages of the Triassic. Uplifts shorten the path of radiation through the atmosphere, enhancing heating and cooling of the land surface, and intensifying high and low pressure systems. The intensified pressure systems and the topography itself guide the winds, and concentrate and localize precipitation. The effects of uplift are best developed at low latitudes.

The effects simulated by GENESIS (1.02a) are primarily the large-scale effects expected from plateau uplift. The intensification of high and low pressure systems by elevation is apparent in the wind patterns. The smaller-scale effects, such as the rain shadow in the lee of mountain ranges, are not apparent in the simulations, even at the 2° X 2° resolution of the surface process model. The smaller-scale effects can be simulated using proxy models.

Acknowledgments

This work was carried out with support from grants EAR 9320136 and EAR 9405737 from the Earth Sciences Section of the U.S. National Science Foundation, by a grant from the Donors of The Petroleum Research Fund administered by the American Chemical Society, and by the Deutsche Forschungsgemeinschaft. CNW was also supported by the Global Change Distinguished Postdoctoral Fellowships Program sponsored by the U.S. Department of Energy, Office of Health and Environmental Research, and administered by the Oak Ridge Institute for Science and Education. All of the illustrations were prepared using GMT (Wessel and Smith, 1991).

References

Barron, E. J., 1985, Explanations of the Tertiary global cooling trend, *Palaeogeogr., Palaeoclimatol., Palaeoecol., 50*, 45–61.

Barron, E. J., and W. M. Washington, 1984, The role of geographic variables in explaining pale-oclimates: Results from Cretaceous climate model sensitivity studies, *J. Geophys. Res., 89*, 1267–1279.

Barry, R. G., 1981, *Mountain Weather and Climate*, Meuthen and Co. Ltd., London.

Barry, R. G., and R. J. Chorley, 1982, *Atmosphere, Weather & Climate*, Meuthen and Co. Ltd., London.

Berner, R. A., 1980, *Early Diagenesis: A Theoretical Approach*, Princeton Univ. Press, Princeton.

Bolin, B., 1950, On the influence of the Earth's orography on the general character of the westerlies, *Tellus, 2*, 1894–1895.

Brock, J. C. D., C. A. McLean, and W. W. Hay, 1992, A southwest monsoon hydrographic climatology for the northwestern Arabian Sea, *J. Geophys. Res., 97*, 9455–9465.

Budyko, M. I., A. B. Ronov, and A. L. Yanshin, 1987, *History of the Earth's Atmosphere*, Springer–Verlag, New York.

Busson, G., 1982, Le Trias comme periode salifere, *Geol. Rundsch., 71*, 857–880.

Charney, J.G., and A. Eliassen, 1949, A numerical method for predicting the perturbations of the middle-latitude westerlies, *Tellus, 1*, 38–54.

Choubert, G., and A. Faure-Muret, 1976, *Geological World Atlas*, 22 sheets with explanations, UNESCO, Paris.

Cloetingh, S., and R. Wortel, 1985, Regional stress field of the Indian plate, *Geophys. Res. Lett., 12*, 77–80.

Covey, C., and S. L. Thompson, 1989, Testing the effects of ocean heat transport on climate, *Palaeogeogr., Palaeoclimatol., Palaeoecol., 75*, 331-341.

Crowley, T. J., 1983, The geologic record of climatic change, *Rev. Geophys. Space Phys., 21*, 828–877.

Dickinson, R. E., A. Henderson-Sellers, P. J. Kennedy, and M. F. Wilson, 1986, *Biosphere-atmosphere transfer scheme (BATS) for the NCAR community climate model*, NCAR Technical Note NCAR/TN-275+STR, 69 pp.

Findlater, J., 1966, Cross-equatorial jet streams at low level over Kenya, *The Meteorological Magazine, 95*, 353–364.

Findlater, J., 1974, The low-level cross-equatorial air current of the western Indian Ocean during the northern summer, *Weather, 29*, 411–416.

Findlater, J., 1977, Observational aspects of the

low-level cross-equatorial jet stream of the western Indian Ocean, *Pageogh, 115,* 1251–1261.

Flohn, H., 1950, Studien zur allgemeinen Zirkulation der Atmosphäre, *Berichte des deutschen Wetterdienstes, 18,* 34–50.

Golonka, J., M. I. Ross, and C. R. Scotese, 1994, Phanerozoic paleogeographic and paleoclimatic modeling maps, in *Pangea; Global Environments and Resources, Canadian Society of Petroleum Geologists Memoir* 17, pp. 1–47, Calgary.

Gordon, W. A., 1975, Distribution by latitude of Phanerozoic evaporite deposits, *J. Geol., 83,* 671–684.

Gregor, C. B., 1985, The mass-age distribution of Phanerozoic sediments, in *The Chronology of the Geological Record*, N. J. Snelling (ed.), pp. 284–289, Geol. Soc. Mem. 10, Blackwell Scientific Publications, Boston, Mass.

Habicht, J. K. A., 1979, Paleoclimate, paleomagnetism, and continental drift, *AAPG Studies in Geology, 9,* 1–32.

Hay, W. W., 1981, Sedimentological and geochemical trends resulting from the breakup of Pangaea, Proceedings, 26th International Geological Congress, Geology of Oceans Symposium, *Oceanol. Acta,* SP, 135–147.

Hay W. W., 1983, Significance of runoff to paleoceanographic conditions during the Mesozoic and clues to locate sites of ancient river inputs, *Proceedings, 5th Joint Oceanographic Assembly, Canadian Department of Fisheries and Oceans,* pp. 9–17, Ottawa, Canada.

Hay, W. W., 1996, Tectonics and climate, *Geol. Rundsch, 85/3,* 409–437.

Hay, W. W., and J. R. Southam, 1977, Modulation of marine sedimentation by the continental shelves, in *The Fate of Fossil Fuel CO_2 in the Oceans, Marine Science Series,* vol. 6, N. R. Anderson, and A. Malahoff (eds.), pp. 569–604, Plenum Press, New York.

Hay, W. W., and C. N. Wold, 1993, Mass-balanced reconstruction of paleogeology, in *Computerized Basin Analysis. Computer Applications in the Earth Sciences,* J. Harff, and D. F. Merriam (eds.), pp. 101–113, Plenum Press, New York.

Hay, W. W., C. A. Shaw, and C. N. Wold, 1989, Mass-balanced paleogeographic reconstructions, *Geol. Rundsch., 78,* 207–242.

Hay, W. W., E. J. Barron, and S. L. Thompson,

1990a, Global atmospheric circulation experiments on an Earth with a meridional pole-to-pole continent, *J. Geol. Soc. London, 147,* 385–392.

Hay, W. W., C. N. Wold, and C. A. Shaw, 1990b, Mass-balanced paleogeographic maps: Background and input requirements, in *Quantitative Dynamic Stratigraphy,* T. Cross (ed.), pp. 261–275, Plenum Press, N. Y.

Hay, W. W., J. F. Behensky, Jr., E. J. Barron, and J. Sloan, 1982, Late Triassic-Liassic paleoclimatology of the proto-Central North Atlantic rift system, *Palaeogeogr., Palaeoclimatol., Palaeoecol., 40,* 13–30.

Hay, W. W., R. M. DeConto, C. N. Wold, K. M. Wilson, S. Voigt, M. Schulz, A. M. Rossby, Wold, W. C. Dullo, A. B. Ronov, and A. N. Balukhovsky, An alternative global Cretaceous paleogeography, in *The Evolution of Cretaceous Ocean/Climate Systems, Geological Society of America Special Publication,* E. Barrera, and C. Johnson (eds.), in press.

Hay, W. W., M. L. Rosol, J. L. Sloan, and D. E. Jory, 1987, Plate tectonic control of global patterns of detrital and carbonate sedimentation, in *Carbonate Clastic Transitions, Developments in Sedimentology,* Vol. 42, L. J. Doyle, and H. H. Roberts (eds.), pp. 1–34, Elsevier Scientific Publishing, Amsterdam.

Hay, W. W., S. L. Thompson, D. Pollard, K. M. Wilson, and C. N. Wold, 1994, Results of a climate model for Triassic Pangaea, *Zentralblatt für Geologie und Paläontologie, Teil I, 11/12,* 1253–1265.

Kasahara, A., and W. M. Washington, 1969, Thermal and dynamical effects of orography on the general circulation of the atmosphere, in *Proceedings of WMO/IUGG Symposium on Numerical Weather Prediction,* pp. IV47–IV56, Japan Meteorological Agency, Tokyo.

Kasahara, A., T. Sasamori, and W. M. Washington, 1973, Simulation experiments with a 12 layer stratospheric global circulation model, I, Dynamical effect of the Earth's orography and thermal influence of continentality, *J. Atmos. Sci., 30,* 1229–1251.

Kinsman, D. J., 1975, Rift valley basins and sedimentary history of trailing continental margins, in *Petroleum and Global Tectonics,* A. G. Fischer, and S. S. Judson (eds.), pp. 83–126, Princeton University Press, Princeton, N. J.

Kutzbach, J. E., and R. G. Gallimore, 1989, Pangaean climates: Megamonsoons of the megacontinent, *J. Geophys. Res., 94*, 3341–3357.

Kutzbach, J. E., P. J. Guetter, W. F. Ruddiman, and W. L. Prell, 1989, Sensitivity of climate to Late Cenozoic uplift in southern Asia and the American West: Numerical experiments, *J. Geophys. Res., 94*, 18,393–18,407.

Kutzbach, J. E., W. L. Prell, and W. F. Ruddiman, 1993, Sensitivity of Eurasian climate to surface uplift of the Tibetan Plateau, *J. Geol., 101*, 177–190.

Manabe, S., and A. J. Broccoli, 1990, Mountains and arid climate of middle latitudes, *Science, 247*, 192–195.

Manabe, S., and T. Terpstra, 1974, The effects of mountains on the general circulation of the atmosphere as identified by numerical experiments, *J. Atmos. Sci., 31*, 3–42.

Parrish, J. M., J. T. Parrish, and A. M. Ziegler, 1986, Permian-Triassic Paleogeography and Paleoclimatology and implications for theraspid distribution, in *The Ecology and Biology of Mammal-Like Reptiles*, N. Hotton, III, P. D. MacLean, J. J. Roth, and E. C. Roth (eds.), pp. 109–131, Smithsonian Institution Press, Washington, D. C.

Pollard, D., and M. Schulz, 1994, A model for the potential locations of Triassic evaporite basins driven by paleoclimatic GCM simulations, *Glob. and Planet. Change, 9*, 233–249.

Pollard, D., and S. L. Thompson, 1995, Use of a land-surface-transfer scheme (LSX) in a global climate model: the response to doubling stomatal resistance, *Glob. and Planet. Change, 10*, 129–161.

Prell, W. L., and J. E. Kutzbach, 1992, Sensitivity of the Indian monsoon to forcing parameters and implications for its evolution, *Nature, 360*, 647–652.

Ronov, A., V. Khain, and A. Balukhovsky, 1986, The global quantitative sedimentation balance for the continents and oceans for the last 150 million years, *Int. Geol. Rev., 28*, 1–9.

Ronov, A., V. Khain, and A. Balukhovsky, 1989, *Atlas of Lithological-Paleogeographical Maps of the World: Mesozoic and Cenozoic of Continents and Oceans*, USSR Academy of Sciences, Leningrad.

Ruddiman, W. F., and J. E. Kutzbach, 1989, Forcing of late Cenozoic Northern Hemisphere climate by plateau uplift in Southern Asia and the American West, *J. Geophys. Res., 94*, 18,409–18,427.

Ruddiman, W. F., and J. E. Kutzbach, 1990, Late Cenozoic plateau uplift and climate change, *Trans. Roy. Soc. Edinburgh, Earth Sci., 81*, 301–314.

Ruddiman, W. F., and J. E. Kutzbach, 1991a, Plateau uplift and climatic change, *Sci. Am., 264*, 66–75.

Ruddiman, W. F., and J. E. Kutzbach, 1991b, Plateaubildung und Klimaänderung, *Spektrum der Wissenschaften, 5*, 114–125.

Shaw, C. A., and W. W. Hay, 1990, Mass-balanced paleogeographic maps: Modeling program and results, in *Quantitative Dynamic Stratigraphy*, T. Cross (ed.), pp. 277–291, Plenum, New York.

Slingo, A., and J. M. Slingo, 1991, Response of the National Center for Atmospheric Research Community Climate Model to improvement in the representation of clouds, *J. Geophys. Res., 96*, 15,341-15,357.

Southam, J. R., and W. W. Hay, 1981, Global sedimentary mass balance and sea level changes, in *The Sea Vol. 7: The Oceanic Lithosphere*, C. Emiliani (ed.), pp. 1617–1684, John Wiley and Sons, New York.

Stein, C. A., and S. Stein, 1992, A model for the global variation in oceanic depth and heat flow with lithospheric age, *Nature, 359*, 123–129.

Thompson, S. L., and D. Pollard, 1995a, A global climate model (GENESIS) with a land-surface transfer scheme (LSX). Part I: Present climate simulations, *J. Climate, 8*, 732–761.

Thompson, S. L., and D. Pollard, 1995b, A global climate model (GENESIS) with a land-surface transfer scheme (LSX). Part II: CO_2 sensitivity, *J. Climate, 8*, 1104–1121.

Thompson, S. L., V. Ramaswamy, and C. Covey, 1987, Atmospheric effects of nuclear war aerosols in general circulation model simulations: Influence of smoke optical properties, *J. Geophys. Res., 92*, 10,942–10,960.

Trümpy, R., 1982, Das Phänomen Trias, *Geol. Rundsch., 71*, 711–723.

Van der Voo, R., 1990, Phanerozoic paleomagnetic poles from Europe and North America and comparisons with continental reconstructions, *Rev. Geophys., 28*, 167–206.

Von Huene, R., and D. W. Scholl, 1991, Observation at convergent margins concerning sediment subduction, subduction erosion, and the growth of continental crust, *Rev. Geophys., 29*, 279–316.

Von Huene, R., and D. W. Scholl, 1993, The return of sialic material to the mantle indicated by terrigenous material subducted at convergent margins, *Tectonophysics, 219*, 163–175.

Wessel, P., and W. H. F. Smith, 1991, Free software helps map and display data, *Eos Trans. AGU, 72*, 441.

Williamson, D. L., 1990, Semi-Lagrangian moisture transport in the NMC spectral model, *Tellus, 42A*, 413-428.

Williamson, D. L., and P. J. Rasch, 1989, Two-dimensional semi-Lagrangian transport with shape-preserving interpolation, *Monthly Weather Rev., 117*, 102-129

Wilson, K. M., 1989, Mesozoic suspect terranes and global tectonics, Ph.D. Sci. thesis, 372 pp., University of Colorado, Boulder, Colo.

Wilson, K. M., M. J. Rosol, and W. W. Hay, 1989, Global Mesozoic reconstructions using revised continental data and terrane histories: A progress report, in *Deep Structure and Past Kinematics of Accreted Terranes*, J. W. Hillhouse (ed.), pp., 1–40, *AGU Geophys. Monogr., 50*.

Wilson, K. M., D. Pollard, W. W. Hay, S. L. Thompson, and C. N. Wold, 1994, General circulation model simulations of Triassic climates: Preliminary results, in *Pangaea: Paleoclimatology, Tectonics and Sedimentation During Accretion, Zenith and Breakup of a Supercontinent*, G. D. Klein (ed.), pp. 91–116, Special Paper 288, Geological Society of America, Boulder, Colo.

Wold, C. N., W. W. Hay, D. Pollard, and M. Schulz, 1994, An evaporite basin experiment using Late Triassic paleoclimate simulation results, *Eos Trans. AGU, 75*, 153.

Wold, C. N., C. A. Shaw, and W. W. Hay, 1993, Mass-balanced reconstruction of overburden, in *Computerized Basin Analysis: The Prognosis of Energy and Mineral Resources*, J. Harff, and D. F. Merriam (eds.), pp. 115–130, Plenum Press, N. Y.

Ziegler, A. M., D. B. Rowley, A. L. Lottes, D. L. Sahagian, M. L. Hulver, and T. C. Gierlowski, 1985, Paleogeographic interpretation: With an example from the Mid-Cretaceous. *Ann. Rev. Earth Planet. Sci., 13*, 385–425.

PART IV

ROLE OF EPEIRIC SEAS

CHAPTER 7

The Vanishing Record of Epeiric Seas, with Emphasis on the Late Cretaceous "Hudson Seaway"

Alfred M. Ziegler and David B. Rowley

The remnants of Cretaceous strata left after erosion are the source of knowledge of these strata, and the zero isopach line, no matter how obtained, often seems to be considered the "margin" to any "basin" of deposition in paleogeographic discussions. Facies studies and recognition of exotic elements in the interior basins commonly indicate that these pseudo-margins are anomalies, and the implications of these anomalies have led to the speculations in this chapter. Williams and Stelck (1975)

The geological past differs from the present in many respects; one of the most obvious is the extent of epeiric, or epicontinental, seas that flooded broad areas of continental crust. Modern examples do exist including Hudson Bay and the Baltic Sea, but most shallow seas today are limited to shelf regions which average about 100 km in width. This means that oceanic salinities and temperatures extend across the shelf to the coast in most regions because there is little impediment to the mixing of water masses. In the Cretaceous, by contrast, the Western Interior Seaway invaded 5000 km across North America with depths in the 200 m range, yielding an aspect ratio of 25,000:1. With such a geometry, water masses within the seaway were dictated by local climate patterns (Slingerland et al., 1996) while at the same time they must have had important feedback effects on atmospheric circulation. Large water bodies, including lakes and seaways, have an ameliorating effect on climate because they retain heat relative to the more reflective land surface. The dimensions of the Western Interior Seaway, and especially its probable extension to the northeast through Hudson Bay and Hudson Strait, are the subject of this paper.

The geologic record is notoriously incomplete, particularly as represented in the deposits of epeiric seas because of vulnerability to erosion during im- mediately subsequent sea level lowstands. Thus, the "Geological Atlas of the Western Canada Sedimentary Basin" summarized a virtually complete Phanerozoic record based on 193,000 wells (Mossop and Shetsen, 1994), while the Canadian Shield immediately to the east has been stripped of most of its sedimentary deposits, probably as recently as the Pleistocene. A good Early to Middle Paleozoic record is present in the center of the Shield in Hudson Bay, and there is every reason to expect that high sea level stands in the Cretaceous would have inundated this area. Indeed, Late Cretaceous marine rocks have recently been drilled on the southwest margin of the Bay, and the purpose of this chapter is to infer the timing and extent of what we call "The Hudson Seaway." This seaway was not quite as extensive as the coeval Western Interior Seaway, but it did divide the continent to the east into subequal land masses, and must have had an ameliorating effect on climate. It also provided a migration pathway to the North Atlantic as late as during the Paleocene, while at the same time it must have constituted a barrier for terrestrial organisms between northern Canada and eastern North America.

The suggestion of a marine connection between the Western Interior Seaway and the Baffin Bay margin of west Greenland during the Late Cretaceous is not new and was first stimulated by similarities in the faunas of the two areas. Teichert pointed this out in 1939, and also noted that the west Greenland faunas differ from those of the eastern seaboard of the United States as well as those of east Greenland, which seem to rule out connections to the south and east. He concluded that a northern seaway extended around the margin of the Canadian Shield through the Arctic and thence to the Western Interior Seaway. He thereby linked all areas with this very distinctive fauna and this suggestion has been echoed in subsequent biogeographic studies (Rosencrantz et al., 1942; Jeletzky, 1971). Jeletzky went further, and tentatively drew a projection of the Western Interior Seaway eastward into Hudson Bay for the Coniacian, Santonian, and Campanian Stages of the middle Late Cretaceous, but considered a further continuation down the Hudson Strait and into the Labrador Sea to be improbable. His eastward projection was based on the fact that marginal marine deposits are absent for these intervals in Manitoba. Jeletzky (1971) thought that the eastern margin was definable by the late Campanian and this oddly seems to be the reason he objected in general to a throughgoing Hudson Seaway. However, contemporary lithofacies maps give no hint of an eastern margin of the Western Interior Seaway for any later Cretaceous interval (Macqueen and Leckie, 1992; Mossop and Shetsen, 1994).

A throughgoing connection was finally proposed by Williams and Stelck (1975) and this was defined using a broad topographic saddle trending northeastward across Manitoba to Hudson Bay and thence down the Hudson Strait. This proposal was adopted by Kauffman (1984), but a number of works show variable acceptance of the concept (Funnell, 1990; Smith et al., 1994; and essays in Caldwell and Kauffman, 1993). In view of this confusion, we have assembled the key topographical, geophysical, sedimentational, geochemical, and biogeographical evidence bearing on the dimensions and duration of the Hudson Seaway. The affect on climate of such a feature has been examined by one modeling study which compared climate simulations with and without the Seaway and found significant differences in the continentality of Late Cretaceous climate (Valdes et al., 1996).

Our approach to the problem is first to review the events affecting Canada since the Cretaceous, and then to try to restore what, if anything, remains of

Mesozoic physiography. Pleistocene icesheets produced a massive downwarp centered on Hudson Bay, and the effects have been incompletely dissipated. Gravity studies reflect this warping and indicate some 300 m of rebound remaining in some areas. We have made corrections for this. In addition, the icesheets significantly modified the relief particularly in the areas of major ice streams, so that areas like Hudson Strait and passages among the Arctic Islands have been over-deepened. Sediment volume studies in surrounding oceans confirm that great volumes of material were eroded and transported to the sea during the Pleistocene. In the Tertiary, considerable amounts of uplift in the Western Interior Basin and elsewhere are indicated by the fact that Upper Cretaceous marine deposits currently rest at elevations of 500–800 m in western Canada, far above reasonable estimates of sea levels for the period. All these post-Cretaceous modifications must first be accommodated before Late Cretaceous paleogeography can be reconstructed. The direct evidence provided by Cretaceous outliers on the Shield will be discussed, followed by a review of the deep ocean connections of the various epeiric seaways to the Gulf of Mexico, Arctic Ocean, and Labrador Sea. These outlets yield evidence on the duration and termination of marine connections. Only then will the available biogeographic evidence be reviewed and the Seaway reconstructed. Finally, the evidence for other possible seaways in the Late Mesozoic will be briefly discussed.

Present Geology and Topography of the Canadian Shield

The North American craton has been a coherent entity since 1.7 Ga (Hoffman, 1989) so it is not surprising that the exposed portion, the Canadian Shield, is topographically reduced. Elevations across this broad region are generally in the 200–400 m range, an exception being a broad upland region in south central Quebec (Fig. 7.1). We have speculated that this upland is isostatically supported by a remnant of crustal thickening formed in the Grenville Orogeny about 1 Ga (Ziegler et al., 1996). Lower Paleozoic sedimentary rocks onlap the Canadian Shield around most of its perimeter except for a few areas where the Cretaceous rests directly on, or is faulted against the Precambrian (Fig. 7.2). Remnants of the Lower Paleozoic that occur

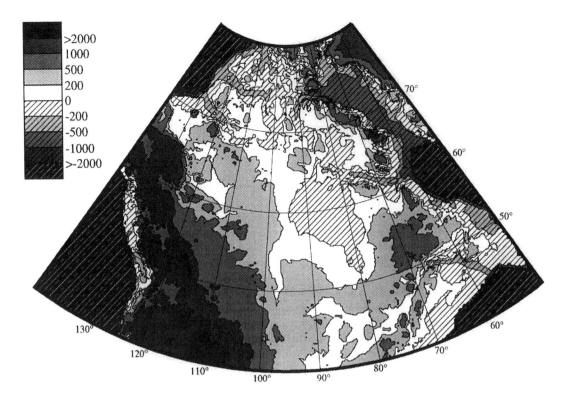

Fig. 7.1. Present topography of Canada and adjacent areas. This and subsequent maps employ the ETOPO5 5-min. gridded topography/bathymetry data from Global Relief CD-Rom, National Geophysical Data Center (1993) resampled at 30 min. resolution. This and subsequent maps employ a Lambert conformal conic projection with standard parallels at 60°N and 49°N, centered at 55°N, 95°W.

are patchily distributed on top of the Shield and range in size from the Hudson Bay Basin, a broad crustal downwarp spanning nearly a million km^2, to much smaller blocks within meteorite impact structures. These Paleozoic rocks lie at elevations below 250 m with the exceptions of the Baffin Bay margin and the Manicouagan impact structure which is surrounded by elevations in the 600 m range. Since most Paleozoic rocks lie within elevations thought to have been subject to sea level inundations of the time (Algeo and Seslavinsky, 1995), their disposition lends credence to the concept of stability within the shield area. Hudson Strait is a fault-bound trough with a sedimentary fill unconformably above the Paleozoic that is thought to be of Cretaceous age (Sanford and Grant, 1990). Sup-

port for the Cretaceous age of this faulting comes from the fact that it is subparallel to known Cretaceous extensional features such as the Labrador Sea and inlets to the north in Baffin Island (Balkwill et al., 1990). Shield rocks are uplifted, presumably as rift shoulders, along the margins of these features.

Post-glacial Rebound of North America

Two different techniques have been used to determine the residual uplift of the area affected by the Wisconsin ice sheet in Canada. Andrews (1970) derived curves for post-glacial uplift for many sites, based on radiocarbon dating of the marine terraces which are extensive in central Canada. He determined the exponential form of these curves and

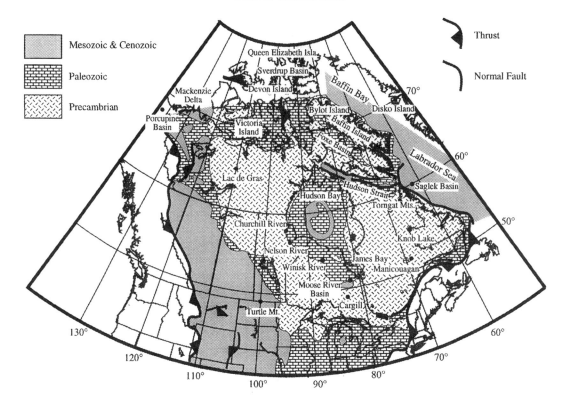

Fig. 7.2. Location map of Canada and adjacent regions, showing general geological relationships.

simply projected them into the future, obtaining estimates ranging up to 160 m for the remaining uplift of southeastern Hudson Bay. Walcott (1970) obtained greater values, in excess of 250 m, based on a study of the free air gravity anomaly which he attributed to crustal warping. His map was constructed by averaging the free air anomalies over 1° (latitude) by 2° (longitude) squares, and this yielded maximum negative anomalies of 50 and 40 mgals for northern Hudson Bay and the Foxe Basin to the north, respectively. He estimated a conversion factor of 7 m/mgal to calculate the residual uplift (Fig. 7.3), but as far as we know, our map (Fig. 7.4) is the first restoration of the topography of Canada to its equilibrium value using these data (Fig. 7.4). Barr

(1972) performed a similar exercise using Andrews' map, but because even Andrews (1970) seemed to have had doubts about his decay model, we have opted for the geophysical data. Our map eliminates the surviving downwarp and also shows the marine coverage if the sea-level were higher by 65 m, an appropriate figure for an ice-free world (Rowley and Markwick, 1992). In such a configuration, the sea would still flood Hudson Strait, but only the northern portion of Hudson Bay. The northern Arctic Islands would remain as distinct features, but Baffin Island and Victoria Island, as well as many smaller features, would be united to the mainland. As Barr pointed out, the amount of rebound makes Canada the fastest growing country in the world!

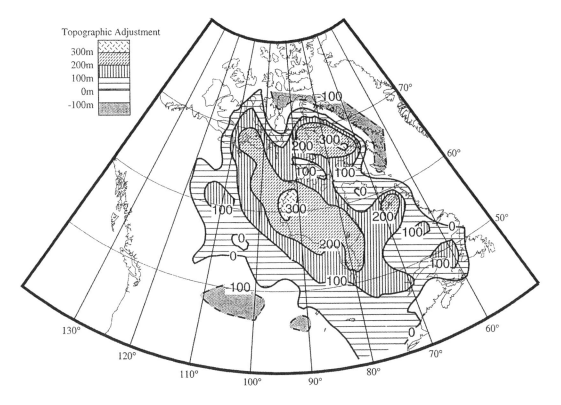

Fig. 7.3. Elevational correction for Canada and adjacent regions derived by applying a correction of 7 m/mgal to the free air gravity anomaly map (after Walcott, 1970).

Pleistocene Physiographic Modifications of Canada

An average of 120 m of rock has been physically eroded from the Laurentide region, to judge by Pleistocene-age sediments in marine basins surrounding North America (Bell and Laine, 1985). A considerable amount of this must have come from the regolith which has been removed in all but a few sites (White, 1972; Dyke and Dredge, 1989). The original thickness of the regolith is unknown, but it must have been considerable because the time available for its formation must be reckoned in 10s or even 100s of millions of years. Two sites within the Precambrian shield have preserved remarkable examples of the regolith—Knob Lake, Quebec and Cargill, Ontario—and both have been dated as Cretaceous, by macro- and microfloras, respectively

(see Fig. 7.2 for localities). The iron ores of Knob Lake, although hosted in Precambrian rock, owe their enrichment to 150 m deep chemical weathering in Mesozoic and later times (Gross, 1968), while the Cargill Township Carbonatite Complex has a karstic residuum 170 m thick (Sage, 1988). These sites are obviously exceptional, but they do confirm that weathering of the Canadian Shield was deep and extremely irregular, and that at least local relief was considerable. Accounting for the general removal of this material is extremely difficult.

On the local scale, greater thicknesses of material were eroded by the major ice streams that deepened valleys like the Hudson Strait and the St. Lawrence estuary, and created depressions like the Great Lakes (Hughes, 1987). Still smaller features like fjords and lakes are also sites of considerable

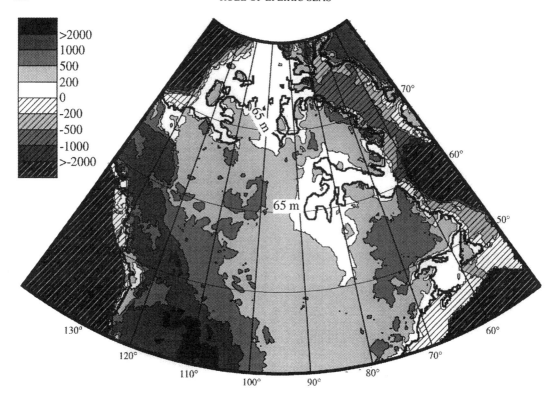

Fig. 7.4. Topography "corrected" for the residual effects of glacial loading. This map was derived from Figures 7.1 and 7.3. The ice-carved channels have been filled to sea level. The 65 m contour shows the position of the shoreline if the world's present ice-caps were melted.

erosion. The arctic Queen Elizabeth Islands are thought to have formed as a result of overdeepening and straightening of river valleys by ice streams (Trettin, 1991). The depths of the channels range up to 900 m, and the volume of these channels below present sea level is 2.4 x 10^5 km^3 corresponding to removal of 6.7 x 10^5 km^3 of rock. It has been proposed that Hudson Bay originated by glacial scouring (White, 1972), and indeed it is floored by soft rocks like those underlying the Great Lakes. However, Pelletier (1969, p. 130) described "steep-walled valleys up to 30 metres in height, which extend more than 130 km from present estuaries" and proposed that these represent a submerged river system. These V-shaped valleys are clear on more recent bathymetric maps (Sanford, 1987) and, if correctly identified, indicate very little modification of much of Hudson Bay during at least the last glacial

advance. Hughes (1987) has assumed a frozen bed of the icesheet in the Bay which would imply no erosional modification. We note that the submarine valleys are not seen in the deepest parts of the bay, the parts that would still be submerged in a fully re-bounded Canada (Fig. 7.4).

Additions to the topography of the glaciated area have also occurred. Till thicknesses up to 250 m are not uncommon over the Mesozoic rocks of the Great Plains of western Canada (Fenton et al., 1994) and over the Paleozoic rocks of Michigan (Sommers, 1977). Interestingly, the greatest thicknesses seem to occur on pre-existing topography, so the general result of Pleistocene physiographic modification, both positive and negative, has been to increase relief. We have made a crude attempt to restore the topography of Canada to its pre-glacial state by simply filling the obvious ice carved channels to sea level (Fig. 7.4). We conclude that a considerable number of these small- to intermediate-scale physiographic features in Canada are of extremely young origin.

Tertiary Uplift of Western Canada

Marine Late Cretaceous and Early Tertiary rocks lie at elevations of 500–800 m in western Canada (Mossop and Shetsen, 1994), far above reasonable estimates of sea levels for those intervals (Hallam, 1992). These elevations apply to unfolded regions well inboard of the fold and thrust belt; in fact, Paleocene marine fossils occur above 661 m at Turtle Mountain, Manitoba (Bamburak, 1978) which is 800 km from the thrust front. Obviously, adjacent parts of the Canadian Shield have also been involved in this apparent uplift, so some understanding of the process is necessary if shoreline positions are to be inferred. The Late Cretaceous or Early Tertiary shoreline and shallow water deposits to the east have been eroded, and Mitrovica et al. (1989) proposed that the continent has been tilted about a north-south axis resulting in the uplift of the Great Plains to a height of 500–1000 m. This, they proposed, is simply the termination of a down-then-up cycle which allowed the westward thickening Cretaceous wedge to accumulate in the first place. Mitrovica et al. (1989) model the downward deflection as a result of subsurface loading from the subducting slab; an effect that is independent of the surface loading of the lithosphere by the thrusts and sediment wedges. The mass anomaly of the subducted slab in the viscous mantle results in mantle flow pulling the surface of the Earth downward above heavy subducting slabs (Gurnis, 1993). The wavelength of the horizontal deflection is sensitive to subduction angle (Mitrovica et al., 1989) while the amplitude of the vertical deflection depends largely on the age of the subducted lithosphere (Gurnis, 1993). Termination of subduction, as in the case of western Canada, or changes in dip or reduction in age of the subducted lithosphere result in subsequent recovery and uplift of the basin and consequent erosion which would, in turn, stimulate a positive feedback through isostatic effects.

We stop short of trying to model this tilting of western Canada because of uncertainties in the position of a hingeline. The subduction angle could be expected to vary in time as well as along strike and

in any case is unknown, making the modeling of the hinge position difficult. We made an attempt to determine the hinge position by projecting the dips of successive Cretaceous marine units eastward across strike in western Canada, but given the width of the belt, east of the obvious effects of structural loading, the uncertainties proved to be too great for this purpose. Combining the Turtle Mountain locality with the presence of marine Cretaceous rocks near Lac de Gras, Northwest Territories, discussed later, we can, however, infer that the hinge line lay more than 800 km east of the front of the fold-thrust belt.

Tertiary Uplift Around Baffin Bay and the Labrador Sea

Another region of considerable post-Cretaceous uplift parallels the Late Cretaceous to Early Tertiary rifted margins of northeastern Canada and west Greenland adjacent to the Labrador Sea and Baffin Bay. The region plays a part in biogeographic links between the Arctic Ocean and Baffin Bay, and ultimately, between the Western Interior Seaway and the Labrador Sea. At the northwest corner of Baffin Bay, outliers of Late Cretaceous marine rocks occur on Devon and Bylot Islands and are assigned formation names derived from the Sverdrup Basin of the Arctic Ocean margin. The Devon Island examples lie within a 900 m high plateau of unfolded Siluro-Devonian rocks and are preserved as jumbled blocks in karstic features (Thorsteinsson and Mayr, 1987). Bylot Island contains the fault-bound Eclipse Trough which is parallel to Baffin Bay immediately to the east, and like it must have been subject to crustal stretching in the Late Cretaceous (Roest and Srivastava, 1989; Miall et al., 1980). The Byram Martin Mountains between these features today reach elevations of 1900 m but provided "minor quantities of coarse detritus" to the Eclipse Trough as late as the Eocene (Miall et al., 1980). This, together with the fact that the partially marine stratigraphic sequence of the Eclipse Trough rests at 760 m (Trettin, 1991), indicates considerable post-depositional uplift. The continuation of this uplift of Precambrian basement is seen along strike to the north on Ellesmere Island where sedimentary provenance indicates uplift and erosion of basement beginning in middle to late Paleocene time (Riediger and Bustin, 1987). On the Greenland margin of Baffin Bay, Cretaceous marine deposits have been uplifted to 900 m (Henderson et al., 1981).

This type of rift margin-parallel uplift is observed in the topography today along both margins of Baffin Bay and the Labrador Sea and was referred to by Trettin (1991) as a "delayed shoulder uplift" because the main activity occurs entirely after rifting and late in the seafloor spreading phase or after it altogether. The across-strike dimension of the uplift is considerable since the Devon Island occurrence is part of it and is 350 km from Baffin Bay. A number of mechanisms for the uplifts have been proposed employing kinematic and dynamical models (Keen and Beaumont, 1990). Whatever the explanation, the uplifts seem not to have been much of a factor in Late Cretaceous paleogeography and a broad epeiric connection of the Arctic Ocean and Baffin Bay is envisaged (Balkwill et al., 1983; Gradstein et al., 1990; but see Embry, 1991 and Miall, 1991 for alternative interpretations). The uplift along Ellesmere Island has been partly overprinted by the contractional Eurekan Orogeny (Trettin, 1991) which began in the Late Paleocene when sea floor spreading studies show that Greenland began a northward trajectory with respect to northeast Canada (Rowley and Lottes, 1988; Roest and Srivastava, 1989).

Cretaceous Outliers on the Laurentian Platform

Marine Cretaceous rocks were recently identified southeast of Hudson Bay by M. A. Miller (personal communication, 1996) of the Amoco Exploration and Production Technology Group, Houston TX, and he generously provided the following information (see Fig. 7.2 for localities and Figure 7.5 for paleogeography). The material was penetrated by drilling near the Winisk River, Ontario (54° 18' 30" North, 87° 2' 30" West) at a depth of 73 m, and we reconstruct the elevation of the deposit at about 37 m as calculated from the local quadrangle map. The fossils are still being studied but include a diverse dinoflagellate assemblage attributed to the early Campanian (~80 Ma) or slightly earlier, as well as an admixture of Albian terrestrial pollen similar to the Moose River Basin in the James Bay Lowlands to the south (Norris, 1993), and even some Devonian microfossils. The fossils were found in grey-green shales overlying Ordovician carbonates, and

Miller is confident that the Cretaceous forms are indigenous to these shales.

About 500 km to the northwest, Cretaceous erratic blocks have been discovered in several areas in the drainages of the Nelson and Churchill Rivers of Manitoba by E. Nielsen (unpublished manuscript) of the Manitoba Department of Energy and Mines. The following information comes from Nielsen's draft report. The erratic blocks are of interest because they are near the Hudson Bay Paleozoics, and at least 550 km from the nearest Cretaceous outcrop in the Western Interior Basin. The most significant sites are in glacial tills northeast of the town of Gillam (about 56° 30' North, 94° West). Ice flow directions imply provenance from the north-northwest and the east in separate tills, but not from the direction of the Western Interior Basin. The pebbles are described as soft shale, less than 2 cm in diameter; one has yielded Turonian (~90 Ma) dinoflagellates and the other Santonian (~85 Ma) dinoflagellates. In view of their fragility, Nielsen concluded that they must have had a "relatively local undiscovered source".

Cretaceous rocks have recently been dated from beneath Hudson Bay and their distribution in the Bay and in Hudson Strait has been inferred using seismic data (Sanford and Grant, 1990). Fossils recovered so far are nonmarine and suggest an age in the range Aptian to Cenomanian (R. Fensome, personal communication, 1996). In these respects they are similar to the Mattagami Formation fossils from the Moose River Basin of the James Bay Lowlands to the south (Norris, 1993). Lamprophyric dikes and sills dated as Early Cretaceous intrude older rocks along the southern margin of the Moose River Basin. All that these observations prove is that the Hudson Platform was a depression in the Cretaceous as indeed it was in the early to mid-Paleozoic, and that subsidence occurred after the mid-Cretaceous. The extension of marine Cretaceous to Hudson Strait is tenuous but seems a reasonable inference because of that structure's parallelism with dated extensional features to the northeast.

Blocks of marine Cretaceous rocks have lately been discovered in a number of kimberlite pipes near Lac de Gras, Northwest Territories (64° 34' North, 110° 7' West). These sites are 390 km east of the main outcrop belt and indicate that the Western Interior Seaway was much wider at this latitude than has been generally thought. Marine and marginal-

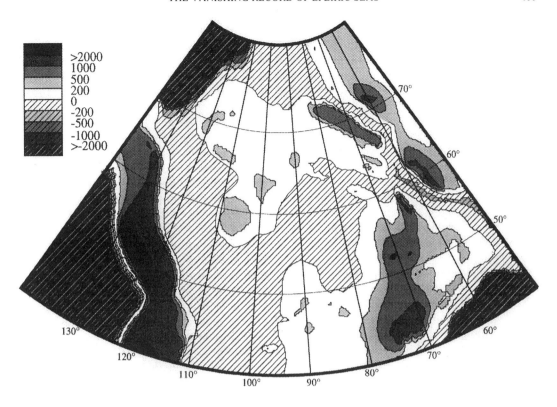

Fig. 7.5. Late Cretaceous (Coniacian Stage) paleogeography of North America. The Hudson Seaway is seen in the center of the map and would presumably have had this configuration for all Late Cretaceous intervals, as well as the Paleocene until the mid-Thanetian.

marine environments are represented and include horizons from the Albian (~100 Ma) to the Maestrichtian (~65 Ma) (Nassichuk and McIntyre, 1995). The blocks have been recovered from depths of 50–150 m in the "diatreme facies" and from as shallow as 25 m in the "pyroclastic crater facies," while there is an overlying "epiclastic lake facies." Wood samples recovered in lignitic peat in the crater facies yield low vitrinite reflectance values indicating that temperatures never exceeded 30°C (Stasiuk and Nassichuk, 1995). All this suggests that relatively little erosion of the landscape has occurred since the kimberlites were intruded at about 52 Ma in the Eocene, except of course for the overlying Cretaceous strata. Present elevations in the area range from the 416 m lake level to about 500

m, so this area appears to have been uplifted like other areas in western Canada. Lac de Gras is about 600 km east of the fold-thrust belt.

Deep Ocean Connections of the Epeiric Seaways

The Western Interior Seaway was progressively in-filled by clastics generated by the overthrust belt to its west as well as by uplift of the Laramide Mountains which rose from within. The Seaway's demise must have been hastened by lowering sea levels in the Cenozoic, placing the Late Cretaceous and Early Tertiary record in an erosive regime. Accordingly, the paleogeographic details of this time interval are sketchy and must be reconstructed from widely dispersed outcrops. Some of the most useful information can be garnered from the outlets of the epeiric seas to the surrounding ocean basins. These outlets were transformed into river systems with major deltas, and the timing of the change is relatively clear in the stratigraphic record. There appear to be three outlets including the Gulf Coast of Texas, the Mackenzie Delta of the Northwest Territories, and the entry point of Hudson Strait into the Labrador Sea. A connection north of the Ozarks to the lower Mississippi valley has been proposed (Williams and Stelck, 1975), but shoreline deposits are reasonably well defined there, making this an unlikely outlet for an epeiric sea.

Marine conditions persisted in the Western Interior Seaway into the Late Paleocene (58 Ma) as seen in the Cannonball Formation of North Dakota (Cherven and Jacob, 1985). Connections to the Gulf of Mexico had probably been ended by infilling of Laramide clastic deposits at some time earlier. The Laramide Orogeny began about the beginning of the Late Cretaceous Maestrichtian Stage (~75 Ma), and by that time coarse continental clastics were prograding east across the Denver and Raton Basins of Colorado and New Mexico (Dickinson et al., 1988). This clastic sheet arrived along the Gulf Coast of Texas in the form of the deltaic Wilcox Formation just after the beginning of the Late Paleocene, and by that time the shoreline can be traced across the entire state subparallel with, and inland about 150 km from the present shoreline (Galloway, 1989). This cannot be said with certainty for the underlying Midway Formation, but it seems likely that this formation was the vanguard of the clastic wedge, and that a marine connection to the interior was severed by the beginning of the Cenozoic (see the following section on biogeography).

The outlet of the Western Interior Seaway to the Arctic seems to have terminated before the end of the Cretaceous. During the Late Maestrichtian (~70 Ma), there was a 225 km northward shift of the main coarse-clastic depocenter to the continental margin of the Canada Basin in the Mackenzie Delta area (Dixon et al., 1992). These authors described this as a new "tectonostratigraphic phase" related to compressional tectonics and prograding deltaic sediments, which had followed the extensional sea floor spreading phase in the Canada Basin. We interpret the arrival of this clastic wedge as evidence for the termination of the northern end of the Seaway in the mid-Maestrichtian when the outlet to the Arctic Ocean effectively became a river mouth.

The outlet of Hudson Strait to the Labrador Sea is the remaining part of the epeiric network to be considered, and was in fact proposed as the only marine connection by Paleocene times on biogeographic grounds (Feldman, 1972). Like the other outlets, it was then transformed into a river system, the "Bell River System," which is thought to have drained much of Canada from the Shield to the Rocky Mountains (McMillan, 1973; McMillan and Duk-Rodkin, 1995). This system was proposed to account for the derivation of some 2.5 x 10^6 km^3 of Tertiary sediments in the Saglek Basin of the eastern Labrador Sea, as well as the introduction of reworked Late Cretaceous palynomorphs of western Canadian aspect. The evidence for the initiation of the river system seems not to have been addressed in the literature, although McMillan and Duk-Rodkin state its duration to be from Late Paleocene to Pliocene. Presumably, the origin of the Bell River system coincides with a major formational boundary in the Saglek Basin, the Markland to Cartwright formational transition which is assigned to the early Late Paleocene (Balkwill et al., 1990). The Cartwright Formation is much thicker and more extensive, and the clay mineral suite, interpreted as continent-wide in derivation, diversifies in this formation (Hiscott, 1984). The Markland Formation is described as a starved basin deposit, although the amount of fine clastic material increased about the Campanian. The interpretation of such an event is uncertain, but it could mean that conditions at the outlet of the Hudson Seaway alternated between marine and fluvial, prior to the final termination of the seaway.

In summary, the outlets of the Western Interior Seaway were transformed into river mouths by a combination of clastic input and lowered sea levels in the mid-Maestrichtian, in the case of the Arctic Ocean, and near the end of the Maestrichtian, in the Gulf of Mexico. However, the outlet of the Hudson Seaway remained open until the middle of the Late Paleocene, when it became the mouth of the Bell River System.

Cretaceous and Paleogene Biogeography of the Boreal Province

Biogeography may be used to test paleogeographic

concepts, particularly the continuity of seaways which often provide migration pathways for marine organisms, or indeed migration barriers for land-based forms. Interpretation problems arise because regional temperature and salinity gradients may mimic geographic barriers in their effect on organisms. Also, water masses, which are usually defined by continuity in these parameters, typically transport faunas and floras to different latitudes on opposite sides of the same ocean basin. Accordingly, the obvious aspects of paleogeography—shoreline positions and latitude—may not always provide the primary control on biotic distributions. In this section it is necessary to consider both North America and other northern continents in the Late Mesozoic and Cenozoic in order to trace all the seaways and water masses affecting boreal biogeography.

The Late Cretaceous faunas of Canada have been referred to by Kauffman (1984) as the "Northern Interior Subprovince" with a "cool temperate" climate, and by Jeletzky (1971) as the "North American Boreal Province." This Boreal Province became distinct from the Pacific Province in the Early Cretaceous by the development of a major tectonic barrier, the Cordilleran fold-thrust belt. Deformation began in the mid-Jurassic (~170 Ma) but some mixing of Pacific and Boreal faunas is indicated in the Yukon region as late as the Barremian Stage of the Early Cretaceous (Jeletzky, 1971). This is about the time of collision of the North Slope-Chukotka Block with Siberia and central Alaska (Green et al., 1986; Rowley and Lottes, 1988) which completed the mountainous land barrier along the Pacific margin of North America and Siberia, a feature which remained unbroken until late in the Tertiary by the Bering Strait. The continuity of this barrier has not been recognized by all paleogeographers (Smith et al., 1994) but it did serve to isolate the deep (>~200m) Arctic Ocean from the world ocean, and only tenuous links through epeiric seaways (Fig. 7.6), until the early Cenozoic (~53 Ma) opening of the Norwegian Sea, or even later with the submergence of the Greenland-Scotland Ridge in the Miocene (Theide and Eldholm, 1983).

By the middle of the Cretaceous (Late Albian Stage, ~100 Ma) marine conditions extended along the entire length of the Western Interior Seaway from the Gulf of Mexico to the Arctic Ocean, and this seaway existed for the rest of the Cretaceous Period. We assume a similar duration for the Hudson Seaway based on general sea level considerations (Hallam, 1992) and sedimentation patterns in the Labrador Sea. We admit that these arguments are not compelling, but better information is unlikely to emerge until a drilling program is undertaken in Hudson Bay and Hudson Strait. It cannot be assumed that marine connections were continuously maintained throughout the epeiric network during intervals of lowered sea level or delta-building. We suggest that dinosaur distribution patterns might be used to test whether the epeiric seas were pervasive barriers over long intervals.

Tropical faunas indicate that warm waters invaded north along the Western Interior Seaway from the Gulf of Mexico during a number of high sea level stands in late Early and Late Cretaceous times (Kauffman, 1984). The area affected varied from the Gulf to Nebraska and even to mid-Alberta, or from one quarter to over one half of the 5000 km long seaway during limited intervals. The notion of Kauffman (1984) that lowered salinities were otherwise the norm in the Seaway is supported by carbonate isotopic studies (Wright, 1987; Pratt et al., 1993) and the fact that high latitude seas today, like the Arctic Ocean, Hudson Bay and the Baltic Sea, have well below average salinities (Parkinson et al., 1987). Also the dimensions of this boreal epeiric network must have precluded much exchange with the deep ocean. The fact that the final invasion of tropical waters into the western states was mid-Maestrichtian (Kauffman, 1984), indicates that the Gulf connection was still open during the latest Cretaceous stage.

The contact of Atlantic waters with the Boreal Province is observed in the Labrador Sea. Crustal extension occurred here in the Early Cretaceous while the seafloor spreading phase began about the middle of the period (~92 Ma). In Baffin Bay to the north, opening was delayed until late in the period (~69 Ma) (Roest and Srivastava, 1989). Boreal Province faunas are found along the eastern shore of Baffin Bay in the Disko Island area of Greenland beginning in the Late Cretaceous (Upper Turonian, ~90 Ma) and these forms are then joined by Atlantic ammonites (Campanian–Maestrichtian, 83–65 Ma); earliest Tertiary (Danian, 63 Ma) faunas show greater taxonomic diversity and clear relationships with Europe (Henderson et al., 1976; Balkwill et al., 1990). Accordingly, the Boreal faunas appeared early in the stretching phase and probably reflect

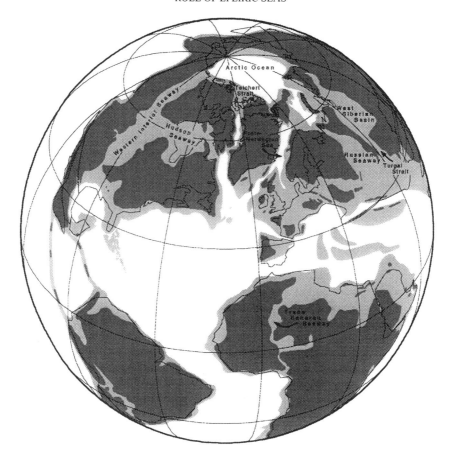

Fig. 7.6. Arctic Ocean connections in the Early Maestrichtian Stage. Deep sea areas are white, shallow seas are light grey, and land areas are dark grey.

connection to the Western Interior Seaway via the Devon Island exposures (mentioned earlier) in the Canadian Arctic (Fig. 7.6). The Hudson Seaway, originally proposed to account for these faunal similarities (Williams and Stelck, 1975), may have provided an additional link but because the intervening Labrador Sea was becoming oceanic, it is likely that normal oceanic salinities and faunas extended to this latitude by mid-Cretaceous times. The only direct evidence for conditions within the Hudson Seaway comes from the dinoflagellate flora from the Winisk borehole near Hudson Bay which compares in some respects with the McIntyre Suite found in the Western Interior Seaway and Canadian Arctic (Lentin and Williams, 1980; Miller, personal communication, 1996).

Boreal-Tethyan connections through Eurasia were as tenuous as they were through North America

(Fig. 7.6). They included the proto-Norwegian Sea (P.A. Ziegler, 1988), the "Russian Seaway" extending from the Barents Sea to the Caspian depression along the west side of the Urals (see later discussion), and the West Siberian Basin connection through the narrow Turgai Strait of Kazakhstan (Vinogradov, 1968). Midway along the proto-Norwegian Sea are Traill Island and adjacent localities of east Greenland where mixtures of Boreal and Tethyan elements are found throughout the Cretaceous sequence (Birkelund and Perch-Nielsen, 1976). The Late Cretaceous outcrops along the Russian Seaway have been eroded, except for a karstic site in the pre-Urals where Santonian Boreal forms are found (Beznosov et al., 1978). The West Siberian Basin, like the Western Interior Basin, experienced low salinities (Teys et al., 1978) with incursions of Tethyan faunas during high sea level stands to paleolatitudes of about 60° (Zakharov et al., 1991; Zakharov, 1994).

The Boreal Province became even more restricted in the Tertiary, and molluscan and foraminiferal assemblages show a high level of endemicity in the Arctic Ocean (McNeil 1990; Marincovich, 1994). Paleogeographic changes include the mid–Maestrichtian termination of the Arctic outlet of the Western Interior Seaway and the hot spot activity to form the Greenland-Scotland Ridge, which became a land bridge between Europe and North America about the Thanetian Stage of the Late Paleocene (Marincovich et al., 1990; P.A. Ziegler, 1988). Of interest here are the connections of the Cannonball Sea remnant of the Western Interior Seaway in the Dakotas and Manitoba. Marine conditions persisted through the Danian and into the Thanetian, to judge by a number of marine groups and interfingering mammal horizons (Cherven and Jacob, 1985; Cvancara and Hoganson, 1993). The marine invertebrate and vertebrate groups seem to be most closely related to European faunas and have affinities with west Greenland and the east coast of the United States, but not with the Gulf Coast (cf. Kurita and McIntyre, 1995). This would point to dispersal through the Hudson Seaway and Labrador Sea, as suggested by Feldman (1972), and indeed this constitutes the best evidence for the continuity and persistence of the Hudson Seaway. Drying out of the Hudson Seaway must have been early Late Paleocene because it was during this stage that deltaic clastics appear at the outlet to the Labrador Sea (McMillan and Duk-Rodkin, 1995), and during which the direct connection to Europe was severed by the emergence above sea level of the Greenland-Scotland Ridge. Alternatively, the European influence could have been made by a route extending around northern Greenland and Baffin Bay to the Hudson Seaway; Marincovich (1994) argued for an Arctic influence on Cannonball mollusks which such a circuitous route would satisfy.

The Hudson Seaway and Other Vanishing Seaways

We adopt the Williams and Stelck (1975) concept of an epeiric extension of the Western Interior Seaway across Manitoba to Hudson Bay and along Hudson Strait to the Labrador Sea, and we follow these authors in defining the channel by existing topo-

graphic and bathymetric lows (Fig. 7.5). The Manitoba sag (Nelson River drainage, Fig. 7.2) formed a link between the Paleozoic Williston and Hudson Basins, while the Hudson Strait probably formed as a graben during the Cretaceous. This concept, originally based on the similarities of west Greenland faunas with those of the Western Interior Basin, is confirmed by new discoveries described earlier in the section on Cretaceous outliers.

We also resurrect the older concept of an epeiric connection from Baffin Bay to the Sverdrup Basin margin of the Arctic Ocean in the vicinity of Devon and Ellesmere Island, a distance of about 500 km. This idea was proposed by Teichert (1939), endorsed by Jeletzky (1971), and confirmed by Miall et al. (1980) and Thorsteinsson and Mayr (1987) with the discovery of Late Cretaceous outliers. We refer to this seaway as the Teichert Strait in honor of Curt Teichert. It was important in Boreal Province biogeography as it provided a link for marine faunas between west Greenland and the Western Interior Seaway.

Seaways whose records have nearly vanished are not limited to North America, and a number of Late Mesozoic examples that have come to our attention are briefly discussed below. They are mentioned here to alert paleogeographers to these interesting problems as well as to warn climatologists of uncertainties in current reconstructions. The "Russian Seaway" is a feature shown on Russian paleogeographic maps for Middle Jurassic through Early Cretaceous intervals (Vinogradov, 1968). This feature (Fig. 7.6) is based on a chain of large cuestas of Jurassic and Cretaceous rock that extends north from the Caspian Depression through Moscow to the Pechora Basin on the Arctic margin. The tops of the sequences have been eroded in such a way that the north-south continuity of marine deposits is broken before the Late Cretaceous. Russian authors have been unwilling to project the shorelines much beyond the existing outcrops, while at the same time admitting that the preserved strata represent deeper water environments (Naidin et al., 1980; Naidin, 1981). An exception is made for Santonian marine strata in karstic depressions on the western slope of the Urals at elevations of 200–300 m (Vinogradov, 1968; Borisevich, 1992). These occurrences are within 500 km of the Caspian Depression but contain Boreal Province fossils (Beznosov et al., 1978), thus establishing a Late Cretaceous link. We believe that the Russian Seaway existed throughout the Cretaceous and may very well have persisted into the Paleogene. Marine rocks of Paleocene and Eocene age exist at elevations above 300 m in the Pre-Volga Upland just north of the Caspian Depression in the vicinity of the city of Penza (Vinogradov, 1967). In the Arctic zone of Finland, Early Tertiary diatomaceous deposits rest on the Precambrian basement at an elevation of 205 m (Tynni, 1982). The present land-surface between these localities is below the 200 m contour, certainly allowing the possibility of an Early Tertiary marine connection across the Russian Platform. Such a seaway, if persistent could have restricted mammal migrations, so terrestrial biogeography might be used to test this possibility.

Late Mesozoic epeiric seaways also occurred in the Southern Hemisphere and some of these are known from very scanty records. The most intriguing example is represented by the Stanleyville Formation of the Zaire Interior Basin from which marine Jurassic bivalves have been described (Cox, 1953, 1960). Support for a marine environment also derives from a geochemical analysis of the organic matter in the Stanleyville Formation (Clifford, 1986). These deposits are centrally disposed in southern Africa and are at least 1500 km from the nearest marine outcrops. They were originally dated as Upper Jurassic, but are now thought to be Middle Jurassic based on microfloral and microfaunal evidence (Colin, 1994). Overlying formations have yielded diverse fish faunas dated as Aptian through Turonian, which Colin characterized as fluvial and lacustrine based on fresh water ostracodes. However, Belgian workers regard the fish as marine with affinities to Tethys and not to the South Atlantic (Lepersonne, 1977). The possibility of a Baltic-like sea, with a gradual transition to brackish and fresh-water conditions, suggests itself. In any case, the Zaire Interior Basin is about 1000 km across, so it is a feature that should be incorporated into climate modeling schemes. We suggest the term, "Zaire Seaway," for this enigmatic feature. The Late Cretaceous–Paleogene Trans-Saharan Seaway, extending from Algeria to Nigeria, has been documented from outcrops and biogeographic connections (Petters, 1991) and so does not constitute a "vanishing seaway," although considerable debate exists as to its extent.

Australia had widespread epeiric seas in the Ap-

tian and Albian Stages of the Early Cretaceous but none are shown on Late Cretaceous maps (Struckmeyer and Totterdell, 1990). This pattern seems contrary to expectations based on sea level curves and invites suspicion that Late Cretaceous marine deposits might have been eroded and epeiric coverage underestimated. Veevers (1984) has recognized this problem, and explains that mid-Cretaceous volcanic activity and uplift in the Eastern Highlands produced thick sediment sequences that overwhelmed the system. This is about the time of opening of the Tasman Sea to the east (~80 Ma), so the uplift could have been related to this major shift in the tectonic regime.

It has been proposed that East Antarctica was inundated during the Late Cretaceous and Paleogene, based on microfossils in a Pliocene till in the Transantarctic Range (Webb et al., 1984). However, the possibility that the fossils represent eolian contamination has also been raised (Kellogg and Kellogg, 1996; Lawver and Gahagan, this volume, chapter 10). In any case, parts of the East Antarctic craton lie below sea level, even when isostatic adjustments for the ice load are made (Drewry, 1983), so it seems likely that epeiric seas existed in this area.

Finally, South America may have been diagonally traversed by an epeiric seaway in latest Cretaceous times (Riccardi, 1987; Uliana and Biddle, 1987). This narrow channel has been reconstructed mostly from subsurface data and skirts the northern margin of Argentina from Buenos Aires to the Altiplano region of Bolivia.

Conclusion

Ancient epeiric seas were more extensive than commonly realized. Evidence for their existence sometimes comes from extraordinary sites, such as diatremes, karstic features and meteorite impact structures. Their extent may be indicated by biogeographic patterns, terrestrial as well as marine. In any case, the "zero edge" of the outcrop should never be assumed to represent a paleoshoreline. We have employed present topography as a guide to inferring the boundaries of the Hudson Seaway, however this has been done only as a last resort. Difficulties in using the present topography arise from the amount of crustal warping evident in "stable continental cratons," and from the uncertainties in the rates of erosion over periods of tens of millions of years. We have referred to standard sea level curves in projecting seaways across continental interiors, but we must admit that virtually all areas of Cretaceous sedimentation in North America were driven by tectonic subsidence. This naturally leads us to question the utility of the sea level curves to project shorelines across areas with a scanty geologic record.

The combined Arctic Ocean and boreal seaway network through Eurasia and North America must represent one of the most extensive in earth history, especially during the Late Cretaceous. Such seaways, as well as large lakes, must have had an ameliorating effect on the climate of the time (Kutzbach and Ziegler, 1993). Climate reconstructions must be based on reliable numerical models and accurate paleogeography, and they must be tested with paleontological, geochemical, and sedimentological data. Model failure to confirm geological data may therefore result from the use of inappropriate paleogeographic boundary conditions (Yemane, 1993; Ziegler, 1993). Specifically, the high latitude warmth indicated by Mesozoic vegetation patterns (Spicer and Parrish, 1990; Ziegler et al., 1993) probably resulted from an epeiric sea coverage more extensive than initially realized (Valdes et al., 1996). Proportionately more effort should be applied to basic paleogeographic research as a means of building a better foundation for climate studies!

References

Algeo, T. J., and K. B. Seslavinsky, 1995, The Paleozoic world: Continental flooding, hypsometry, and sealevel, *Am. J. Science, 295,* 787–822.

Andrews, J. T., 1970, A geomorphological study of post-glacial uplift with particular reference to Arctic Canada, *Institute of British Geographers Special Publication 2,* Institute of British Geographers, London.

Balkwill, H.R. et al., 1983, Arctic North America and northern Greenland, in, *The Phanerozoic Geology of the World II: The Mesozoic, B,* M. Moullade and A.E.M. Nairn (eds.), pp. 1–31, Elsevier, Amsterdam.

Balkwill, H.R., N. J. McMillan, B. MacLean, G. L. Williams, and S. P. Srivastava, 1990, Geology

of the Labrador Shelf, Baffin Bay, and Davis Strait, *The Geology of North America, vol. I-1,* pp. 295–348. Geological Society of America, Boulder, Colo.

Bamburak, J. D., 1978, *Stratigraphy of the Riding Mountain, Boissevain and Turtle Mountain Formations in the Turtle Mountain area, Manitoba.* Manitoba Department of Mines, Resources and Environmental Management, Mineral Resources Division, Geological Survey Rep. 78-2, 47 pp.

Barr, W., 1972, Hudson Bay: the shape of things to come, *Musk Ox, 11,* 64.

Bell, M., and E. P. Laine, 1985, Erosion of the Laurentide region of North America by glacial and glaciofluvial processes, *Quaternary Research, 23,* 154–174.

Beznosov, N. V., T. N. Gorbatchik, I. A. Mikhilova, and M. A. Pergament, 1978, Soviet Union, in *The Phanerozoic Geology of the World II: The Mesozoic,* A.M. Moullade and A.E.M. Nairn (eds.), pp. 5–53, Elsevier, Amsterdam.

Birkelund, T., and K. Perch-Nielsen, 1976, Late Palaeozoic-Mesozoic evolution of central east Greenland, in, *Geology of Greenland,* A. Escher and W.S. Watt (eds.), pp. 305–339, Gronlands Geologiske Undersogelse, Denmark.

Borisevich, D. V., 1992, Neotectonics of the Urals, *Geotectonics, 26,* 41–47.

Caldwell, W. G. E. and E.G. Kauffman (eds.), 1993, *Evolution of the Western Interior Basin,* Geological Association of Canada Special Paper 39.

Cherven, V. B., and A. F. Jacob, 1985, Evolution of Paleogene depositional systems, Williston Basin, in response to global sea level changes, in, *Cenozoic Paleogeography of the West-Central United States,* R. M. Flores and S. S. Kaplan (eds.), pp. 127–170, Rocky Mountain Paleogeography Symposium 3, Rocky Mountain Section SEPM, Denver, Colo.

Clifford, A. C., 1986, African oil—past, present, and future, in, *Future Petroleum Provinces of the World,* M. T. Halbouty (ed.), pp. 339–372, *American Association of Petroleum Geologists Mem. 40.*

Colin, J.-P., 1994, Mesozoic–Cenozoic lacustrine sediments of the Zaïre Interior basin, in, *Global Geological Record of Lake Basins,* E. Gierlowski-Kordesch and K. Kelts (eds.), pp. 31–36, IGCP Project 324.

Cox, L. R., 1953, Lamellibranchs from the Lualaba

beds of the Belgian Congo, *Revue de Zoologie et de Botanique Africaines, 47,* 99–107.

Cox, L. R., 1960, Further mollusca from the Lualaba Beds of the Belgian Congo. *Annales du Musee Royal du Congo Belge,* 37, 1–15.

Cvancara, A.M., and J. W. Hoganson, 1993, Vertebrates of the Cannonball Formation (Paleocene) in North and South Dakota. *J. Vertebrate Paleontol., 13,* 1–23.

Dickinson, W. R., 1988, Paleogeographic and paleotectonic setting of Laramide sedimentary basins in the central Rocky Mountain region, *Geol. Soc. Am. Bull., 100,* 1023–1039.

Dixon, J., J. Dietrich, L. R. Snowdon, G. Morrell, and D. H. McNeil, 1992, Geology and petroleum potential of Upper Cretaceous and Tertiary strata, Beaufort-Mackenzie area, northwest Canada. *Am. Assoc. Petrol. Geologists Bull., 76,* 927–947.

Drewry, D. J. (ed.), 1983, *Antarctica: Glaciological and Geophysical Folio.* Cambridge: Scott Polar Research Institute. 9 plates.

Dyke, A. S., and L. A. Dredge, 1989, Quaternary geology of the northwestern Canadian Shield, in *Geology of Canada and Greenland,* R. J. Fulton (ed.), pp. 189–214, *The Geology of North America, vol. K-1,* Geological Society of America, Boulder, Colo.

Embry, A. F., 1991, Mesozoic history of the Arctic Islands, in *The Geology of the Innuitian Orogen and Arctic Platform of Canada and Greenland,* H. P. Trettin (ed.), pp. 371–433, *The Geology of North America E,* The Geological Society of America, Boulder, Colo.

Feldmann, R. M., 1972, First report of Hercoglossa ulrichi (White, 1882) (Cephalopoda: Nautilida) from the Cannonball Formation (Paleocene) of North Dakota, U.S.A., *Malacologia, 11,* 407–413.

Fenton, M. M., B. T. Schreiner, E. Nielson, and J. G. Paulowicz, 1994, Chapter 24—Quaternary geology of the western plains, in, *Geological Atlas of the Western Canada Sedimentary Basin,* Canadian Society of Petroleum Geologists and Alberta Research Council, Calgary, G. Mossop and I. Shetson (eds.), pp. 413–420.

Funnell, B. M., 1990, Global and European Cretaceous shorelines, stage by stage, in, *Cretaceous Resources, Events and Rhythms: Background and Plans for Research,* R. N. Ginsburg and B. Beaudoin (eds.), Kluwer Academic Publishers, Dordrecht, *NATO ASI Series C: Mathematical and Physical Sciences, 304,* 221–235.

Galloway, W. E., 1989, Genetic stratigraphic sequences in Basin Analysis II: application to northwest Gulf of Mexico Cenozoic basin, *Am. Assoc. Petrol. Geologists Bull., 73,* 143–154.

Gradstein, F. M., L. F. Jansa, S. P. Srivastava, M. A. Williamson, G. B. Carter, and B. Stam, 1990, Aspects of North Atlantic paleo-oceanography, in, *Geology of the Continental Margin of Eastern Canada,* M. J. Keen, and G. L. Williams (eds.) pp. 353–389, *The Geology of North America, vol. I-1,* Boulder, Colorado, Geological Society of America, Boulder, Colo.

Green, A. R., A. A. Kaplan, and R. C. Vierbuchen, 1986, Circum-Arctic petroleum potential, in, *Future Petroleum Prospects of the World,* M. T. Halbouty (ed.), pp. 101–130, *American Association of Petroleum Geologists Mem., 40,* 101–130.

Gross, G. A., 1968, Geology of iron deposits in Canada. Volume III–Iron ranges of the Labrador geosyncline, *Geological Survey of Canada Economic Geology Report 22.*

Gurnis, M., 1993, Depressed continental hypsometry behind oceanic trenches: A clue to subduction controls on sea-level change, *Geology, 21,* 29–32.

Hallam, A., 1992, *Phanerozoic Sea-Level Changes,* Columbia University Press, New York, The Perspectives in Paleobiology and Earth History Series, 266 pp.

Henderson, G., A. Rosenkrantz, and E. J. Schiener, 1976, Cretaceous-Tertiary sedimentary rocks of west Greenland, in, *Geology of Greenland,* A. Escher and W.S. Watt (eds.), pp. 341–362, Gronlands Geologiske Undersogelse, Denmark.

Henderson, G., E. J. Schiener, J. B. Risum, C.A. Croxton, and B. B. Andersen, 1981, The West Greenland Basin, in *Geology of the North Atlantic Borderlands,* J. W. Kerr and A. J. Fergusson (eds.), pp. 399–428, *Canadian Society of Petroleum Geologists Mem. 7.*

Hiscott, R. N., 1984, Clay mineralogy and clay-mineral provenance of Cretaceous and Paleogene strata, Labrador and Baffin shelves, *Can. Petrol. Geol. Bull., 32,* 272–280.

Hoffman, P. E., 1989, Precambrian geology and tectonic history of North America, in, *The Geology of North America–An Overview,* A. W. Bally and A. R. Palmer (eds.), pp. 447–512, *The*

Geology of North America, vol. A, Geological Society of America, Boulder, Colo.

Hughes, T., 1987, Ice dynamics and deglaciation models when ice sheets collapsed, in *North America and Adjacent Oceans During the Last Deglaciation,* W.F. Ruddiman and H. E. Wright, Jr. (eds.), pp. 183-220, *The Geology of North America, vol. K-3,* pp. 183–220, Geological Society of America, Boulder, Colo.

Jeletzky, J. A., 1971, Marine Cretaceous biotic provinces and paleogeography of western and Arctic Canada: Illustrated by a detailed study of ammonites, *Geol. Survey Can. Pap. 70-22.*

Kauffman, E.G., 1984, Chapter 8: The fabric of Cretaceous marine extinctions, in *Catastrophes and Earth History,* W. A. Berggren and J. A. Van Couvering (eds.), pp. 151–246, Princeton University Press, Princeton, NJ.

Keen, C. E., and C. Beaumont, 1990, Geodynamics of rifted continental margins, in, *Geology of the Continental Margin of Eastern Canada,* M. J. Keen, and G. L. Williams (eds.), pp. 393–472, *The Geology of North America, vol. I-1,* Geological Society of America, Boulder, Colo.

Kellogg, D. E., and T. B. Kellogg, 1996, Diatoms in South Pole ice: Implications for eolian contamination of Sirius Group deposits, *Geology, 24,* 115–118.

Kurita, H., and D. J. McIntyre, 1995, Paleocene dinoflagellates from the Turtle Mountain Formation, southwestern Manitoba, Canada, *Palynology, 19,* 119–136.

Kutzbach, J. E., and A.M. Ziegler, 1993, Simulation of Late Permian climate and biomes with an atmosphere-ocean model: Comparisons with observations, *Phil. Trans. R. London, Ser. B, 341,* 327–340.

Lentin, J. K., and G. L. Williams, 1980, Dinoflagellate provincialism with emphasis on Campanian peridinlaceans, *Am. Assoc. Stratigraphic Palynologists Contrib. Ser. 7,* 1–46.

Lepersonne, J., 1977, Structure geologique du bassin interieur du Zaire, *Bulletin de l'Academie Royale de Belge, 63,* 941–965, (in French).

Macqueen, R. W., and D. A. Leckie, 1992, *Foreland Basins and Fold Belts,* American Association of Petroleum Geologists, Mem. 55, Cincinnati, Ohio, 460 pp.

Marincovich, Jr., L., 1994, Earliest Tertiary paleo-

geography of the Arctic Ocean, in, *1992 Proceedings, International Conference on Arctic Margins, Anchorage, Alaska,* D. K. Thurston and K. Fujita (eds.), pp. 45–48, U.S. Department of the Interior, Anchorage, Alaska.

Marincovich, Jr., L., E. M. Brouwers, D. M. Hopkins, and M. C. McKenna, 1990, Late Mesozoic and Cenozoic paleogeographic and paleoclimatic history of the Arctic ocean basin, based on shallow-water marine faunas and terrestrial vertebrates, in, *The Arctic Region,* A. Grantz, L. Johnson, and J. F. Sweeney (eds.), pp. 403–426, The *Geology of North America vol. L,* Geological Society of America, Boulder, Colo.

McMillan, N. J., 1973, Shelves of Labrador Sea and Baffin Bay, Canada, in, *The Future Petroleum Provinces of Canada–Their Geology and Potential,* R. G. McCrossan (ed.), pp. 473–517, *Canadian Society of Petroleum Geologists Mem. 1,* Canadian Society of Petroleum Geologists, Calgary.

McMillan, N. J., and A. Duk-Rodkin, 1995, The Bell River system: Tertiary drainage from the eastern Cordillera to the Labrador Sea, in *Proceedings of the Oil and Gas Forum '95 Energy from Sediments,* J. S. Bell et al. (eds.), pp. 495–496, Geological Survey of Canada Open File 3058.

McNeil, D. H., 1990, Tertiary marine events of the Baeufort–Mackenzie basin and correlation of Oligocene to Pliocene marine outcrops in Arctic North America, *Arctic, 43,* 301–313.

Miall, A. D., 1991, Late Cretaceous and Tertiary basin development and sedimentation, Arctic Islands, in, *The Geology of the Innuitian Orogen and Arctic Platform of Canada and Greenland,* H. P. Trettin (ed.), pp. 437–458, *The Geology of North America vol. E,* The Geological Society of America, Boulder, Colo.

Miall, A. D., H.R. Balkwill, and W.S. Hopkins, 1980, Cretaceous and Tertiary sediments of Eclipse Trough, Bylot Island area Arctic Canada, and their regional setting, *Geol. Survey Can. Pap. 79–23.*

Mitrovica, J. X., C. Beaumont, and G. T. Jarvis, 1989, Tilting of continental interiors by the dynamical effects of subduction, *Tectonics, 8,* 1079–1094.

Mossop, M. M. and I. Shetsen, 1994, *Geological Atlas of the Western Canada Sedimentary Basin,*

Canadian Society of Petroleum Geologists and Alberta Research Council, Calgary.

Naidin, D. P., 1981, The Russian Platform and the Crimea, in, *Aspects of Mid-Cretaceous Regional Geology,* R. A. Reyment and P. Bengtson (eds.), pp. 29–68, IGCP Project 58, Academic Press, London.

Naidin, D. P., et al., 1980, Cretaceous transgressions and regressions on the Russian Platform, in Crimea and central Asia, *Cretaceous Res., 1,* 375–387.

Nassichuk, W. W., and D. J. McIntyre, 1995, Cretaceous and Tertiary fossils discovered in kimberlites at Lac de Gras in the Slave Province, Northwest Territories, *Current Res. 1995-B,* 109–114.

Nielson, E., Geological Survey of Canada, unpublished manuscript, received 1996.

Norris, A. W., 1993, Hudson platform—Geology, in, *Sedimentary Cover of the Craton in Canada,* D. F. Scott and J. D. Aitken (eds.), pp. 655–700, *The Geology of North America, vol. D-1,* Geological Society of America, Boulder, Colo.

Parkinson, C. L., J. S. Comiso, H. J. Zwally, D. J. Cavalier, P. Gloersen, and W. J. Campbell, 1987, *Arctic Sea Ice, 1973–76: Satellite Passive-Microwave Observations,* NASA SP-489, NASA, Washington, D.C.

Pelletier, B. R., 1969, Submarine physiography, bottom sediments and models of sediment transport in Hudson Bay, in, *Earth Science Symposium on Hudson Bay,* P. J. Hood (ed.), pp. 100–135, *Geol. Survey Can. Pap. 68-53.*

Petters, S. W., 1991, *Regional Geology of Africa,* Springer-Verlag, Berlin, 722 pp.

Pratt, L. M., M. A. Arthur, W. E. Dean, P.A. Scholle, 1993, Paleo-oceanographic cycles and events during the Late Cretaceous in the Western Interior Seaway of North America, in, *Evolution of the Western Interior Basin,* W. G. E. Caldwell and E.G. Kauffman (eds.), pp. 333–53, *Geol. Assoc. Can., Spec. Pap. 39.*

Riccardi, A. C., 1987, Cretaceous paleogeography of southern South America, *Palaeogeog., Palaeoclimat., Palaeoecol., 59,* 169–195.

Riedeger, C. I. and R. M. Bustin, 1987, The Eureka formation, southern Ellesmere Island. *Bull. Can. Petrol. Geol., 35,* 123–142.

Roest, W. R. and S. P. Srivastava, 1989, Sea-floor spreading in the Labrador Sea: a new reconstruction, *Geology, 17,* 1000–1003.

Rosencrantz, A., A. Noe-Nygaard, H. Gry, S. Munck, and D. Laursen, 1942, A geological reconnaissance of the southern part of the Svartenhuk Peninsula West Greenland. *Meddelelser om Gronland, 135,* 1–72.

Rowley, D. B., and A. L. Lottes, 1988, Plate kinematic reconstructions of the North Atlantic and Arctic: Late Jurassic to Present. *Tectonophysics, 155,* 73–120.

Rowley, D. B., and P. J. Markwick, 1992, Haq. et al. eustatic sea level curve: implications for sequestered water volumes, *J. Geol., 100,* 703–715.

Sage, R. P., 1988, *Geology of Carbonatite-Alkalic Rock Complexes in Ontario: Cargill Township Carbonatite Complex, District of Cochrane,* Ontario Geological Survey Study 36, 92 pp.

Sanford, B. V., 1987, Paleozoic geology of the Hudson Platform, in, *Sedimentary Basins and Basin-forming Mechanisms,* C. Beaumont and A. J. Tankard (eds.), pp. 483–505, *Canadian Society of Petroleum Geologists Memoir 12.*

Sanford, B. V., and A. C. Grant, 1990, New findings relating to the stratigraphy and structure of the Hudson Platform, in, *Current Research, Part D,* pp. 17–30, *Geol. Survey Can. Pap. 90–1D.*

Slingerland, R., L. R. Kump, M. A. Arthur, P. J. Fawcett, B. B. Sageman and E. J. Barron, 1996, Estuarine circulation in the Turonian Western Interior Seaway of North America, *Geol. Soc. Am. Bull., 108,* 941–952.

Smith, A. G., D. G. Smith, and B. M. Funnell, 1994, *Atlas of Mesozoic and Cenozoic Coastlines,* Cambridge University Press, New York.

Sommers, L. M., 1977, *Atlas of Michigan,* Michigan State University Press, Ann Arbor.

Spicer, R. A., and J. T. Parrish, 1990, Late Cretaceous-Early Tertiary palaeoclimates of northern high latitudes: a quantitative view, *J. Geol. Soc., London, 147,* 329–341.

Stasiuk, L. D., and W. W. Nassichuk, 1995, Thermal history and petrology of wood and other inclusions in kimberlite pipes at Lac de Gras, Northwest Territories, *Cur. Res. 1995-B,* 115–124.

Struckmeyer, H. M. and J. M. Totterdell, 1990, *Australia: Evolution of a Continent,* Canberra, Australian Government Publishing Service.

Teichert, C., 1939, Geology of Greenland, in, *Ge-*

ology of North America, vol. 1, R. Ruedemann and R. Balk, eds., pp. 100–175.

Teys, R. V., M. A. Kiselevskiy and D. P. Naidin, 1978, Oxygen and Carbon isotopic composition of organogenic carbonates and concretions in the Late Cretaceous rocks of northwestern Siberia, *Geochem. Int., 15,* 74–81.

Theide, J., and O. Eldholm, 1983, Speculations about the paleodepth of the Greenland-Scotland Ridge during Late Mesozoic and Cenozoic times, in, *Structure and Development of the Greenland-Scotland Ridge: New Methods and Concepts,* M.H.P. Bott, S. Saxov, and J. Theide, (eds.), pp. 445–456, Plenum Press, New York.

Thorsteinsson, R., and U. Mayr, 1987, The sedimentary rocks of Devon Island, Canadian Arctic Archipelago. *Geol. Survey Can. Mem. 411.*

Trettin, H. P., 1991, Middle and Late Tertiary tectonic and physiographic developments, in, *The Geology of the Innuitian Orogen and Arctic Platform of Canada and Greenland,* H. P. Trettin (ed.), pp. 493–496, The *Geology of North America vol. E,* Geological Society of America, Boulder Colo.

Tynni, R., 1982, The reflection of geological evolution in Tertiary and interglacial diatoms and silicoflagellates in Finnish Lapland, *Geol. Survey Finland Bull., 320,* 5–40.

Uliana, M. A. and K. T. Biddle, 1987, Permian to Late Cenozoic evolution of northern Patagonia: main tectonic events, magmatic activity, and depositional trends, in, *Gondwana Six: Structure, Tectonics and Geophysics,* G. D. McKenzie (ed.), pp. 271–286, *Geophysical Monograph 40,* American Geophysical Union, Washington, D.C.

Valdes, P. J., B. W. Sellwood, and G. D. Price, 1996, Evaluating concepts of Cretaceous equability, *Palaeoclimates, 2,* 139–158.

Veevers, J. J., ed., 1984, *Phanerozoic Earth History of Australia,* Clarendon Press, Oxford.

Vinogradov, A. P., ed., 1967, *Atlas of the Lithological-Paleogeographical Maps of the U.S.S.R., Vol. 4: Paleogene, Neogene and Quaternary,* Akademia Nauk SSSR, Moscow.

Vinogradov, A. P., ed., 1968, *Atlas of the Lithological-Paleogeographical Maps of the U.S.S.R., Vol. 3: Triassic, Jurassic and Cretaceous,* Akademia Nauk SSSR, Moscow.

Walcott, R. I., 1970, Isostatic response to loading of the crust in Canada, *Can. J. Earth Sci., 7,* 716–734.

Webb, P.N., D. M. Harwood, B. C. McKelvey, J. H. Mercer, and L. D. Stott, 1984, Cenozoic marine sedimentation and ice-volume variation on the east Antarctic craton. *Geology, 12,* 287–291.

White, W. A., 1972, Deep erosion by continental ice sheets, *Geol. Soc. Am. Bull., 83,* 1037–1056.

Williams, G. D., and C. R. Stelck, 1975, Speculations on the Cretaceous palaeogeography of North America, in, *The Cretaceous System in the Western Interior of North America,* W. G. E. Caldwell (ed.), pp. 1-20, *Geol. Assoc. Can. Spec. Pap. 13.*

Wright, E. K., 1987, Stratification and paleocirculation of the Late Cretaceous Western Interior Seaway of North America, *Geol. Soc. Am., Bull., 99,* 480–490.

Yemane, K., 1993, Contribution of Late Permian palaeogeography in maintaining a temperate climate in Gondwana. *Nature, 361,* 51–54.

Zakharov, V. A., 1994, Climatic fluctuations and other events in the Mesozoic of the Siberian Arctic, in, *1992 Proceedings International Conference on Arctic Margins, Anchorage, Alaska,* Thurston, D. K., and K. Fujita (eds.), pp. 23–28, U.S. Department of the Interior, Anchorage, Alaska.

Zakharov, V. A., A. L. Beizel, N. K. Lebedeva, and O. V. Khomentovskii, 1991, Evidence of Upper Cretaceous world ocean eustacy in the north of Siberia. *Soviet Geology and Geophysics, 32,* 1–6.

Ziegler, A.M., 1993, Models come in from the cold, *Nature, 361,* 16–17.

Ziegler, A.M., M. L. Hulver, and D. B. Rowley, 1996, Permian world topography and climate, in, *Late Glacial and Postglacial Environmental Changes,* I. P. Martini (ed.), pp. 111–146, Oxford University Press, Oxford.

Ziegler, A.M., J. M. Parish, J. P. Yao, E. D. Gyllenhaal, D. B. Rowley, J. T. Parrish, S. Y. Nie, A. Bekker, and M. L. Hulver, 1993, Early Mesozoic phytogeography and climate, *Phil. Trans. R. Soc. London, B341,* 297–305.

Ziegler, P.A., 1988, Evolution of the Arctic North Atlantic and the Western Tethys, *American Association of Petroleum Geologists Mem. 43,* 198 pp.

PART V

ROLE OF OCEAN GATEWAYS

CHAPTER 8

Caribbean Constraints on Circulation Between Atlantic and Pacific Oceans Over the Past 40 Million Years

André W. Droxler, Kevin C. Burke, Andrew D. Cunningham, Albert C. Hine,
Eric Rosencrantz, David S. Duncan, Pamela Hallock, and Edward Robinson

Openings and closings of those narrow passageways between neighboring oceans which are called straits by geographers and gateways by paleoceanographers have had a powerful influence on oceanic circulation and, therefore, on climate change over the past approximate 50 million years (see, e.g., Kennett, 1977; Lawver and Gahagan, this volume, chapter 10). The closing of the Panama passageway (also referred to as the Central American Seaway) by blocking the transport of oceanic waters between the low latitudes of the western North Atlantic and the eastern Pacific oceans and by strengthening interhemispheral oceanic transport from low to high latitudes in the Atlantic Ocean, has long been recognized to have strongly influenced both oceanic circulation and climatic change (see, e.g., Warren 1983).

The establishment of the Caribbean Current and the subsequent strengthening of the western boundary current in the western North Atlantic Ocean are direct consequences of the closing of the Panama passageway. The Caribbean Current today originates as a mixture of the Northern Atlantic Equatorial Current, the Southern Atlantic Cross-Equatorial Current, and the Guiana Current (Fig. 8.1a). The Caribbean Current consists of surface and thermocline waters combined with lower thermocline waters identified as an upper part of the Antarctic Intermediate Water (AAIW). These waters fill the Caribbean basins at a rate of about 20 Sv (or 20×10^6 m³/sec) (Schmitz and McCartney, 1993) (Fig. 8.1a). According to Schmitz and McCartney (1993), 13 of the Caribbean Current's 20 Sv have their origin in the southern Atlantic. By its linkage with the Gulf Stream via the Loop and Florida currents, the Caribbean Current becomes an efficient transporter of equatorial, saline, and warm waters into the high latitudes of the North Atlantic, where in the Labrador and Norwegian seas these waters ultimately sink at rates of 7 Sv and 6 Sv, respectively, to form the North Atlantic Deep Water (NADW) (Gordon, 1986; Gordon et al., 1992) (Fig. 8.1a). The 13 Sv of the Southern Atlantic Cross-Equatorial Current, therefore, compensate for the volume of water sinking in the Labrador and Norwegian seas and exiting the North Atlantic Ocean as NADW (Schmitz and McCartney, 1993; Fig. 1a). Broecker et al. (1985) explained that the NADW production, and, therefore, the driving mechanism of the global thermohaline conveyor belt, is linked to the excess salt left behind in the Atlantic as the result of vapor export toward the Pacific (as much as 0.3 Sv [or 0.3×10^6 m³ per sec]), especially at the latitudes of Central America (Fig. 8.1b).

Recent studies on the timing of the Central American Seaway closure (e.g., Coates et al., 1992; Farrell et al., 1995; Collins et al., 1996; Burton et al., 1997) and modeling of the effects of the closure (Maier-Reimer et al., 1990; Mikolajewicz et al., 1993; Mikolajewicz and Crowley, 1997) have served to show the importance of the isthmus for heat and salt contrasts between the Caribbean and the eastern equatorial Pacific, for moisture transfer across this oceanic barrier, and finally for related global climate change. However, plate tectonic reconstructions (e.g., Pindell and Dewey, 1982) have also illustrated that even before the establishment of the Isthmus of Panama, the oceanic passage between the North and South America was far from being a simple deep water environment. Figure 8.2 shows why.

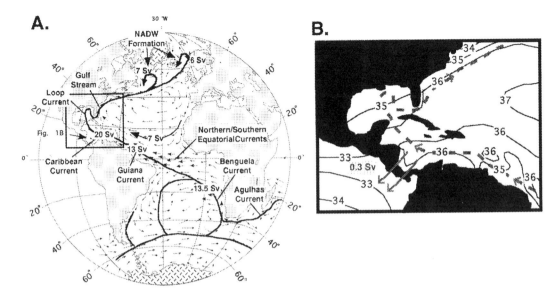

Fig. 8.1. (a) Modern Atlantic surface and thermocline circulation. Surface water circulation is shown by small arrows; the thick, black, and continuous lines indicate the movement of the Antarctic Intermediate Water (AAIW) also identified as lower thermocline waters in the Atlantic [(adapted by Haddad (1994), Haddad and Droxler (1996) from Gordon (1986) and Gordon et al. (1992)]. The volume of water (13 Sv = 13 x 10^6 m^3/second) transported from the Southern Atlantic through the Caribbean as surface, thermocline, and lower thermocline circulation is equivalent to the volume of water (13 Sv) sinking in the Labrador Sea (7 Sv) and the Norwegian/Greenland seas (6 Sv) at the origin of the North Atlantic Deep Water (NADW) (Schmitz and McCartney, 1993). (b) Map of the Caribbean Sea and neighboring oceanic areas showing the contrast between the high salinity (contour lines in per thousand) of the Caribbean compared with the relatively low salinity of the eastern equatorial Pacific Ocean. This contrast can be at least partially explained by the transport of water vapor by the tropical easterlies (Trade Winds) from the Caribbean/western Central Atlantic into the eastern Equatorial Pacific Ocean (0.3 Sv) (Broecker, 1991).

At about 80 Ma, the Great Arc of the Caribbean whose surviving active segment is represented by the Lesser Antilles (Burke, 1988)(Figs. 8.2 and 8.3) made its initial entry from the Pacific Ocean into the Atlantic Ocean. Ever since that time, shallow water environments have occurred in various areas of the Caribbean Sea and have surely influenced circulation within and between the two great ocean basins. The area of about one million km^2 occupied by the Caribbean Oceanic Plateau, although irregular in relief, lies now at an average depth of about 4 km and since its formation at ~ 90 Ma (Duncan et al., 1994) has subsided thermally by an estimated average of ~2 km. Much shallower areas, at times possibly including large surfaces of land, have been associated with the two arcs that bound the eastern and western margins of the oceanic plateau as well as with the northern Nicaragua Rise

and Greater Antilles which mark the boundary of the Caribbean Plate on its northern side (Fig. 8.3).

Here, we review the tectonic history of the Caribbean over the past 40 million years to show how it has influenced the paleobathymetry of the Caribbean seafloor. We attempt to show how changes of the general Caribbean configuration, developed over time scales of tens of millions of years, modified the interoceanic and interhemispheric surface, mid-depth, and deep ocean circulation. During the Cenozoic, the Caribbean Sea has been a major conduit for water exchange between the Atlantic and the Pacific oceans. Tectonic processes have constructed and destroyed barriers to that exchange in three critical regions of the Caribbean Sea: (1) the Aves Swell and Lesser Antilles, (2) the northern Nicaragua Rise, and (3) the Central American Seaway, later to be occupied

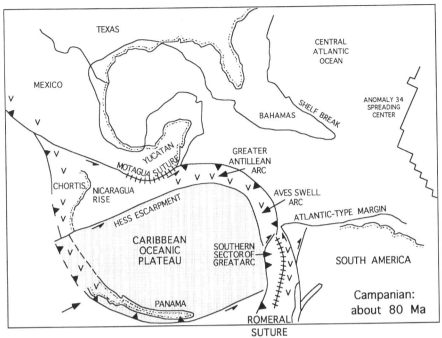

Fig. 8.2. Since the Caribbean oceanic plateau began to enter the Atlantic about 80 Ma, areas of shallow water capable of influencing the circulation of waters between the Atlantic and the Pacific Oceans have occupied various parts of the Caribbean Sea. (Modified from Burke, 1988.)

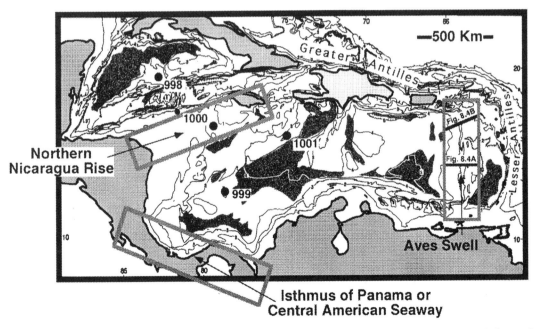

Fig. 8.3. Map of the Caribbean showing the three critical areas discussed in this paper: the Aves Swell, the Isthmus of Panama, and the northern Nicaragua Rise. Tectonic activity has modified the bathymetry of these three areas. These modifications have influenced the general Atlantic Ocean circulation and in particular the Atlantic/Pacific ocean interchange at various times during the past 40 m.y. Four of the five ODP Sites (998, 999, 1000, and 1001) drilled during Leg 165 are also located on the map. The locations of the two profiles illustrating the tentative Cenozoic geologic evolution of the Aves Swell in Fig. 8.4 are also shown.

by the Isthmus of Panama (Fig. 8.3). We focus our exercise on these three critical regions, which are similar to a set of valves that have regulated oceanic circulation through time. Prior to the final closure of the Panama passageway or Central American Seaway, there were at least two earlier episodes along the Aves Swell and the northern Nicaragua Rise capable of strongly modifying oceanic circulation. Our Caribbean example suggests to us that the evolution of gateway areas may generally be quite complex. Processes as straightforward as the opening or the closing of a gate, which are suitable for controlling such activities as the passage of sheep, may prove too simple to be very useful analogs of the opening and the closing of straits between oceans. We will treat the evolution of the northern Nicaragua Rise in more detail than the history of the Aves Swell, because the northern Nicaragua Rise has been the object of recent research programs whose results are for the first time included in a general model of the evolution of the Caribbean.

Aves Swell

The Aves Swell (Fig. 8.3) is a broad N-S trending structure which extends almost all the way across the eastern part of the Caribbean Sea from the Venezuelan continental borderland to the Anegada Trough. The swell lies at depths ranging from sea level to 2000 m. The Aves Swell is a part of the Great Arc of the Caribbean which, except at its northern extremity has become separated from the Lesser Antillean volcanic arc by seafloor spreading which led to the formation of the Grenada Basin. Bird et al., (1993) have shown that the seafloor spreading of the Grenada Basin took place about a N-S trending spreading center but they were not able to establish exactly when the Grenada Basin formed. Their best estimate was that the basin formed over a few million years at some time during the latest Cretaceous or Paleocene (~70–55 Ma). From that time on, the Aves Swell has behaved as an inactive or remnant arc. It is, therefore, likely to have experienced thermal subsidence as it has aged and to have provided, where its surface has been in the photic zone, a platform for carbonate rock accumulation.

Seismic surveys, drilling, and dredging have shown that during the Eocene, Oligocene, and early Miocene (~50–15 Ma) neritic conditions existed on most of the topographic highs along the Aves Swell. Deepwater clastic sedimentation was restricted to intervening troughs and to adjacent basins (Figs. 8.4a and 8.4b; Bouysse et al., 1985; Pinet et al., 1985; Holcombe et al., 1990). There is thus some likelihood, given its tectonic and sedimentary environment, that the Aves Swell might have been shallow enough for at least part of a 35 m.y. long interval to have modified the circulation of oceanic waters in the western North Atlantic and to have formed a partially or fully developed barrier to circulation between the Atlantic and the Pacific oceans, somewhat similar in setting to the modern Central American barrier (or Isthmus of Panama).

Very strong supporting evidence for this possibility comes from the islands of the Greater Antilles where fossil skeletal remains of early Miocene land mammals with South American affinities including sloths, have been discovered (McPhee and Iturralde-Venant 1994, 1995; Iturralde-Venant et al., 1996). These authors have postulated the existence of an Oligocene land bridge between the Americas along the Aves Swell (Figs. 8.5a, 8.5b, and 8.5c). There are no other places in the Caribbean where the establishment of a land bridge seems as likely (see Fig. 8.3). The Lesser Antillean volcanic arc might have fulfilled a similar role but, by analogy with today, that arc seems more likely to have been interrupted by deepwater passages. Because the early and middle Oligocene (~35–30 Ma) was a time of at least two major world-wide sea level low stands (Haq et al., 1987; Abreu and Haddad, in press), it would seem to have been the most likely interval for the Aves Swell to have been subaerially exposed and to have provided a land bridge between South America and the area on the northern side of the Caribbean now represented by the islands of the Greater Antilles (Figs. 8.5b and 8.6a).

We see some analogy between the mid Oligocene Aves Swell land bridge and the Isthmus of Panama land bridge joining today North and South America. Establishment of the Panama land bridge has done more than simply block the flow of ocean waters between the Atlantic and the Pacific oceans. It has, for example, enhanced the transport of water, heat, moisture, and salt from low to high latitudes in the Atlantic (Broecker et al., 1985; Gordon, 1986;

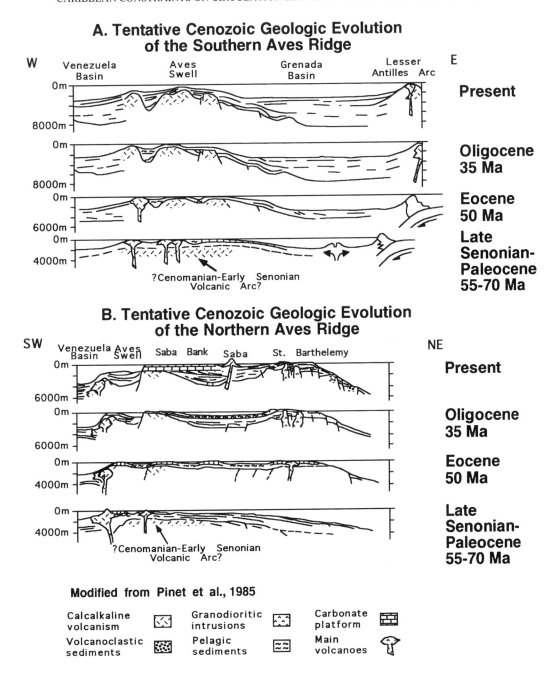

Fig. 8.4. Tentative Cenozoic evolution of the Aves Swell. The Grenada Basin isolated the Aves Swell from the active arc of the Lesser Antilles. This Fig. shows the Grenada Basin as forming between ~ 70 and 55 Ma, although the timing of the Grenada Basin formation is poorly known. The four upper cross sections in Fig. 8.4a show the Aves Swell as having formed part of the Great Arc of the Caribbean until Paleocene times (~60 Ma) and becoming separated from the active arc by the formation of the Grenada Basin. Fig. 4b shows four northern sections. Extension split the northern part of the arc but did not attain the formation of ocean floor such as underlies the Grenada Basin. Since the Aves swell became separated from the active arc, it has experienced thermal subsidence. Only at Aves Island and on Saba Bank at the northern end of the swell have shallow water conditions persisted until today.

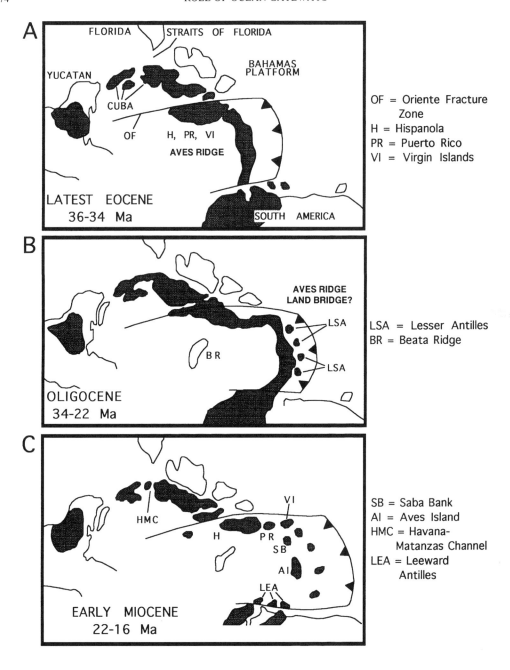

Fig. 8.5. Sketch maps from McPhee and Iturralde-Vinent (1994, 1995) indicating how mammals may have reached Cuba and Hispaniola by way of a land bridge formed along the crest of the Aves Swell during several early Oligocene intervals characterized by major sea level falls (Fig. 8.5b). Since at least the early Eocene, the Aves Swell, over most of its length, has been separated from the active arc of the Lesser Antilles and has experienced thermal subsidence. Only at Aves Island and on Saba Bank at the northern end of the swell has shallow water environment persisted until today.

A. Early Oligocene, 36-30 Ma

B. Early Miocene, 17 Ma

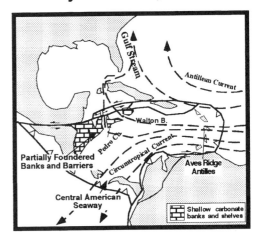

C. Late-Middle Miocene, 12-8 Ma D. Last 4 Ma

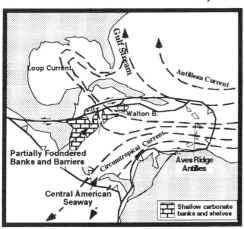
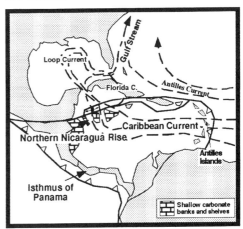

Fig. 8.6. Simplified reconstruction sketches of the Caribbean (after Pindell, 1994) illustrating several scenarios for closure and opening of gateways in the Caribbean. Fig. 8.6a illustrates the possible establishment of a land bridge along the Aves Swell during the early Oligocene. The resulting strengthening of the Western Boundary Current in the western North Atlantic could explain the first production of a North Component Deep Water during the early Oligocene. During late Oligocene times and the early part of the early Miocene, continuous carbonate banks and barrier reefs enhanced the water exchange between the eastern equatorial Pacific and the western tropical North Atlantic, and minimized the development of a western boundary current within the Caribbean. The partial foundering and subsidence of the continuous shallow carbonate system along the northern Nicaragua Rise may have occurred as late early to middle Miocene (20-12 Ma); this partial collapse may have initiated the Caribbean Current farther downstream in the Caribbean (Fig. 8.6b). In the late early Miocene (Fig. 8.6b), the Caribbean Current bypassed the Gulf of Mexico by flowing through the Havana/Matanzas Channel in western Cuba (Iturralde-Vinent et al., 1996) into the Straits of Florida and strengthened the Gulf Stream. The late early Miocene corresponds to an interval of newly established production of North Component Deep Water in the North Atlantic (see Fig. 8.10; Wright and Miller, 1996). During the late middle Miocene (Fig. 8.6c), the Havana/Matanzas Channel closed and the Central American Seaway was still opened, though possibly quite restricted since the middle to late Miocene transition (12-10 Ma) (Duque-Caro, 1990; Farrell et al., 1995). At that time, the Caribbean Current was flowing for the first time through the Gulf of Mexico where it became the Loop Current and strengthened the Gulf Stream even more. Strong North Component Deep Water is well established in the late middle Miocene (Fig. 8.10), interval characterized by the "Carbonate Crash" on both sides of the Isthmus of Panama (Fig. 8.12). The final closure of the Central American Seaway occurred only between 4.0 and 3.5 Ma according to Coates et al. (1992) and is expected to be the cause of the salinity contrast observed between the western North Atlantic and the East Pacific oceans (see Figs. 8.1b and 8.13), and the clear establishment of the modern North Atlantic Deep Water (Crowley and North, 1991) in the early late Pliocene.

Schmitz and McCartney, 1993). Low latitude, warm and salty water masses moving into high latitudes today stimulate the production of the NADW in the Labrador Sea and in the Norwegian Sea (Broecker et al., 1985; Gordon, 1986; Schmitz and McCartney, 1993, their Fig. 12). Perhaps therefore, it is no coincidence that a brief pulse of Northern Component Water (NCW or Proto-NADW), which has been interpreted as the first episode of simultaneous production of deep water from both poles during the Cenozoic (the past 65 m.y.), has been detected using global benthic marine carbon isotopic data for the early Oligocene (~36.5–34 Ma) (Miller and Fairbanks, 1985; Miller et al., 1991; Miller, 1992; Wright and Miller, 1993). Even if the Aves Swell was not a continuous subaerially exposed land bridge, it still may have acted as an important "valve," as was the rising Isthmus of Panama during the late Miocene–earliest Pliocene (see below section on "The Central American Seaway and the Isthmus of Panama).

Pinet et al. (1985) have shown that subsidence of the Aves Swell suddenly accelerated during the middle Miocene to depths of 600 m to 1200 m. This sudden change resulted in pelagic and hemipelagic deposition along most of the length of the swell. Only at Aves Island and on Saba Bank at the northern end of the swell have shallow water conditions persisted until today (Figs. 8.4b and 8.5c). Saba Bank is an active carbonate bank similar to the active banks of the northern Nicaragua Rise and like them is underlain by a thick shallow water Neogene carbonate platform sequence (Figs. 8.4b and 8.5c) (Speed et al., 1984; Bouysse et al., 1985; Despretz et al., 1985; Pinet et al., 1985).

Northern Nicaragua Rise

Modern northern Nicaragua Rise physiography has been assumed in the past to reflect the general morphology of the northern Nicaragua Rise since the Eocene, when the major basins and channels formed as rifted depressions and the intervening isolated banks established themselves on topographic highs (Mann and Burke, 1984). This model implies that Neogene tectonic activity along the northern boundary zone of the Caribbean Plate has exerted little or no influence on the evolution of the northern Nicaragua Rise carbonate system. As a

result, it has been assumed that the northern Nicaragua Rise has not interfered with or modified the pattern of the Caribbean oceanic surface circulation since the late Eocene. However, several lines of evidence point toward an overall stepwise strengthening of the Western Boundary Surface Current downcurrent from the northern Nicaragua Rise in the Gulf of Mexico/Straits of Florida and North Atlantic gyre since the middle Miocene. Closure of the Central American Seaway alone may not explain this complex evolution of the North Atlantic Western Boundary Current. In this chapter for the first time, results of several bathymetric and high resolution seismic surveys, and dredging operations on the R/V *Cape Hatteras* between 1988 and 1992, in Walton Basin, Pedro Channel, Rosalind, and Diriangen Channels, and the northern part of Rosalind Bank are compiled. Results of our research show that the basin and channel sub-seafloor consists at least partially of a series of foundered, faulted and folded shallow carbonate banks and barrier reefs as young as early/middle (?) Miocene in age that have been buried under fairly recent periplatform sedimentary cover.

Sedimentary and Environmental Record

The northern Nicaragua Rise, stretching from the east coast of Honduras and Nicaragua to the island of Jamaica, morphologically consists of several isolated carbonate banks and shelves, covered on average by 30 m of water depth (Glaser and Droxler, 1991; Triffleman et al., 1992) (Figs. 8.7a, 8.7b). The rise was continuously covered by shallow carbonate banks and barrier reefs until early Miocene time. Partial foundering of these banks and reefs in the middle Miocene (~15–12 Ma), possibly as early as the late early Miocene (20–15 Ma) permitted the initiation of the Caribbean current and thus contributed to the intensification of the Western Boundary Current in the central North Atlantic and ultimately to the strengthening of the Gulf Stream (Droxler et al., 1992). We summarize the geological history of the rise central area to show how the changes happened.

Since the Cayman Trough began to open ~ 40 million years ago (Rosencrantz et al., 1988), the Nicaragua Rise has rotated away from the position adjacent to North America that it occupied at 80 Ma (see Fig. 8.2). The roots of a Late Cretaceous

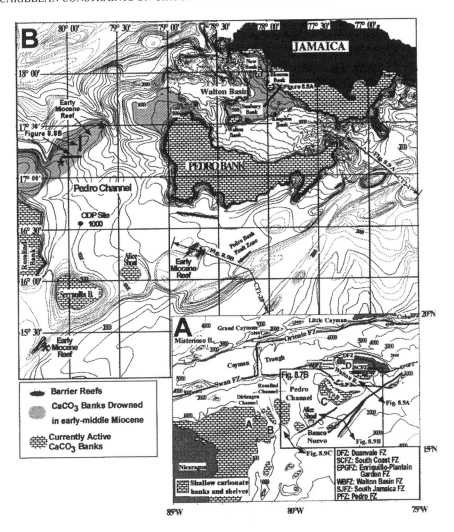

Fig. 8.7. (a) Modern physiography of the northern Nicaragua Rise. Seaways separated by carbonate banks/shelves are identified and major fault zones within the northern Caribbean plate boundary zone are indicated. Our main study area, including Pedro Channel and Walton Basin is blown up in Fig. 8.7b. UTIG MCS line segments across the Pedro Bank fault zone, interpreted in Fig. 8.9, are also indicated in this figure. (b) Detailed bathymetry in Pedro Channel and Walton Basin (Robinson, 1976; Cunningham, unpublished) represents the complexity and the segmented character of the sea floor morphology within the seaways along the northern Nicaragua Rise. Three segments of high resolution seismic profiles from Pedro Channel and Walton Basin are reproduced in Fig. 8.8. UTIG MCS line segments across the Pedro Bank fault zone, interpreted in Fig. 8.9, are also indicated. Parts of the Pedro Channel and Walton Basin (light gray pattern) were deposited, at least until 20 Ma, as carbonate banks and barrier reefs. In association with the current carbonate banks that have remained areas of neritic carbonates since the late Eocene, drowned banks and reefs observed in the Pedro Channel and Walton Basin formed along the northern Nicaragua Rise an east-west barrier, where continuous shallow water environments prevailed from late Eocene to the early part of the early Miocene (~35-20 Ma). Some of the carbonate banks and barriers (light gray pattern) subsided and drowned as early as the late early Miocene, but most probably deep water conditions became prevalent in these areas of Pedro Channel and Walton Basin in the late middle Miocene. Growth rates of currently active banks, indicated with brick pattern, have kept pace with subsidence.

volcanic arc that formed part of an Andean-type arc margin to North America occupy the northern Nicaragua Rise and have been overlain above a regional unconformity by an Early Tertiary section characterized by continental sediments. The clastic rocks which resemble those of the Wagwater trough in Jamaica (Robinson, 1994) are succeeded by middle to late Eocene carbonate sediments similar to the middle and late Eocene "Yellow Limestones" of Jamaica (Robinson, 1994) in being of widely varying thicknesses. Widespread, uniform late Eocene to early Miocene limestones, comparable to the "White Limestones" of Jamaica (Robinson, 1994) overlie the "Yellow Limestones" and are themselves overlain discontinuously by Miocene through Holocene bank carbonates and periplatform sediments.

High resolution seismic profiles in interbank channels across the northern Nicaragua Rise (Droxler et al., 1989, 1992; Hine et al., 1992, 1994) (Figs. 8.8a, 8.8b) generally show in most parts of the subseafloor a distinct sequence of flat, sub-parallel, and high amplitude reflections typical of those generated by neritic carbonate sediments and rocks. The top of this unit is marked by a major unconformity (Fig. 8.8). Rocks below the unconformity are extensively block faulted, tilted, and gently folded. Indications of strike-slip motion are strong. Using the geological evolution of Jamaica (e.g., Robinson, 1994; Leroy et al., 1996) and the interpretation of these recently acquired high resolution seismic reflection profiles from the major northern Nicaragua Rise basins and channels as guides, we develop a model in which the entire length of the northern Nicaragua Rise, including the island of Jamaica and part of the Honduras and Nicaragua shelves, was covered by a series of relatively continuous carbonate banks and barrier reefs during late Eocene to early Miocene times (~38–20 Ma) (see Fig. 8.7b) (Horsfield and Roobol, 1974; Droxler et al., 1989, 1992; Lewis and Draper, 1990; Hine et al., 1992, 1994). Beginning in late early Miocene and perhaps as early as late Oligocene time based upon the recent drilling of ODP Site 1000 in Pedro Channel (Sigurdsson et al., 1997) (Fig. 8.7b), this series of relatively continuous carbonate banks and barrier reefs progressively broke up and partly drowned. The main phase of segmentation and partial drowning probably occurred in the middle Miocene. A set of smaller banks, separated by a series of basins and seaways began to develop (Fig. 8.6). Shallow water organisms recovered in dredge samples from the top of the foundered carbonate banks and barrier reefs include larger foraminifera such as *Miogypsina gunteri* and corals such as: *Montastrea costata*, *Stylophora cf. imperatoris*, and *Porites trinitatis* all of which yield early Miocene ages (~20 Ma). The surface on which these organisms had lived has subsided by as much as 1200 m since the time it lay in the photic zone (Droxler et al., 1992). At the same time as differential subsidence continued over much of the shallow carbonate areas during the middle Miocene (~12 Ma), the most easterly area of the shallow carbonate system began to be uplifted and is now exposed in central and southern Jamaica (Lewis and Draper, 1990). The presently active carbonate banks of the Nicaragua Rise such as Pedro, Rosalind, and Serranilla banks (Fig. 8.7) are remnants of a larger neritic carbonate system which remain perched on faulted blocks. Most of the older neritic area has been characterized by subsidence rates fast enough to outpace the growth potential of carbonate banks. Only on top of those blocks has carbonate deposition as on Pedro, Rosalind, and Serranilla banks, been able to keep pace with relatively slower subsidence rates.

Neogene Tectonic History

Foundering of the carbonate banks and barrier reefs along the northern Nicaragua Rise took place at a time of substantial tectonic change (~8–20? Ma) along much of the length of the northern Caribbean Plate Boundary Zone (PBZ) (Burke et al., 1980; Leroy et al., 1996). This was the time when the island of Jamaica began to rise from below sea level and it was also the time of the initiation of many new faults and fault systems. Among the most prominent of new fault zones that developed within the northern Caribbean PBZ at ~12–20? Ma was the Enriquillo-Plantain Garden-Walton fault zone which has been traced from Hispaniola through Jamaica to the mid-Cayman spreading center and which bounds the north side of the extensive Oligocene–early Miocene neritic carbonate system on the northern Nicaragua Rise (Fig 8.7a) (Rosencrantz et al., 1988; Mann et al., 1995 and references therein). Individual faults of the Enriquillo-Plantain Garden-Walton fault zone system, such as the Walton Basin

A. High Resolution Seismic Line Across the Central Plateau in Walton Basin

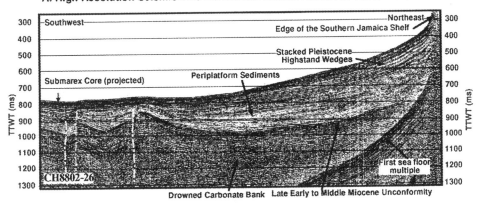

B. High Resolution Seismic Lines Across the Northern Spur in Pedro Channel

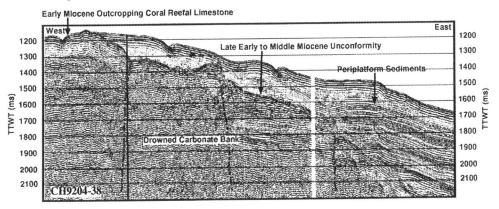

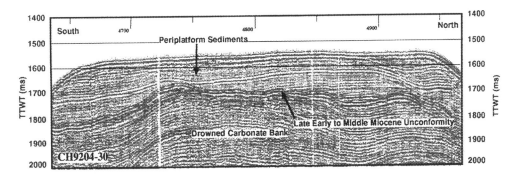

Fig. 8.8. Seismic profiles from the northern Nicaragua Rise illustrating the areas in Walton Basin and Pedro Channel, where carbonate banks and barriers (light gray pattern in Fig. 8.7b) were deposited until the early Miocene (~20 Ma), then drowned in the middle Miocene (12-15 Ma) and subsided to their current upper bathyal water depths. Fig. 8.8a from Walton Basin and Fig. 8.8b from Pedro Channel shows how periplatform deposits have come to overlie a drowned carbonate bank, characterized by flat, sub-parallel, and high amplitude reflections typical of those generated by neritic carbonate sediments and rocks. These drowned banks are usually broken by subvertical faults. Early Miocene corals and diagnostic larger foraminifers were dredged from the western end of profile CH9204-38, where a drowned carbonate bank is cropping out.

fault zone (Rosencrantz and Mann, 1991) are seismically active or show bathymetric evidence of being active today. Plate reconstructions such as those of Pindell (1994, Fig. 2.6n) and Stephan et al. (1990) emphasize the tectonic importance of the Enriquillo-Plantain Garden-Walton fault zone, but the timing of the fault formation differs in the two reconstructions. This fault zone already appears in the early Miocene reconstruction of Stephan et al. (1990, their plate 12 (anomaly 6), but only in the late Miocene reconstruction of Pindell (1994, Fig. 2.6n). In a recent study of the eastern Cayman Trough by Leroy et al. (1996), the northern margin and central part of Jamaica formed during the Eocene, then subsided during the Oligocene–early Miocene, and finally was reactivated along the Duanvale Enriquillo-Plantain Garden-Walton fault zone in the middle Miocene.

Farther south near the limit of the northern Caribbean PBZ, the Pedro Bank fault zone (also known as the Baja Nuevo fracture zone) is much less well-known but is tectonically significant because it lies close to the southern margin of the shallow carbonate system on the northern Nicaragua Rise (Figs. 8.7a, 8.7b). The important question is: Did the Pedro Bank fault zone become active in the middle/early(?) Miocene at the time of the partial foundering of the carbonate banks and barrier reefs and thus play a part in the initiation of the Caribbean Current? Although topographically distinct (see Figs. 8.7b, 8.9), the Pedro Bank fault zone has not been well mapped or closely studied. Pindell (1994) did not include it on his Miocene plate reconstruction map and Stephan et al. (1990) only included the fault on their 10 Ma and younger reconstructions. Edgar et al. (1971) briefly discussed the acoustic cross section of the Pedro bank fault zone and dated the fault as of Miocene age. Holcombe et al. (1990) identified the structure without discussion. Bathymetric maps indicate that the fault zone extends to the north-east to join the Enriquillo–Plantain Garden-Walton fault zone east of Jamaica and that the fault zone includes at least three individual faults trending ENE. These structures could possibly be splays from a single main fault (see Fig. 8.9 as well as Mann and Corrigan, 1990; Mann et al., 1995).

Interpretation of University of Texas Institute of Geophysics multichannel seismic reflection lines (Figs. 8.7 and 8.9) shows that the individual faults

of the Pedro Bank fault zone are separated by turbidite filled basins. The geometry of the fault zone is suggestive of left-lateral displacement which is consistent with the overall sense of movement in the northern Caribbean PBZ. Whether the perched basins formed by the down-dropping of crust between the faults or by the uplift of the flanks is not clear. The southern flank of the Pedro Bank fault zone appears elevated and some of the faults bisect what can be interpreted as push-up structures (Fig. 8.9). Seafloor displacements seen on 3.5 kHz profiles in the area of the push-up structures indicate continuing tectonic activity although, in contrast to the Enriquillo-Plantain Garden-Walton fault zone, earthquakes have not been reported from the Pedro Bank fault zone. We tentatively identify elongate narrow ridges imaged by swath bathymetry (Droxler, 1995) and high resolution profiling (Droxler et al., 1989, 1992) as drowned banks and barrier reefs (Fig. 8.7b). The reefs had formed on blocks of the Pedro Bank fault zone and aggraded until their growth was interrupted. The blocks continued to survive with the drowned reefs on their crests. Early Miocene neritic limestone blocks have been dredged from one of these ridges (Droxler et al., 1992) (Figs. 8.7b, 8.9b, 8.9c). In summary: there are indications that the Pedro Bank fault zone, which bounds the southern margin of the extensive Oligocene–early Miocene neritic carbonate system, developed, as did the fault zone of the northern margin of that shallow carbonate system, at the time of the collapse and partial foundering of the carbonate banks and barrier reefs. However, analysis of the structural and stratigraphic record of the collapse of the carbonate system is still less complete for the southern area.

Caribbean Current Establishment and Its Influence on the Production of North Atlantic Deep Water

Estimated timing for the partial foundering of the carbonate banks and barrier reefs of the northern Nicaragua Rise and the resulting initiation of the Caribbean Current corresponds to the general development and strengthening of the North Atlantic Western Boundary Current in the middle Miocene. Mullins et al. (1987) have shown that the Loop Current of the Gulf of Mexico became established in the middle Miocene at ~12 Ma (Fig.

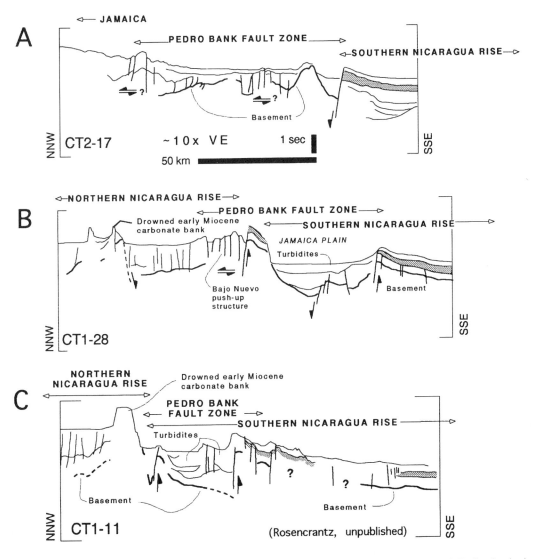

Fig. 8.9. Simple line drawing interpretation of three UTIG MCS profile segments (located in Fig. 8.7) showing basic morphological and tectonic features observed across and along the Pedro Bank fault zone (see text for more details). These features clearly illustrate recent and active tectonic and subsidence along this fault zone.

8.6c). Contemporaneous strengthening of the Florida Current and Gulf Stream has been established in the Straits of Florida by Denny et al. (1994), Gomberg (1974), Mullins and Neumann (1979), Austin et al. (1988), Mullins et al. (1980), and finally on the Blake Plateau by Popenoe (1985). We suggest that the establishment of the Loop and Caribbean currents, the variations in their strengths over the past 12–15 m.y. and the related downstream strengthening of the Florida Current and Gulf Stream, were direct consequences of the

extensive foundering of shallow carbonate banks and barrier reefs on the northern Nicaragua Rise.

The establishment and strengthening of the Caribbean Current in the late middle Miocene, as a phenomenon caused by the partial foundering of carbonate banks and barrier reefs on the northern Nicaragua Rise, appears to correspond, in terms of timing, to a significant enhancement of a northern component for deepwater formation (NCW, equivalent to a proto-NADW) in the high latitudes of the North Atlantic (Woodruff and Savin, 1989,

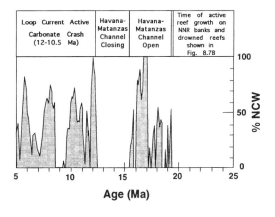

Fig. 8.10. Record of the Northern Component Water flux over the past 25 m.y. estimated by interbasinal differences of benthic carbon isotope records (North Atlantic minus Pacific oceans) (Wright and Miller, 1996). The North Atlantic was a source for deep to intermediate water masses (referred to as North Component Water or NCW), first to a certain degree from 18 to 16 Ma, and more clearly since 12.5 Ma. NCW was minimum, or the Southern Component (deep) Water (SCW) was maximum, between 16 and 12.5 Ma. North Atlantic component deep water (or NCW) was well established at about 10 Ma, time of the "carbonate crash" optimum in the eastern equatorial Pacific (see Fig. 8.12 or Lyle et al., 1995) and in the Caribbean (Sigurdsson et al., 1997). The timing of NCW establishment in the middle Miocene could relate to either one or both of the following tectonic events: the partial drowning and subsidence of an almost continuous shallow carbonate system along the northern Nicaragua Rise and/or an early episode in closure of the Central American Seaway in Panama.

1991; Crowley and North, 1991; Wright et al., 1992; Wright and Miller, 1996) (Fig. 8.10). According to Wright and Miller (1996), an earlier Neogene interval of NCW production had already occurred in the late early Miocene (16–19 Ma with peak production around 17 Ma) (Fig. 8.10) well before the establishment of the Loop Current in the Gulf of Mexico. This first Neogene interval of NCW production is itself younger than the dredged reefal limestones capping the drowned banks in Pedro Channel and other seaways along the northern Nicaragua Rise, but older than the interval of Loop Current initiation (Fig. 8.10). This first Neogene (late early Miocene) interval of NCW production can be tied to an early establishment of the Caribbean Current through seaways created by partially foundered banks and reefal barriers along

the northern Nicaragua Rise. During this interval, the Caribbean Current would have flowed downstream from the northern Nicaragua Rise through a seaway in western Cuba, the Havana-Matanzas Channel, into the Straits of Florida, bypassing the Gulf of Mexico (Figs. 8.5c, 8.6b). According to Iturralde-Vinent et al. (1996), the Havana-Matanzas Channel was fully open during the early Miocene and closed sometime in the early middle Miocene. Two late early and early middle Miocene significant hiatuses in ODP Site 1003, recently drilled on the lower slope of Great Bahama Bank in the Straits of Florida (Swart et al., 1997), may have been produced by some strengthening of the Florida Current related to the upstream flow of the Caribbean Current through the Havana-Matanzas Channel. Prior to the early late Miocene, during Oligocene times and early part of the early Miocene, from ~35–~20(?) Ma, comparable currents could not have developed because the extensive east to west trending series of continuous carbonate banks and barrier reefs on the northern Nicaragua Rise impeded the development of a strong northwestward flowing Caribbean Current (Fig. 8.6a; Fig. 8.10).

Because the stepwise closing of the Central American Seaway overlaps in time the partial foundering of the carbonate banks and barrier reefs on the northern Nicaragua Rise, we suggest that the strengthening of the Western Boundary Current in the western North Atlantic was also linked to this closing. The stepwise closing of the Central American Seaway may have led to the partial isolation of deep and possibly intermediate water masses in the eastern Pacific Ocean from those of the western Central Atlantic Ocean (Duque-Caro, 1990; Collins et al., 1996). This isolation may have triggered the rapid and relatively short-lived shoaling of the carbonate compensation depth in the eastern Equatorial Pacific Ocean and caused the observed eastern Equatorial Pacific "carbonate crash" at the middle to late Miocene transition, between 11 and 9 Ma (Lyle et al., 1995; Farrell et al., 1995) (Fig. 8.12). However, recent shipboard results of ODP Leg 165 (Sigurdsson et al., 1997) have clearly documented that the "carbonate crash," at the middle to late Miocene transition, also occurred in the Caribbean basins. It is the dominant feature observed between 12.5 and 10.5 Ma at Sites

999 and 1001 in the Colombian Basin, Site 1000 in Pedro Channel, and Site 998 in the Yucatan Basin, in water depths ranging from 900 to nearly 3300 m (see Fig. 8.3 for Leg 165 site locations). The "carbonate crash" is, therefore, not limited to the eastern equatorial Pacific Ocean but corresponds to a more widespread low latitude oceanographic phenomenon. The "carbonate crash," because of its occurrence on both sides of the Isthmus of Panama and in the northwest Caribbean, is related to a major reorganization of the ocean circulation during the late Neogene and possibly to the first establishment of a pattern of global thermohaline ocean circulation approaching that of modern times. The association of both the stepwise closing of the Central American Seaway and the somewhat penecontemporaneous opening of the major seaways on the northern Nicaragua Rise would have caused a major reorganization of the deep and intermediate oceanic circulation, and perhaps the initial production of North Atlantic Deep Water. The influx into the Caribbean basins of AAIW, characterized today by low carbonate concentration (Haddad and Droxler, 1996), would have been initiated at this time to replenish the waters sinking in the northern latitudes of the North Atlantic.   This influx of southern source intermediate waters, corrosive toward carbonate sediments, would explain the occurrence of the systematic carbonate dissolution observed in the Caribbean basin at the middle to late Miocene transition (Sigurdsson et al., 1997).

The establishment of a north component deepwater mass in the high latitudes of the North Atlantic in the middle Miocene could have caused the 10–12 Ma shift of the silica deposition from the North Atlantic to the North Pacific ocean basins (Woodruff and Savin, 1989). Another probable or possible consequence of the partial foundering of a series of carbonate banks and barrier reefs on the northern Nicaragua Rise, the contemporaneous stepwise closing of the Central American Seaway, and the resulting acceleration of the North Atlantic western boundary current is the formation of major phosphate deposits on Florida and along the southeast U.S. continental margin as well as widespread submarine erosion farther offshore (Riggs, 1984; Scott, 1988; Snyder et al., 1990; Allmon et al., 1996).

The Central American Seaway and the Isthmus of Panama

Analysis of the evolution of this classic example of a gateway closing has become an oft-told tale but publication of splendid new syntheses of relevant data in the scientific results of ODP 138 (Pisias et al., 1995) and in GSA Special Paper 295 (Mann, 1995) has encouraged a brief review. The ODP report deals with the results of a drilling leg in the eastern Pacific west of Panama and the Special Paper addresses the geologic and tectonic evolution of Costa Rica and Panama making use of both on and offshore geological and geophysical observations. What is absent at present is a comparable overview of the paleoceanography of the Caribbean to the east of Panama. The recent completion of ODP Leg 165 in Caribbean waters is an assurance that something of that sort will be available within the next few years (Sigurdsson et al., 1997). For that reason, if no other, our present review must be regarded as a progress report and in no way definitive. Discussion is divided into sections on: (1) tectonic evolution, and (2) sedimentary and environmental record. It is important to emphasize, however, that observations of both kinds are commonly used together.

Tectonic Evolution of the Isthmus of Panama

Subduction at the Andean margin of South America led to the collision of Panama with Colombia and resulted in closure of the Central American Seaway. As a result of the collision, Central, North, and South America formed a continuous landmass. Consequently, the collision isolated the Caribbean Sea from the Pacific. Restoration of plate geometry by conventional rotational methods (see, e.g., Pindell and Barrett, 1990) shows how the collision may have happened. Because post-collisional and continuing convergence of Panama with South America (see, e.g., Mann, 1995) have led to deformation of both Panama and the part of Colombia with which it collided, it is not possible to restore the areas that were involved in the collision to their precise paleogeographic positions at the time of collision by simply using rigid body rotations.

Many tectonicists have been impressed by evidence of the timing of uplift of the Cordillera of northern Colombia, called locally "the Andean Orogeny," which they have inferred to have been related to the Panama collision (see, e.g., Van der Hammen et al., 1973; Irving, 1975; Pindell and Dewey, 1982; Duque-Caro, 1990; Dengo and Covey, 1993). This observation leads to ages for the collision ranging from about 16 Ma to about 6 Ma with a concentration around 10 Ma. Although it is not easy to attach an exact physical significance to the event indicated by the uplift of the Colombian Cordillera, comparison with current collisional events, suggests that bathymetric depths of two or more kilometers might have existed in a trench between Panama and Colombia at the time when the uplift became perceptible in the Colombian Cordillera.

In an attempt to improve understanding of the tectonics, timing, and environmental changes associated with the closing of the Central American Seaway, Wadge and Burke (1983) tried to restore the shape of peninsular Panama to that which it had before the collision with Colombia (Fig. 8.11). That was done by estimating the former length of the peninsula. The present length of the isthmus when added to the length of the collided arc in Colombia gives about 500 km (see, e.g., Kellogg and Vega, 1995, p. 86). Wadge and Burke (1983) (Fig. 8.11) chose to draw pre-collision Panama as a straight peninsula trending to the ENE because they interpreted the absence of volcanoes from the area to indicate that Panama was being carried toward South America along a transform boundary. The more northwesterly region of Central America, where there were active volcanoes at the time, was drawn with a northwesterly trend (Fig. 8.11).

Silver et al. (1990) and Mann and Corrigan (1990) expanded on the simple model of Wadge and Burke (1983) and concluded that deformation or oroclinal bending by processes including folding, thrusting and strike-slip motion of an originally straight peninsula is a likely way to have generated the present peculiar shape of Panama with its curious resemblance to the letter S lying on its back. More recently, Kolarsky and Mann (1995, especially, Fig. 21) have shown that oblique subduction of the Nazca plate beneath the Panama arc and collision of the Cocos ridge with the Azuero peninsula may induce additional forces that act together with those generated by the collision between Panama and Colombia. Pindell (1994, p.

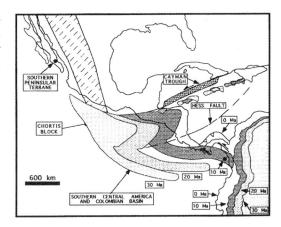

Fig. 8.11. Map showing how the former Panama peninsula has moved over the past 30 m.y. to generate the present Panama Isthmus. In this figure, North America has been held fixed and, as the Cayman Trough has opened, the Caribbean Plate has moved eastward. Panama, forming the southern boundary of the Caribbean plate, collided with the slowly westward moving South American continent. The present S-shape of Panama has developed from internal deformation rather than rigid body rotation largely within the past 10 m.y. Modified from Mann's (1995) representation of the model of Wadge and Burke (1983).

33 and Fig. 2.6 k-n) introduced a radically different suggestion in which Panama collided with Colombia as early as middle Eocene times (49 Ma), but this idea appears incompatible with paleoceanographic constraints on the history of the Central American Seaway.

The foregoing discussion shows that, although tectonic analysis has been helpful in showing what has happened and where it happened during the closing of the Central American Seaway, our understanding of high resolution temporal and spatial pattern is still poor. The main problem involves the difficulty in restoring precollisional shapes. Continued post-collisional convergence has deformed, distorted, and even bodily removed precollisional rock. For these reasons, paleobiological and paleoenvironmental indicators yield a much better high-resolution understanding of the closure of the seaway (e.g. Farrell et al., 1995, p. 748).

Sedimentary and Environmental Record of the Closing of the Central American Seaway

Farrell et al. (1995, Table 5 and pp. 748 and 749)

reviewed the geologic history of the closure of the Panamanian Seaway in the light of results obtained from ODP Leg 138. They emphasized that, although the history of sill uplift and separation of Caribbean from eastern equatorial Pacific waters have been addressed by using sedimentary records from the deep-sea, near-shore and terrestrial environments, reconstruction has proved difficult. Apart from the acknowledged problems of interpreting tectonic information, there is also the possibility that emergence of the Panamanian isthmus may have been episodic (e.g., Duque-Caro, 1990; Collins et al., 1996). The interchange of some terrestrial mammals several million years before the final closure (Marshall, 1985) opens up the additional alternative that the seaway after having been temporarily closed, may have later reopened.

Compilation of published results enabled Farrell et al. to generate a chronology of 10 events marking the closure of the Isthmus of Panama over the interval from ~13 Ma to ~2 Ma (Farrell et al., 1995, Table 5 and p. 744). The results of the various studies do not agree completely, especially in such matters as sill depths and timing of flow limitation but a general consensus does emerge from the data. This chronology is quite consistent with major results of ODP Leg 138. For example the earliest indication of seaway closure occurred between 12 and 13 Ma at a time which coincided with the onset of an interval characterized with dramatic increase of carbonate dissolution recorded in Leg 138 cores that is referred to as "the carbonate crash" (Lyle et al., 1995) (Fig. 8.12). During this episode much less carbonate was deposited on the ocean floor and the Carbonate Compensation Depth (CCD) rose dramatically (Fig. 8.12). Lyle et al. (1995) showed that the crash could not have been related to an abrupt increase in productivity or a loss of organic carbon from the continental shelves and suggested that the crash could have resulted from a reduction in deepwater exchange between the Atlantic and Pacific oceans through the Panama Gateway prior to the emergence of the isthmus. The timing of the carbonate crash coincides with the timing of the main phase of the carbonate bank partial foundering along the northern Nicaragua Rise and the consequent initial flow of the Caribbean Current. For that reason and because the carbonate crash has been observed on both sides of

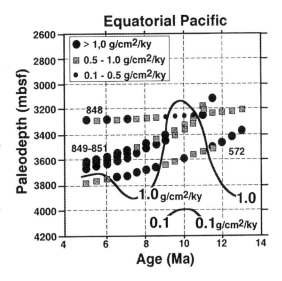

Fig. 8.12. Variations of carbonate mass accumulation rates at several ODP and DSDP sites in the eastern Equatorial Pacific clearly illustrate the "carbonate crash" of Lyle et al. (1995). Shoaling of the Carbonate Compensation Depth (or CCD) at the middle to late Miocene transition is illustrated by the 700 m shallowing excursion of the 1.0 g/cm^2/ky line between 12 to 9 Ma. A contemporaneous interval of low carbonate deposition was also observed in several ODP sites recently drilled in the Caribbean during Leg 165. This time of low carbonate deposition in the eastern Equatorial Pacific Ocean and Caribbean Sea corresponds to an interval during which the North Atlantic component Deep water (or NCW) was well estblished (see Fig. 8.10; Wright and Miller, 1996), possibly triggered by the partial drowning and subsidence of an almost continuous shallow carbonate system along the northern Nicaragua Rise and/or an early episode in closure of the Central American Seaway in Panama.

the Isthmus of Panama, we suggest that the partial collapse of the northern Nicaragua Rise may have been a substantial contributor to if not the dominant cause of the carbonate crash (see above section "Caribbean Current Establishment and its Influence on the Production of North Atlantic Deep Water", Sigurdsson et al., 1997).

The first passage of mammals between the continents at 9.5-10.5 Ma "coincides with the nadir of the carbonate crash" (Farrell et al., 1995, p. 749) and a time of well established proto-NADW or NCW (Figs. 8.10). The interval from 9-5 Ma shows differences between Caribbean and Pacific in foraminiferal assemblages and in benthic foraminiferal isotopic compositions (Duque-Caro, 1990; Farrell et al., 1995, Table 5 and p. 744).

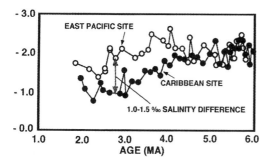

Fig. 8.13. Caribbean and East Pacific planktic oxygen isotopic records diverge from about 4.2 Ma (Keigwin, 1982). Keigwin suggested that the divergence of $\delta^{18}O$ values beginning at 4.2 Ma may reflect increasing western North Atlantic surface-water salinity contrasting with relatively lower salinity in the East Pacific (see Fig. 8.1b for modern situation). This may have resulted from the final closing of the Isthmus of Panama.

Recent studies of benthic foraminifers on both side of the Isthmus of Panama by McDougall (1996) demonstrated that the flow of deep and intermediate water across the Central American sill stopped about 5.6 Ma. The same table shows that a large body of evidence is suggestive of final closure between 4.5 and 2.6 Ma. The planktic $\delta^{18}O$ records are identical on both sides of the Isthmus of Panama prior to 4.5 Ma (Keigwin, 1982) (Fig. 8.13). Because sea surface temperatures are currently substantially lower in the eastern equatorial Pacific than in the western Caribbean, the 0.6–1.0‰ $\delta^{18}O$ enrichment of the Caribbean mixed-layer waters since about 4.5 Ma is interpreted by Keigwin (1982) to correspond to significantly saltier Caribbean waters than those in the eastern equatorial Pacific. In addition differences in planktonic foraminiferal assemblages (Keller et al., 1989) developed between 3.5 and 2.6 Ma. Recently, Burton et al. (1997) demonstrated that the Nd, Pb, and Sr isotopic signals, recorded by hydrogenous ferromanganese crusts and associated with NADW, strengthened about 3–4 Ma. These events also coincide with a southeastward shift in the locus of maximum opal accumulation in the eastern equatorial Pacific at about 4.4 Ma indicating major changes in wind-driven circulation (Farrell et al., 1995, p. 749).

Conclusion

Emergence of the Aves Swell at ~35–30 Ma, subsidence of a series of carbonate banks and barrier reefs on the northern Nicaragua Rise at ~12 Ma, and emergence of the Isthmus of Panama at ~3–4 Ma have been bathymetric and tectonic changes in the Caribbean that took place at times which roughly coincided with changes in oceanic circulation on both local and global scales as well as with other major environmental changes (see Fig. 8.6 for schematic summary diagrams). Because of the varied and different kinds of observation that have been made and because of the very high spatial and temporal resolution of the data generated from those observations, reconstruction of the tectonic history is best for the closure of the Isthmus of Panama (see, e.g., Farrell et al., 1995, Table 5) and less good for the subsidence of the Aves Swell and the northern Nicaragua Rise. Nevertheless, Farrell and his co-workers can only conclude that "a large body of evidence suggests that the final closure occurred between 4.6 and 2.6 Ma" (Farrell et al., 1995, p. 749). One reason for this is that "final closure" does not necessarily mean the same thing for all the processes involved which range from the cessation of the flow of an ocean current to the passage of a dry-footed mammal.

What part the other two events, the exposure of the Aves Swell and the partial foundering of a series of carbonate banks and barrier reefs along the northern Nicaragua Rise, may have played in modifying the oceanic circulation and the regional and global environment is much more speculative.

Making the picture clearer requires:

(1) Better temporal resolution of the timing of two tectonic events.

(2) Better spatial and temporal resolution for the ~30–35 Ma, ~17 Ma and ~12 Ma intervals of the record of environmental change in the sediments of the Caribbean and neighboring ocean basins.

(3) The need for better resolution of the environmental changes on a global scale. For example: Was ~30–35 Ma the time of establishment of the East Antarctic Ice-Sheet ?

(4) It will also be useful to know with better resolution when other gateway events, such as the opening of the Drake Passage, took place. This should reduce the risk of confusion as to the cause of a particular environmental change. At present, there are too many tectonic events which appear to have happened at about the same time for clear pictures of the causes of environmental changes to be drawn.

Requirement (1) can be addressed by determining with better resolution exactly when the Grenada Basin opened. This will show when the Aves Swell ceased to be an active arc and allow better modeling of its thermal subsidence. For the northern Nicaragua Rise a better understanding of events in the Cayman Trough, the timing of faulting in the northern Caribbean Plate Boundary Zone (the Enriquillo-Plantain Garden-Walton and Pedro Bank fault zones) should help. Requirement (2) calls for more information from the Neogene sediments of the Caribbean ocean floor. The results of ODP Leg 165 should help to some degree but there are extensive areas for which no information presently exists. Requirements (3) and (4) are no more than a reiteration of the general need to address the problems of tectonics and climate together on a global scale. We have outlined changes in the Caribbean but to understand their significance it is, as always, essential to fit them in to a global understanding of climatic and environmental change.

Acknowledgments

The main financial support for this study was provided by several National Science Foundation research grants OCE-8715922, -8900040, and -9116323 to A. W. Droxler, A.C. Hine, and P. Hallock. Special thanks are due to Captain Richard Ogus and the crew of the R/V Cape Hatteras. David Mucciarone and Timothy W. Boynton's help was invaluable in the seismic data acquisition and processing from Walton Basin and Pedro Channel. Comments and suggestions from Tom Crowley, Mark Leckie, Larry Lawver, John Farrell, and an anonymous reviewer have helped us to clarify our model and a first version of this manuscript.

References

Abreu, V.S., and G.A. Haddad, in press, Glacioeustatic fluctuations: The mechanism linking stable isotope events and sequence stratigraphy from the early Oligocene to middle Miocene. *SEPM Special Publication #57.*

Allmon, W.D., S.D. Emslie, D.S. Jones, and G.S. Morgan, 1996, Late Neogene Oceanographic change along Florida's West Coast: Evidence and Mechanisms, *J. Geol.*, 104, 143–162.

Austin, J.A., W. Schlager, A.A. Palmer, and ODP Leg 101 Scientific Party, 1988, *Proceedings of the Ocean Drilling Program, Initial Reports (Part A)*, vol. 101, Ocean Drilling Program, College Station, Tex.

Bird, D., S.A. Hall, J.F. Casey, and P.S. Millegan, 1993, Interpretation of magnetic anomalies over the Grenada Basin, *Tectonics*, 12, 1267–1279.

Bouysse, P., P. Andreieff, M. Richard, J.C. Baubron, and A. Mascle, 1985, Aves Swell and northern Lesser Antilles Ridge: Rock dredging results from ARCANTE 3 Cruise, in *Géodynamique des Caraïbes*, A. Mascle (ed.), pp. 65–76, Editions Technip., Paris.

Broecker, W.S, 1991. The great ocean conveyor, *Oceanography*, 4, 79–89.

Broecker, W.S., et al., 1985. Sources and flow patterns of deep-ocean waters as deduced from potential temperature, salinity, and initial phosphate concentration, *J. Geophys. Res.*, 90, 6925–6939.

Burke, K.C., 1988, Tectonic evolution of the Caribbean, *Annu. Rev. Earth Planet. Sci.*, 16, 201–230.

Burke, K.C., J. Grippi, and A.M.C. Sengör, 1980, Tectonic style of the northern boundary of the Caribbean, *J. Geol.,* 88, 375–386.

Burton, K.W, H.-F. L., O'Nions, R. K., 1997, Closure of the Central American Isthmus and its effect on deep-water formation in the North Atlantic, *Nature*, 386, 382–385.

Coates, A.G., et al., 1992, Closure of the Isthmus of Panama: The near-shore marine record of Costa Rica and western Panama. *Geol. Soc. Am. Bull.*, 104, 814–828.

Collins, L.S., A.G. Coates, W.A. Berggren, M.-P. Aubry, and J. Zhang, 1996, The late Miocene Panama isthmian strait, *Geology*, 24(8), 687–690.

Crowley, T.J., and G. R. North, 1991, *Paleoclimatology*, Oxford University Press, New York, 339 pp.

Dengo, C.A. and M.C. Covey, 1993, Structure of the Eastern Cordillera of Colombia: Implications for trap styles and regional tectonics, *AAPG Bull*, 77, 1315–1337.

Denny, W.M., J.A. Austin, and R.T. Buffler, 1994, Seismic stratigraphy and geologic history of mid-Cretaceous through Cenozoic rocks, Southern Straits of Florida, *AAPG Bull.*, 78, 461–487.

Despretz, J.-M., T. E. Daly, and E. Robinson, 1985, Saba Bank Petroleum Geology, NE Caribbean, *Oil and Gas Journal, Nov. 4,* 112–118.

Droxler, A.W., 1995, Caribbean Drilling Program: R/V Maurice Ewing Site Survey preliminary Report, JOI-USSAC, 16 pp.

Droxler, A.W., S.A Staples, E. Rosencrantz, R.T. Buffler., and P.A. Baker, 1989, Neogene tectonic disintegration of a carbonate Eocene-early Miocene megabank along the northern Nicaragua Rise, Caribbean Sea, *Twelfth Caribbean Geological Conference St. Croix, USVI, Program and Abstracts*, 41.

Droxler, A.W., A. Cunningham, A.C. Hine, P. Hallock, D. Duncan, E. Rosencrantz, R. Buffler, and E. Robinson, 1992, Late middle (?) Miocene segmentation of an Eocene-early Miocene carbonate megabank on the northern Nicaragua Rise, *EOS, Suppl.*, 73, S299.

Duncan, R.A., C.W. Sinton, and T.W. Donnelly, 1994, The Caribbean Cretaceous basalt province: An oceanic LIP. *EOS Trans., Amer. Geophys. Union*, 75(44), 594.

Duque-Caro, H., 1990, Neogene stratigraphy, paleoceanography and paleobiogeography in northwest South America and the evolution of the Panama Seaway, *Paleogeogr., Paleoclimat., Paleoecol.*, 77, 203–234.

Edgar, N.T., J.I. Ewing, and J. Hennion, 1971, Seismic refraction and reflection in the Caribbean Sea, *Amer. Assoc. Petrol. Geol. Bull.*, 55, 833–870.

Farrell, J.W., et al., 1995, Late Neogene sedimentation patterns in the eastern equatorial Pacific Ocean, in *Proceedings of the Ocean Drilling Program, Scientific Results, vol. 138*, N.G. Pisias, L.A. Mayer, T.R. Janecek, A. Palmer-Julson, and T.H. van Andel (eds.), 717–753, Ocean Drilling Program, College Station, Tex.

Glaser, K.S. and A.W. Droxler, 1991, Holocene high stand shedding, producing a periplatform wedge in the surroundings of "drowned" shallow carbonate Bank and Shelf. Walton Basin, Northern Nicaragua Rise, *J. Sedimentary Petrology*, 61(1), 126–142.

Gomberg, D., 1974, Geology of the Portales Terrace, *Florida Science*, 37, (Suppl. 1), 15.

Gordon, A.L., 1986, Inter ocean exchange of the thermocline water, *J. Geophys. Res.*, 91, 5037–5046.

Gordon, A.L., F.W. Ray, W.M. Smethie Jr., and M.J. Warner, 1992, Thermocline and intermediate water communication between the South Atlantic and Indian oceans, *J. Geophys. Res.*, 97, 7223–7240.

Haddad, G.A., 1994, Calcium carbonate dissolution patterns at intermediate water depths of the Tropical oceans during the Quaternary, Ph.D. thesis, 494 pp., Rice University, Houston, Tex.

Haddad, G.A. and A.W. Droxler, 1996, Metastable $CaCO_3$ dissolution at intermediate water depths of the Caribbean and western North Atlantic: Implications for intermediate water circulation during the past 200,000 years, *Paleoceanography*, 11, 701–716.

Haq, B.U., J. Hardenbol, and P.R. Vail, 1987, Chronology of fluctuating sea levels since the Triassic, *Science,* 235, 1156–1167.

Hine, A.C., et al., 1992, Megabreccia shedding from modern, low-relief carbonate platforms, Nicaraguan Rise, *Geol. Soc. Amer. Bull.*, 104, 928–943.

Hine, A.C., et al., 1994, Sedimentary infilling of an open seaway: Bawihka Channel, Nicaraguan Rise, *J. Sediment. Res.*, B64, 2–25.

Holcombe, T.L., J.W. Ladd, G. Westbrook, N.T. Edgar, and C.L. Bowland, 1990, Caribbean marine geology: Ridges and basins of the plate interior, in *The Geology of North America, vol. H, The Caribbean Region.*, G. Dengo and J.E. Case (eds.), pp. 231–260, The Geological Society of America, Boulder. Colo.

Horsfield, W.T. and M.J. Roobol, 1974, A tectonic model for the evolution of Jamaica, *Geol. Soc. Jamaica J.*, 14, 31–38.

Irving, E.F., 1975, Structural evolution of the

northernmost Andes. *U.S. Geological Survey Professional Paper*, 846, 47 pp.

Iturralde-Vinent, M., G. Hubbell, and R. Rojas, 1996, Catalogue of Cuban fossil Elasmobranchii (Paleocene to Pliocene) and paleogeographic implications of their Lower to Middle Miocene occurrence, *J. Geological Soc. Jamaica*, 31, 7–21.

Keigwin, L., 1982, Isotopic paleoceanography of the Caribbean and East Pacific: Role of Panama uplift in late Neogene time. *Science*, 217, 350–353.

Keller, G., C.E. Zenker, and S.M. Stone, 1989, Late Neogene history of the Pacific-Caribbean gateway, *J. South Am. Earth Sci.*, 2, 73–108.

Kellogg, J.N. and V. Vega, 1995, Tectonic development of Panama, Costa Rica, and the Colombian Andes: Constraints from Global Positioning System Geodetic Studies and Gravity, in Geologic and Tectonic Development of the Caribbean Plate Boundary in Southern Central America, P. Mann (ed.), pp. 75–86, Geol. Soc. America Special Paper 295.

Kennett, J.P., 1977, Cenozoic evolution of Antarctic glaciation, the circum-Antarctic Ocean, and their impact on global paleoceanography. *J. Geophys. Res.*, 82, 3843–3859.

Kolarsky, R.A., and P. Mann, 1995, Structure and neotectonics of an oblique-subduction margin, southwestern Panama, in *Geologic and Tectonic Development of the Caribbean Plate Boundary in Southern Central America*, P. Mann (ed.), pp. 131–158, Geol. Soc. America Special Paper 295.

Leroy, S., B. Mercier de Lépinay, A. Mauffret and M. Pubellier, 1996, Structural and tectonic evolution of the eastern Cayman Trough (Caribbean Sea) from seismic reflection data, *AAPG Bull.*, 80(2), 222–247.

Lewis, J.F. and G. Draper, 1990, Geology and tectonic evolution of the northern Caribbean margin, in The *Geology of North America, vol. H, The Caribbean Region.*, G. Dengo and J.E. Case (eds.), pp. 77–140, The Geological Society of America, Boulder, Colo.

Lyle, M., K. Dadey, and J.W. Farrel, 1995, The late Miocene (11-8 Ma) eastern Pacific carbonate crash: Evidence for reorganization of deep-water circulation by the closure of the Panama gateway, in *Proceedings of the Ocean Drilling Program, Scientific results, vol. 138*, N.G.

Pisias, L.A. Mayer, T.R. Janececk, A. Palmer-Julson, and T. H. van Andel (eds.), pp. 821–838, Ocean Drilling Program, College Station, Tex.

Maier-Reimer, E., U. Mikolajewicz, and T. Crowley, 1990, Ocean general circulation model sensitivity experiment with an open Central American Isthmus, *Paleoceanography*, 5, 349–366.

Mann, P., F.W. Taylor, R. E., Edwards, and T.L. Ku, 1995, Actively evolving microplate formation by oblique collision and side ways motion along strike-slip faults: An example from the northeastern Caribbean plate margin, *Tectonophysics*, 246, 1–69.

Mann, P. (ed.), 1995, *Geologic and Tectonic Development of the Caribbean Plate Boundary in Southern Central America*, Geol. Soc. of America Special Paper 295, Geological Society of America, Boulder, Colo. 349 p., 4 folding maps and figures.

Mann, P., and K. Burke, 1984, Transverse intra-arc rifting: Paleogene Wagwater Belt, Jamaica, *Marine and Petroleum Geology*, 7, 410–427.

Mann, P. and J. Corrigan, 1990, Model of late Neogene deformation in Panama, *Geology*, 18, 558–562.

Marshall, L.G., 1985, Geochronology and land-mammal biochronology of the transamerican faunal interchange, in The *Great American Biotic Interchange*, F. G. Stehli, and S. D. Webb (eds.), pp. 303–324, Plenum, New York.

McDougall, K., 1996, Benthic foraminiferal response to the emergence of the Isthmus of Panama and coincident paleoceanographic changes, *Marine Micropaleontology*, 28, 133–169.

McPhee, R. and M. Iturralde-Vinent, 1994, First Tertiary land mammal from Greater Antilles: An early Miocene Sloth (Xenarthra, Megalonychidae) from Cuba, *Amer. Mus. Novitates*, 3094, 1–13.

McPhee, R. and M. Iturralde-Vinent, 1995, Origin of the Greater Antillean land mammal fauna, 1: New Tertiary fossils from Cuba and Puerto Rico, *Amer. Mus. Novitates*, 3141, 1–31.

Mikolajewicz, U., E. Maier-Reimer, T.J. Crowley, and K.-Y. Kim, 1993, Effect of Drake and Panamanian gateways on the circulation of an ocean model, *Paleoceanography*, 8, 429–441.

Mikolajewicz, U., and T.J. Crowley, 1997,

Response of a coupled ocean energy balance model to restricted flow through the central American isthmus, *Paleoceanography*, 12, 409–426.

Miller, K.G. and R.G. Fairbanks, 1985, Oligocene to Miocene carbon isotope cycles and abyssal circulation changes, in *The Carbon Cycle and Atmospheric CO2: Natural variations Archean to Present*, E.T. Sundquist and W.S. Broecker (eds.), pp. 469–486, Geophys. Monogr. Ser., 32, AGU, Washington, DC.

Miller, K.G, J.D. Wright, and R.G. Fairbanks, 1991, Unlocking the icehouse: Oligocene–Miocene oxygen isotope, eustasy, and margin erosion, *J. Geophys. Res.*, 96, 6829–6848.

Miller, K.G., 1992, Middle Eocene to Oligocene stable isotopes, climate, and deep-water history: The terminal Eocene Event? in *Eocene-Oligocene Climatic and Biotic Evolution*, D. Prothero and W.A. Berggren (eds.), pp. 160–177, Princeton University, N.J.

Mullins, H.T., and A.C. Neumann, 1979, Geology of the Miami Terrace and its paleoceanographic implications, *Marine Geology*, 30, 205–232.

Mullins, H.T., A.C. Neumann, R.J. Wilber, A.C. Hine, and S.J. Chinburg, 1980, Carbonate sediment drifts in the northern Straits of Florida, *Am. Assoc. Petrol. Geologists Bull.*, 64, 1701–1717.

Mullins, H.T., A.F. Gardulski, S.W. Wise, and J. Applegate, 1987, Middle Miocene oceanographic event in the eastern Gulf of Mexico: Implications for seismic stratigraphic succession and Loop Current/Gulf stream circulation, *Geol. Soc. Amer. Bull.*, 98, 702–713.

Pindell, J.L., 1994, Evolution of the Gulf of Mexico and Caribbean, in *Caribbean Geology: An Introduction*, S. Donovan and T.A. Jackson (eds.), pp. 13–39, University of the West Indies Publishers Association, Kingston, Jamaica.

Pindell, J.L., and J.F. Dewey, 1982, Permo-Triassic reconstruction of western Pangea and the evolution of the Gulf of Mexico/Caribbean region, *Tectonics*, 1, 179–212.

Pindell, J.L., and S.F. Barrett, 1990, Geological evolution of the Caribbean region: A plate tectonic perspective, in The *Geology of North America, vol. H, The Caribbean Region*, G.A. Dengo and J. Case (eds), pp. 405–432. Geological Society of America, Boulder, Colo.

Pinet, B., D. Lajat, P. Le Quellec, and P. Bouysse, 1985, Structure of Aves Ridge and Grenada Basin from multichannel seismic data. in *Geodynamique des Caraïbes*, A. Mascle (ed.), pp. 53–64, Editions Technip., Paris.

Pisias, N.G., L.A. Mayer, T.R. Janecek, A. Palmer-Julson, and T.H. van Andel (eds.), 1995, *Proceedings of the Ocean Drilling Program, Scientific Results*, vol. 138, Ocean Drilling Program, College Station, Tex.

Popenoe, P., 1985, Cenozoic depositional and structural history of the North Carolina margin from seismic stratigraphic analyses, in *Geologic evolution of the United States Atlantic margin*, C.W. Poag (ed.), pp. 125–187, Van Nostrand Reinhold, New York.

Riggs, S.R., 1984, Paleoceanographic model of Neogene phosporite deposition, U.S. Atlantic continental margin, *Science*, 223, 123–131.

Robinson, E., Walton Basin and Pedro Bank bathymetric map, published for a UN meeting on "Marine resources of the Caribbean" and printed by the Survey Department of Jamaica, 1976.

Robinson, E., 1994, Jamaica. in *Caribbean Geology: An Introduction*, S.K. Donovan and T.A. Jackson (eds.), pp. 111–127, University of the West Indies, Kingston, Jamaica.

Rosencrantz, E. and P. Mann, 1991, SeaMARC II mapping of transform faults in the Cayman trough, Caribbean Sea, *Geology*, 19, 690–693.

Rosencrantz, E., M.I. Ross, and J.G. Sclater, 1988, Age and spreading history of the Cayman Trough as determined from depth, heat flow, and magnetic anomalies, *J. Geophys. Res.*, 93(B3), 2141–2157.

Schmitz, W.J., Jr. and M.S. McCartney, 1993, On the North Atlantic circulation, *Rev. Geophysics*, 31, 29–49.

Scott, T.M., 1988, The lithostratigraphy of the Hawthorn Group (Miocene) of Florida, *Florida Geological Survey Bulletin*, vol. 59, 148 pp.

Sigurdsson, H., M. Leckie, G. Acton, and Shipboard Party of ODP Leg 165, 1997. *Proc. ODP, Init. Repts., 165*, College Station, Tex. (Ocean Drilling Program).

Silver, E.A., D.L. Reed, J.E. Tagudin, and D.J. Heil, 1990. Implications of the North and South Panama thrust belts for the origin of the Panama orocline, *Tectonics*, 9, 261–281.

Snyder, S.W., A.C. Hine, and S.R. Riggs, 1990,

Carolina continental margin: Part 2 - The seismic stratigraphic record of shifting Gulf Stream flow paths in response to Miocene glacio-eustasy: Implications for phosphogenesis along the North Carolina continental margin, in *World Phosphate Deposits, Volume 3, Neogene phosphorites of the southeastern United States*, W.C. Burett and S.R. Riggs (eds.), p. 396–423, Cambridge University Press, New York.

Speed, R.C., 1984, et al., Lesser Antilles Arc and adjacent terranes, Ocean Drilling Program, Regional Atlas Series, Atlas 10, 27 plates, Marine Science International, Woods Hole, Mass.

Stephan, J.-F., B. Mercier de Lépinay, E. Calais, M. Tardy, C. Beck, J.-Ch. Carfantan, J.-L. Olivet, J.-M. Vila, Ph. Bouysse, A. Mauffret, J. Bourgois, J.-M. Thiéry, J. Tournon, R. Blanchet, and J. Dercourt, 1990, Paleogeodynamic maps of the Caribbean: 14 steps from Lias to Present. *Bull. Soc. géol. Fr.*, t. VI (no 6), 915–919.

Swart, P., G. Eberli, M. Malone, and Shipboard Party of ODP Leg 166, 1997. *Proc. ODP, Init. Repts., 165*, College Station, Tex. (Ocean Drilling Program).

Triffleman, N. J., P. Hallock, A. C Hine, and M. Peebles, 1992, Morphology, sediments, and depositional environments of a small, partially drowned carbonate platform: Serranilla Bank-southwest Caribbean Sea, *J. of Sedimentary Petrology*, 21, 39–47.

Van der Hammen, T., J. H. Werner, and H. Van Dommelen, 1973, Palynological record of the upheaval of the northern Andes: A study of the Pliocene and lower Quaternary of the Columbian eastern Cordillera and the early evolution of its high-Andean biota, *Rev. Palaeobot. Palyn.*, 16, 1–122.

Wadge, G. and K. Burke, 1983, Neogene Caribbean plate rotation and associated central American tectonic evolution, *Tectonics*, 2(6), 633–643.

Warren, B.A., 1983, Why is no deep water formed in the North Pacific? *J. Mar. Res.*, 41, 327–347.

Woodruff, F. and S. Savin, 1989, Miocene deep-water oceanography, *Paleoceanography*, 4, 87–140.

Woodruff, F. and S. Savin, 1991, Mid-Miocene isotope stratigraphy in the deep sea: High-resolution correlations, paleoclimatic cycles, and sediment preservation, *Paleoceanography*, 6, 755–806.

Wright, J.D. and K.G. Miller, 1993, Southern ocean influences on the late Eocene to Miocene deep-water circulation, in *The Antarctic paleoenvironment: A Perspective on Gobal Change, Part II*, J.P. Kennett and D.A. Warnke (eds.), pp. 1–25, *Antarctic Res. Ser.*, 60.

Wright, J. D. and K.G. Miller, 1996, Control of the North Atlantic Deep Water circulation by the Greenland-Scotland Ridge, *Paleoceanography*, 11, 157–170.

Wright, J.D., K.G. Miller, and R.G. Fairbanks, 1992, Early and middle Miocene stable isotopes: Implications for deep water circulation and climate, *Paleoceanography*, 7, 357–389.

CHAPTER 9

Role of the Greenland-Scotland Ridge in Neogene Climate Changes

James D. Wright

Cenozoic climates are marked by long-term cooling since the early Eocene that evolved into the large amplitude glacial/interglacial cycles of the late Pleistocene. Circulation models (both atmospheric and oceanic) have simulated these changes by altering the concentration of atmospheric trace gases (i.e., increased CO_2) or by changing the meridional heat transports (increased delivery of heat to the poles via oceanic circulation) (e.g., Barron, 1983, 1987; Manabe and Bryan, 1985; Rind, 1987; Covey and Barron, 1988; Covey and Thompson, 1989; Manabe et al., 1990; Rind and Chandler, 1991). There is evidence in support of both mechanisms, leaving unclear the fundamental cause of the long-term climate change during the Cenozoic. Identifying mechanisms that triggered Cenozoic cooling has focused on tectonic changes, both marine and terrestrial, that fundamentally altered the ocean-atmosphere system. Arrangements of continents and marine gateways dictate surface and deepwater circulation patterns, and therefore, how water masses communicate between the ocean basins. Opening of the Drake Passage and uplift of the Central American Isthmus are two gateways that are often associated with climate change. The Drake Passage gateway is thought to be a critical "valve" in the development of a circumpolar circulation that led to the thermal isolation of Antarctica (Kennett, 1977). The closure of the Central American Isthmus has been implicated as a potential causes for large-scale Northern Hemisphere glaciation (Keigwin, 1982) and the formation of deepwater in the North Atlantic (Keigwin, 1982; Maier-Reimer et al., 1990). The Greenland-Scotland Ridge is such a gateway, separating the cold, polar water masses in the Greenland–Norwegian Seas and Arctic Ocean from the open Atlantic. Thus, the opening and

closing of this gateway may provide a critical control on long-term climates.

Today, much of the Greenland-Scotland Ridge is less than 500 m deep, making the water depths across this marine connection sensitive to small vertical changes through either tectonic or eustatic processes. At present, surface and intermediate/deep waters are exchanged between the Nordic seas and open North Atlantic. Vogt (1986) termed the Greenland-Scotland Ridge a "watergate" that acts as a valve for the exchange of water across the ridge. Warmer waters from the North Atlantic Current flow into the Norwegian Sea, while cold polar surface waters flow to the south along the Greenland coast as part of the East Greenland Current (Worthington, 1970; Aagaard, 1982; Swift, 1984, 1986). This sets up an asymmetric circulation pattern from the southwest to the northeast. Intermediate to deep waters flow from the Norwegian-Greenland Seas over the Greenland-Scotland Ridge mainly through three passages: Denmark Straits with a sill depth of ~620 m; Iceland-Faeroe Ridge with a sill depth of ~500 m, and Faeroe Bank Channel with a sill depth between 800 and 900 m (Fig. 9.1) (Worthington, 1970). Once in the North Atlantic, these waters eventually combine with Labrador Sea Water to form North Atlantic Deep Water (NADW). This deep current in the modern ocean flows south to the Southern Ocean where its waters are distributed throughout the world's ocean (Gordon, 1981).

Changes in North Atlantic thermohaline circulation have been an integral part of many climate change hypotheses (e.g., Broecker and Denton, 1990). Indeed, there is a good correlation between deepwater circulation patterns in the North Atlantic and climate change (e.g., Boyle and Keigwin, 1987; Raymo et al., 1990, 1992; Jansen

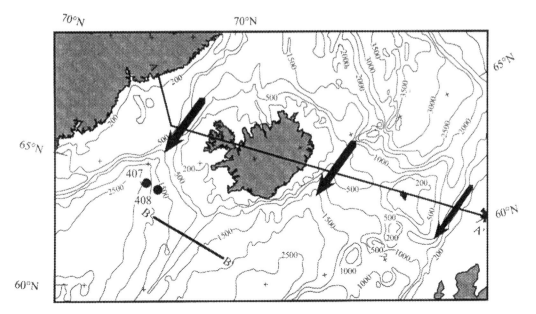

Fig. 9.1. Bathymetry of the northern North Atlantic and Greenland and Norwegian Seas surrounding Iceland (redrawn from *Tucholke et al.*, 1986). The contour interval is 500 m, except for a 200 m contour. Note the shallow depths associated with features proximal to Iceland. Profiles A to A' and B to B' mark cross sections across the Greenland-Scotland and Reykjanes Ridges, respectively shown in Figure 9.2. The large black arrows mark channels where deep water masses overflow into the North Atlantic.

and Veum, 1990; Charles and Fairbanks, 1992; Keigwin et al., 1991, 1994; Lehman and Keigwin, 1992; Oppo and Lehman, 1993; Keigwin and Jones, 1994; Keigwin and Lehman, 1994). One possible link between North Atlantic thermohaline circulation and climate is that higher latitude regions are strongly influenced by the heat released from the ocean during convection. In particular, there is a significant amount of heat released with the formation of NADW (Broecker and Denton, 1990) which is responsible for the relatively warm climates of western Europe and Scandinavia. Furthermore, the upwelling of NADW in the Southern Ocean is an important source of heat for the seasonal meltback of sea ice in the Southern Ocean (Gordon, 1981; Jacobs et al., 1985).

Long-term changes in Northern Component Water (NCW, analogous to modern NADW)[1] production may be related to sill depth changes along the Greenland-Scotland Ridge. The Greenland-Scotland

Ridge system is ~3000 km in length, but less than one half of this length is presently deeper than 200 m, and only 300 km is deeper than 500 m (Fig. 9.1). The relatively shallow depths of the Greenland-Scotland Ridge are associated with the Iceland mantle plume now centered under east central Iceland (Vogt, 1971, 1983; Vogt and Avery, 1974; Nunns, 1983). A large regional swell from the mantle plume extends radially away from Iceland for about 1000 km (Anderson et al., 1973; Vogt et al., 1981; Schilling, 1986; White and McKenzie, 1989; Sleep, 1990). This regional swell results from the injection of heat and buoyant mantle material into the asthenosphere below the crust (White and McKenzie, 1989; Sleep, 1990). The low density and thick crust of the Greenland-Scotland Ridge also contributes to the shallow depths of this gateway. Crustal thicknesses of 30 km are thought to underlie the Iceland-Faeroe Ridge (Bott, 1983). Recent evidence indicates that crustal thicknesses under Iceland are greater than 20 km (White et al., 1996). The isostatic adjustment to this thick, but low density crust (relative to the mantle below) causes the Greenland-Scotland Ridge to be a bathymetric high, sitting well above the surrounding seafloor.

---

[1] Because NADW by definition has specific water mass properties, the term Northern Component Water is used to designate the North Atlantic Deep Water mass that may have formed by similar processes in the past, but with different physical–chemical properties (Broecker and Peng, 1982).

The mantle plume and the crustal thickness and density affect the long-term depths of the Greenland-Scotland Ridge in different ways. Accordingly, there are two general models for Greenland-Scotland Ridge subsidence. One model suggests that the Greenland-Scotland Ridge behaved like normal ocean crust throughout its history and subsided at a rate proportional to the square root of the age of the crust (e.g., Thiede and Eldholm, 1983). Even though the crust is anomalously thick, thermal subsidence will continue as the newly formed material moves laterally away from the mantle plume. These simple thermal contraction subsidence models minimize any changes in the mantle plume activity, favoring long-term thermal subsidence as the dominant control on ridge depths. In contrast, others have suggested that the Greenland-Scotland Ridge has had a more variable history that was tied to changes in the mantle plume flux through time (Vogt, 1972, 1983; Shor and Poore, 1979; Wright and Miller, 1996). These studies indicate that the regional swell has varied in horizontal extent and elevation during the Neogene, and hence, has affected the exchange of water masses between the Greenland-Norwegian Seas and the northern North Atlantic. In this chapter, the evidence and implications for both the simple and more dynamic models of thermal subsidence on the Greenland-Scotland Ridge are reviewed to evaluate the role that this marine gateway has played in Neogene climate change.

Greenland-Scotland Ridge Tectonics

Plume volcanism on what is now Iceland produced the thick crust which forms the Greenland-Scotland Ridge today. Lateral movement from seafloor spreading has carried this crust away from the spreading center, producing the ridge. As product of the oceanic spreading system, the Greenland-Scotland Ridge should have a predictable thermal subsidence (Sclater et al., 1971; Berger and Winterer, 1974):

$$d = id + k\sqrt{t} - s \qquad (9.1)$$

where d is the predicted depth (m), id is the initial depth (m) at the ridge, k is 300, empirically derived for the North Atlantic by Miller et al. (1987), t is the age of the crust, and S is a correction for sedi-

ment thickness. For the Greenland-Scotland Ridge, simple thermal subsidence models indicate that the ridge reached a critical depth as early as the late early Miocene (~17 Ma) (Vogt, 1972) or during the middle Miocene (~14 Ma) (Schnitker, 1980; Thiede and Eldholm, 1983). The "critical" depth is defined as the depth that allowed both surface and deep water to exchange between the northern North Atlantic and Norwegian-Greenland Seas. Vogt (1972) and Schnitker (1980) cited paleoceanographic evidence supporting their interpretation of when subsidence of the ridge reached the critical depth.

The dynamic mantle plume hypothesis was first postulated by Vogt (1971) based on the step-like appearance of several bathymetric (basement) features along the Reykjanes Ridge (Talwani et al., 1971). Vogt identified five major features on each side of the Reykjanes Ridge, Escarpments E and A and Ridges B, C, and D, that were coherent along much of the ridge's length (Fig. 9.2). Escarpments E and A are the largest of these features and have 500–800 m of vertical relief. The observed crustal depths in this region, particularly for the features identified by Vogt (1971), are generally less than those predicted by simple thermal subsidence. The higher topography can be produced thermally through a hotter mantle and/or volumetrically through increased melting and crustal thickening. These features were attributed to fluctuations in plume activity under Iceland that propagated along the ridge axis (Vogt, 1971, 1983). The flux of mantle material along the Reykjanes Ridge is time-transgressive, creating a V-shaped signature (Vogt, 1971, 1983; White et al., 1995). Large variations in water depths indicate that the mantle plume activity under Iceland has not been constant during the Neogene. Vogt used the age versus distance relationships of the escarpments and ridges to estimate that the propagation rate of mantle material along the ridge was ~20 cm/yr. Using this rate, the mantle plume events that formed Escarpments E and A originated under Iceland at ~17 and 7 Ma, respectively.

Two studies (Johansen et al., 1984; Wright and Miller, 1996) have re-analyzed the geophysical data from cruise Vema 23-03 (Talwani et al., 1971). Wright and Miller (1996) estimated that the mantle plume event that created Escarpment E originated around 16.3 Ma, similar to Vogt's estimate. However, their age estimate for the base of Escarpment A was closer to 4 Ma, not 7 Ma as Vogt

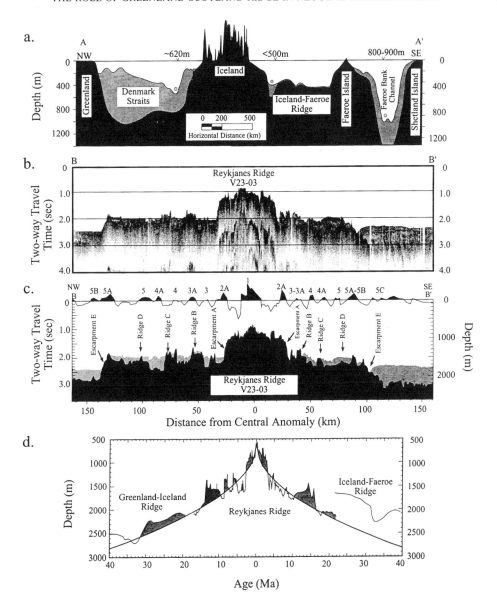

Fig. 9.2. The A-A' cross section shows the Greenland-Scotland Ridge (a, adapted from Miller and Tucholke, 1983). Sediment coverage is shaded in gray. The main conduits of water across the GSR are marked by circles: the Denmark Strait (~600 m), on the Greenland–Iceland segment of the GSR, the Iceland-Faeroe overflow where water depths are ~500 m, and the Faeroe Bank Channel between the Faeroe Islands and Shetland Island. The GSR overflow to the east and west of Iceland is roughly equal. (b) Seismic line across the Reykjanes Ridge shown in Figure 9.1 (B - B'). (c) Interpretation of V23-03 seismic transect across the Reykjanes Ridge. Vogt (1971) identified several coherent bathymetric features along the Reykjanes Ridge (Escarpments E and A; Ridges D, C, B) attributed these to changes in the Icelandic mantle plume discharge. The polarity patterns across the Reykjanes Ridge are shown above the cross section. Numbers above the polarity patterns represent anomaly numbers. (d) Crustal depths versus age for the Reykjanes and GSR were adjusted to account for the effects of sediment loading. The smooth curve radiating from 0 Ma is the predicted thermal subsidence curve which used a decay of $300\sqrt{t}$. The initial starting depth used for this section across the Reykjanes Ridge was 600 m. The initial depth used for the GSR and Iceland-Faeroe Ridge was 1100 m above sea level. The difference between the GSR/Iceland Faeroe Ridge and Reykjanes Ridge initial depths is 1700m and this value was subtracted from the GSR and Iceland Faeroe cross sections to compare to the Reykjanes Ridge cross section. Time when the crust was higher than the predicted depth are shaded in gray.

suggested, indicating a much higher propagation rate of the anomaly along the ridge axis for the younger event. Wright and Miller (1996) proposed that the plume discharge for the older events encountered a colder asthenosphere, slowing the progression of the anomaly away from the source. The discharge from each successive event encountered a hotter asthenosphere and less resistance, allowing faster propagation of the thermal or melting anomaly along the Reykjanes system. To estimate the chronologies for the plume events on Iceland, Wright and Miller (1996) applied a linear regression to the age versus distance (from Iceland) relationship for each event on each Reykjanes Ridge crossing (see Fig. 3 in Wright and Miller, 1996). The zero-distance intercept for each regression should reflect the timing of each mantle plume event under Iceland (e.g., Vogt, 1971).

Depths along the Reykjanes Ridge may provide the most complete proxy record for mantle plume variations; however, additional evidence from Iceland is consistent with variations in mantle plume fluxes. The Iceland Plateau is covered by successive and thick (3–4 km) sequences of lava flows that have accumulated during the last 15 million years. Rates of accumulation of lava sequences on Iceland are well-constrained by K-Ar dates (Saemundsson et al., 1980; McDougall et al., 1984; Saemundsson, 1986) and indicate higher accumulation rates (> 2 km/m.y.) in some lava flows. Intervals of higher rates of lava accumulation occurred between 14 and 12 Ma, 9.5 and 9.0 Ma, and 7.5 and 6.5 Ma (Fig. 9.3). These intervals correspond to intervals of higher mantle plume activity inferred from the Reykjanes Ridge. While the timing between the changes on the Reykjanes Ridge and high lava accumulation is compelling, these records of volcanism on Iceland are incomplete, making it difficult to characterize the volcanic history on Iceland from a succession of lava flows.

Regional mantle plume activity might also be recorded by the volcanic ash layers in marine sediments proximal to the Icelandic Plateau. Deep Sea Drilling Project Sites (DSDP) 407 and 408 were drilled to the south of the Iceland Plateau, while Ocean Drilling Program (ODP) Site 907 was drilled to the north of the plateau (Fig. 9.1) with each site containing substantial volcanic material (Luyendyk et al., 1979; Myhre et al., 1995). The origin of the volcanic sediments is not constrained but the over-all pattern of high ash accumulation indicates relatively high regional volcanism through much of the Miocene (Fig. 9.3). Several ash layers deposited during the middle and late Miocene appear to correspond with the development of Escarpment E and Ridges D and C; however the overall pattern of ash deposition is equivocal concerning times of increased mantle plume activity.

Deep Water Circulation

First-order deepwater circulation patterns during the Neogene have been well-established through various proxies (Jones et al., 1970; Ruddiman, 1972; Shor and Poore, 1979; Schnitker, 1980; Miller and Tucholke, 1983; Miller and Fairbanks, 1985; Vogt and Tucholke, 1989; Woodruff and Savin, 1989; Wright et al., 1991, 1992; Wold, 1994; Wright and Miller, 1996). There is general agreement concerning long-term NCW production from the late middle Miocene (12 Ma) to Present. Many studies have recognized that the most recent phase of NCW production began in the late middle Miocene and has continued with minor interruptions until the present (e.g., Shor and Poore, 1979; Schnitker, 1980; Miller and Fairbanks, 1985; Woodruff and Savin, 1989; Wright et al., 1991; 1992; Wold, 1994; Wright and Miller, 1996). The interpretation of early to early middle Miocene deepwater circulation patterns in the North Atlantic is more controversial with some studies questioning the existence of NCW prior to the middle Miocene (cf., Woodruff and Savin, 1989; Wright et al., 1992).

One of the best tools for inferring past deepwater circulation patterns is carbon isotope records from the deep ocean. In the modern ocean, there are considerable differences in dissolved inorganic carbon (DIC) $\delta^{13}C$ values, reflecting the basin-to-basin fractionation caused by deepwater circulation patterns (Kroopnick, 1985). This fractionation occurs because the $\delta^{13}C$ value and nutrient content of deep/bottom waters is a function of the time the water mass has been isolated from the surface. As water masses sink and move away from their source regions, organic matter falling from the surface ocean collects at depth. Oxidation of this organic matter releases CO_2 with low $\delta^{13}C$ values (-25‰) and high nutrients, lowering the DIC $\delta^{13}C$ value in the deep waters. Therefore, deepwater masses proximal to

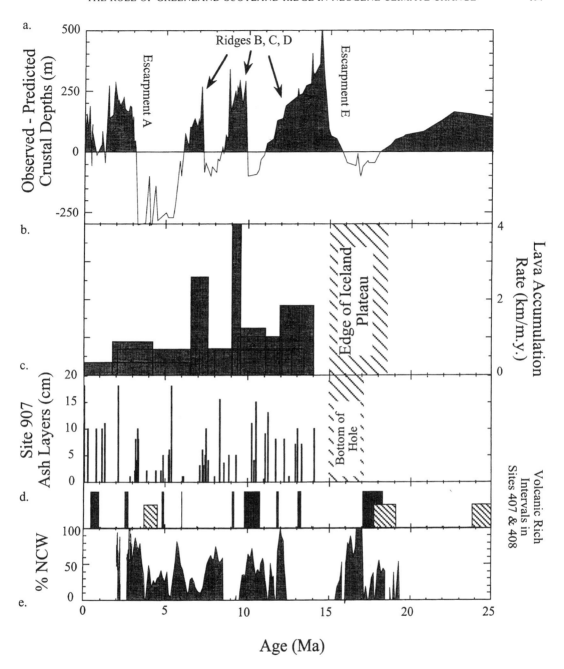

Fig. 9.3. (a) Difference curve between the observed and predicted crustal depths for the Reykjanes and GSR cross sections shown in Figure 9.2d. The chronology of the Reykjanes portion of this record has been adjusted so that Escarpments E and A and Ridges D, C, and B have the ages inferred by Wright and Miller (1996). Positive variations through time represent increased mantle plume activity under Iceland. (b) Rate of lava accumulation for several of the thick sections of radiometrically dated lava sequences on Iceland (Saemundsson et al., 1980; MacDougall et al., 1984; Saemundsson, 1986). The data are summarized in Saemundsson (1986). (c) Ages and thicknesses of ash layers found at ODP Site 907 which lies to the north of Iceland (Mehyre et al., 1995). (d) Age of intervals with volcanic components at DSDP Site 407 (diagonal pattern) and 408 (gray shading) (Luyendyk et al., 1979). (e) History of Northern Component water inferred from carbon isotope records (after, Wright and Miller, 1996).

source regions have high DIC $\delta^{13}C$ values and low nutrients, while more distal water masses have low DIC $\delta^{13}C$ values and high nutrients. Benthic foraminiferal $\delta^{13}C$ records can be used to reconstruct these patterns because certain benthic foraminifera accurately record $\delta^{13}C$ variations in the DIC reservoir (Belanger et al., 1981; Graham et al., 1981). By comparing benthic foraminiferal $\delta^{13}C$ values at various points in the deep ocean, one can determine deepwater patterns.

The late middle Miocene to Recent phase of NCW production is established by several lines of independent evidence. Interbasinal comparisons of benthic foraminiferal $\delta^{13}C$ values show that the Atlantic recorded higher values than those in the Pacific since ~12 Ma, indicating that the North Atlantic was proximal to a deepwater source (Fig. 9.4) (Miller and Fairbanks, 1985; Woodruff and Savin, 1989; Wright et al., 1991, 1992; Wright and Miller, 1996). The beginning of this large interbasinal $\delta^{13}C$ difference was associated with the development of seismic Reflector Merlin throughout the western North Atlantic region (Mountain and Tucholke, 1985). Similarly, Shor and Poore (1979) noted a prominent hiatus between 13 and 10 Ma at DSDP sites drilled to the south of Iceland and on the Rockall Plateau (Legs 48 and 49) (Fig. 9.4). Sedimentation on the Hatton and Snorri drifts increased between 13 and 12 Ma, in response to increased deepwater circulation (Miller and Tucholke, 1983; Wold, 1994). Northern Component Water fluxes have varied over the past 12 m.y. Brief interruptions during the late Miocene occurred around 9 and 7 Ma (Wright et al., 1991; Wright and Miller, 1996) and an unusually high NCW production existed during the Pliocene prior to the development of glacial-interglacial cycles (Fig. 9.4) (Shor and Poore, 1979; Raymo et al., 1992; Wright and Miller, 1996).

The controversy over the early to early middle Miocene NCW history (24–15 Ma) may be related more to the region of emphasis in each study. Woodruff and Savin (1989) concentrated their efforts on reconstructing the Pacific and Indian Ocean circulation patterns, while Wright et al., (1991, 1992) focused much of their attention in the Atlantic basins. Accordingly, Woodruff and Savin (1989) demonstrated the existence of an intermediate to upper deepwater mass that appears to have originated in the northern Indian Ocean, but found

little evidence for NCW production prior to 12 Ma. Those authors postulated that the Indian Ocean water mass was warm and very saline, transporting heat to the high southern latitudes during the early and middle Miocene. Conversely, Wright et al., (1991, 1992) focused on deepwater changes in the North Atlantic during the Neogene. Except for a brief interval of NCW production during the earliest Oligocene (Miller, 1992), the first major phase of NCW production began in the early Miocene around 20 Ma (Wright et al., 1992). This interval lasted for ~4 m.y., reaching peak production around 17 Ma (Fig. 9.4). Seismic Reflector R2 has been traced to the Greenland-Scotland Ridge (Miller and Tucholke, 1983) and was attributed to this phase of NCW (Miller and Fairbanks, 1985). Age estimates for Reflector R2 vary, but fall in the interval between 20 and 16 Ma (Miller and Tucholke, 1983; Mountain and Tucholke, 1985; Bauldauf, 1987). Sites 406 and 407 also recorded a hiatus between 18 and 16 Ma that corresponds to this interval of NCW production (Fig. 9.4) (Shor and Poore, 1979). An important phase of drift accumulation in the North Atlantic began at this time (Jones et al., 1970; Ruddiman, 1972; Wold, 1994). Vogt (1972) cited this sedimentary evidence as support for his hypothesis that a critical depth along the Greenland-Scotland Ridge was reached during the early Miocene.

The phase of long-term NCW production that began during the late middle Miocene is consistent with both the simple thermal subsidence and dynamic mantle plume models (e.g., Schnitker, 1980; Wright and Miller, 1996). Both models predict that the Greenland-Scotland Ridge would subside to a critical depth to allow NCW to flow into the northern North Atlantic during the late middle Miocene. In contrast, the early to early middle Miocene phase of NCW production requires that the "critical" depths along the sill were reached several million years before the timing predicted by simple thermal subsidence models or that sill depths along the Greenland-Scotland Ridge have been more variable.

Wright and Miller (1996) argued that the correlation of inferred variations on the Greenland-Scotland Ridge with NCW production during the Neogene implied that sill depths on the ridge are an important control for deepwater circulation patterns. However, sediments recovered from the Norwegian-Greenland Seas on DSDP Leg 38 and

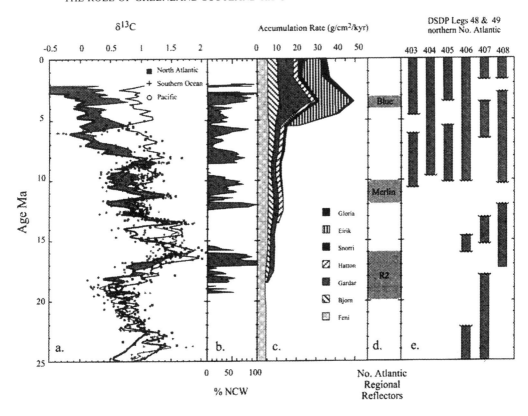

Fig. 9.4. (a) Composite $\delta^{13}C$ curves of the North Atlantic, Southern, and Pacific Oceans for the Neogene (after Wright and Miller, 1996). The shaded area represents times when the Southern Ocean $\delta^{13}C$ value was higher relative to the Pacific $\delta^{13}C$ value, indicating NCW production. (b) The record of Northern Component Water flux over the past 25 Ma estimated by recording the benthic foraminiferal $\delta^{13}C$ changes in the Southern Ocean relative to change in the North Atlantic and Pacific records. (c) Summary of the accumulation of drift sediments in the North Atlantic taken from Wold (1994). (d) Reflectors R2, Merlin, and Blue are three major seismic reflectors in the North Atlantic and their age ranges are shown on the side of the graph (Mountain and Tucholke, 1985). (e) Summary of sedimentation at sites drilled on DSDP Legs 47 and 48 (Shor and Poore, 1979). Note the major hiatuses in the early an middle Miocene which correspond to the major phases of NCW production.

ODP Legs 151 and 152 are equivocal about deep ventilation in this region prior to 10 Ma (Talwani et al., 1976; Larsen et al., 1994; Thiede and Myhre, 1996). Another consideration is that Labrador Sea Water is an important component in modern NADW that forms south of the ridge; thus the formation of Labrador Sea Water may be invoked to explain this discrepancy. However, the distribution of sediment drifts and seismic disconformities implicates the Norwegian and Greenland Seas as a source of NCW during the Miocene (Ruddiman, 1972; Roberts, 1975; Shor and Poore, 1979; Miller and Tucholke, 1983; Vogt and Tucholke, 1989).

Another potential control on the Greenland-Scotland Ridge overflow is glacio-eustatic sea level

changes which would effectively raise or lower sill depths along the ridge relative to sea level. Thiede and Eldholm (1983) noted that sea level changes could influence the exchange of water across the ridge particularly when sill depths along the ridge were not as deep. Haq et al. (1987) estimated that Miocene sea level events ranged between 50 and 150 m. More recent estimates suggest that the maximum sea level lowerings were no larger than 100 m with most Neogene sea level events ranging between 30 and 80 m below the present level (Greenlee and Moore, 1988; Miller et al., 1996). Wright et al. (1992) and Wright and Miller (1996) noted that NCW production during the early and middle Miocene was not constant and that flux changes could be related to sea level changes affecting the overflow across the Greenland-Scotland Ridge.

Tectonic uplift/subsidence along the Greenland-Scotland Ridge and glacio-eustatic sea level changes focus on changes at the gateway that may isolate the probable source region for NCW (Greenland-Norwegian Seas) from the open North Atlantic. However, another potential influence on NCW production relates to the pre-conditioning of surface waters that sink to form NCW. High salinity is necessary for the formation of NCW (Reid, 1979). The surface and thermocline waters that flow into the Nordic Seas acquire high salinities as they flow through tropical and subtropical Indian and Atlantic Oceans. This inter-ocean exchange of surface and thermocline waters is part of a global system that circulates surface and deep water among all of the ocean basins (e.g., Gordon, 1986; Broecker and Denton, 1990). A critical gateway in this circulation system may be the Central American Isthmus. Results from ocean modeling experiments indicate that with an open isthmus, low salinity water from the Pacific enters the North Atlantic, making it difficult for NCW to form (Maier-Reimer et al., 1990; Mikolajewicz et al., 1993). This author recognizes the importance of this aspect, but assumes for this discussion that the surface waters that flowed into the Greenland-Norwegian Seas were sufficiently pre-conditioned to sink with heat loss to the atmosphere and form a deepwater mass.

Deepwater and Climate Change

The most recent glacial to interglacial transition provides the best illustration for how changes in deepwater circulation patterns in the North Atlantic may have affected climate. At present, the NADW production is high and temperatures in the circum-North Atlantic (both marine and terrestrial) are relatively warm. In contrast, NCW production was much lower during the most recent glacial maximum at ~20 to 18 ka (e.g., Oppo and Fairbanks, 1987; among many) and sea surface temperature (SST) estimates indicate 6 to 10°C cooler waters accompanied by large ice sheets on the adjacent continents (CLIMAP, 1981). The relationship of high NCW during interglacials and low NCW during glacials has been extended into the Pliocene (Oppo et al., 1990; Raymo et al., 1990, 1992). The argument

that climate responds to deepwater circulation changes is compelling; however, whether NCW is a primary cause for climate change or just a consequence of climate change is still in doubt. For example, Broecker and Denton (1990) argue that the late Pleistocene climate cycles are a result of thermohaline reorganizations. In contrast, Raymo et al. (1992) attribute a decrease in NCW production during the late Pliocene to the increase in Northern Hemisphere glaciation.

Neogene Climates

Neogene climates are most often and best characterized by benthic foraminiferal $\delta^{18}O$ records (e.g., Savin et al., 1975; Shackleton and Kennett, 1975; Miller et al., 1987). The most prominent feature is the middle Miocene $\delta^{18}O$ increase of > 1 ‰ that is recorded throughout the intermediate and deep oceans between 15 and 12.8 Ma (Fig. 9.5). This increase has been attributed to the development of a permanent ice sheet on Antarctica (Shackleton and Kennett, 1975) with accompanying deepwater cooling (Savin et al., 1975; Shackleton and Kennett, 1975; Miller et al., 1987; Wright et al., 1992). The timing of the permanent ice cap on Antarctica is in question as there is extensive evidence for Antarctic glaciation during the late Eocene and Oligocene (Barron et al., 1991; Miller et al., 1991; Schlich et al., 1992; Zachos et al., 1993). However, it remains unclear whether these Paleogene ice sheets were permanent or intermittent.

Northern Component Water has been implicated as a possible cause for the middle Miocene $\delta^{18}O$ increase. Schnitker (1980) related the middle Miocene $\delta^{18}O$ increase to the initiation of NCW formation during the middle Miocene. He proposed that the upwelling of warm NCW around Antarctica increased the moisture flux to Antarctica, causing large ice sheets to grow. Wright and Miller (1996) noted the coincidence in timing of the NCW shutdown between 16 and 15 Ma and the beginning of the middle Miocene $\delta^{18}O$ increase at 15 Ma. The timing of these deepwater and climate changes are well-constrained with paleomagnetic data (Wright et al., 1992). However, they concluded that the delay between NCW shutdown (16 to 15 Ma) and culmination of the climate change (12.8 Ma) was too long for the reduction in NCW to have been the primary cause. Wright and Miller (1996) suggested that the

Atlantic Composite
(*Cibicidoides* spp.)

$\delta^{18}O$

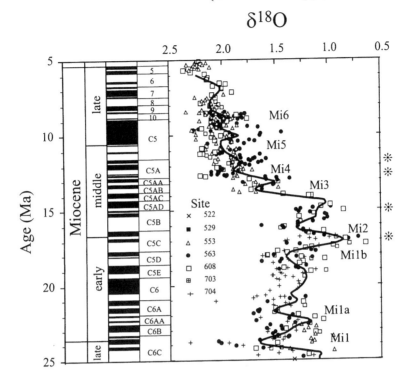

Fig. 9.5. Benthic foraminiferal $\delta^{18}O$ curve for the Neogene modified from Wright and Miller (1993). The solid line was generated by interpolating the data to constant intervals of 0.5 million years and smoothing with an 11 point gaussian filter. The "Mi" glacial events identified by Miller et al. (1991) are shown. The middle Miocene $\delta^{18}O$ increase began around 15.0 Ma and culminated with a $\delta^{18}O$ maximum at 12.8 Ma. The Berggren et al. (1995) time-scale was used. Asterisks denote times used for equator to pole reconstructions.

NCW production warmed the high-latitude climates, but its shutdown was not primarily responsible for the long-term climate changes. Other hypotheses to explain the middle Miocene $\delta^{18}O$ increase include: the Monterey hypothesis of carbon sequestering in isolated basins removing CO_2 from the atmosphere (Vincent et al., 1985); and continent–continent collisions and arc-continent collisions disrupting the carbon cycle which also led to the removal of CO_2 from the atmosphere [e.g., Himalayas (Raymo and Ruddiman, 1992; Raymo, 1994), Indonesia-New Guinea (Reusch and Maasch, this volume, Chapter 13)]. One problem with these carbon cycle models may be the delay in the removal of carbon as indicated by the increase oceanic $\delta^{13}C$ values from 20 to 16 Ma and the beginning of the middle Miocene $\delta^{18}O$ increase around 15 Ma (Hodell and Woodruff, 1994).

Surface Water Reconstructions

One potential measure of the effects of deepwater circulation change on climate during the Neogene may lie in the SST gradients measured at times of contrasting deepwater circulation patterns. Gradients based on planktonic foraminiferal $\delta^{18}O$ values provide the best information regarding paleo-SST gradients (Shackleton and Boersma, 1981; Keller, 1985; Savin et al., 1985; Zachos et al., 1994; among others). However, temperature is not the sole influence on the modern $\delta^{18}O_{calcite}$ gradient. All paleotemperature equations include the $\delta^{18}O_{water}$ term (e.g., Epstein et al., 1953).

$$T = 16.5 - 4.3 \, (\delta^{18}O_{calcite} - \delta^{18}O_{water})$$
$$+ 0.14 \, (\delta^{18}O_{calcite} - \delta^{18}O_{water})^2 \qquad (9.2)$$

where T and $\delta^{18}O_{water}$ are the temperature (°C) and oxygen isotope value of the water in which the organism lived, and $\delta^{18}O_{calcite}$ is the oxygen isotope value measured in the calcite. From this equation, there is a one-to-one relationship between changes in $\delta^{18}O_{calcite}$ and $\delta^{18}O_{water}$ values. In contrast, there is an inverse relationship between $\delta^{18}O_{calcite}$ and temperature such that for every 1°C temperature increase there is a 0.23‰ decrease in the measured $\delta^{18}O_{calcite}$ value.

The key to using planktonic foraminiferal $\delta^{18}O$ values as indicators of past climates is understanding the hydrographic parameters which produce the modern $\delta^{18}O_{calcite}$ gradient. In the modern ocean, the predicted meridional $\delta^{18}O_{calcite}$ gradient (equator to pole) is greater than 5.0 ‰ and is dominated by temperature changes (~28°C). Therefore, it is tempting to ascribe planktonic foraminiferal $\delta^{18}O$ gradients to temperature differences alone. The importance of the $\delta^{18}O_{water}$ term is highlighted when we consider that surface water $\delta^{18}O_{water}$ values vary by ~1.5 ‰ (equivalent to a 6–7°C temperature effect) with high values in the tropics and subtropics and the lowest values in the polar regions. This variation is due to the general transport of water vapor from low (evaporation) to high latitudes (precipitation). During this process, the $\delta^{18}O$ value of the water vapor decreases through a distillation process (Craig and Gordon, 1965). Hence, the precipitation in the polar/subpolar regions has lower $\delta^{18}O$ values than in the tropics/subtropics. These differences affect the oceanic environment because the $\delta^{18}O$ values at various points in the ocean represent mixing between the more saline, high $\delta^{18}O$ waters in the evaporative regions of the oceans (subtropics) and the fresh, low $\delta^{18}O$ riverine water (Craig and Gordon, 1965). As a result, planktonic foraminiferal $\delta^{18}O$ gradients must be evaluated as a gradient which incorporates both temperature and $\delta^{18}O_{water}$ values (e.g., Savin et al., 1985).

Zachos et al. (1994) minimized the errors associated with not knowing the $\delta^{18}O_{water}$ term for the early Cenozoic oceans by using the modern $\delta^{18}O_{water}$ gradient. However, the $\delta^{18}O_{water}$ variations in the ocean are linked to climate and certainly differed in the past. First-order comparisons must be made based on $\delta^{18}O$ values alone. Any subsequent interpretation requires a firm knowledge of either the temperature or $\delta^{18}O_{water}$ term. In this chapter, $\delta^{18}O$ measurements of coretop planktonic foraminifera, sampled along a meridional transect in the North Atlantic, are compared with similar transects representing four Miocene time slices. This method is similar to that of Savin et al. (1985) and provides first-order approximation for surface hydrographic changes.

Planktonic Foraminiferal Equator to Pole Transects in the North Atlantic

Stable isotopic values were determined for the most abundant planktonic foraminiferal taxa at four stratigraphic levels in seven DSDP sites. These sites formed a south to north transect near the Mid-Atlantic Ridge that spans 5–63°N in latitude (Table 9.1). These equator-to-pole transects in the North Atlantic were selected to characterize the surface water hydrographic changes that may have resulted from changes in NCW production. To ensure time equivalence, each transect selected was associated with a distinct benthic foraminiferal $\delta^{18}O$ event, which are assumed to be synchronous (Miller et al., 1991). The Berggren et al. (1995) time-scale is used to determine the age of the surface water $\delta^{18}O$ transects. All analyses for the Miocene transects were run in the Stable Isotope Laboratory at the University of Maine. Precision on the NBS-20 standards during the analyses was 0.06 and 0.05‰ for $\delta^{18}O$ and $\delta^{13}C$, respectively.

The presence of high-frequency climate cycles (40 kyr) during the middle Miocene (Pisias et al., 1985; Flower and Kennett, 1995) can potentially introduce error into the comparisons used in this study because the downcore records, from which the time slices were chosen, were under-sampled. The sample resolution used in this study varies between 20 and 50 kyrs depending on the site. Unlike the late Pleistocene when glacial/interglacial cycles record an amplitude of ~2.0 ‰, the middle Miocene cycles are approximately one third to one-quarter of this amplitude (0.5–0.6‰), minimizing the magnitude of potential errors associated with low sample resolution.

Middle Miocene $\delta^{18}O$ Transects The potential for NCW to influence middle Miocene climates is evaluated by comparing the meridional $\delta^{18}O$ gradients reconstructed from planktonic foraminifera for time intervals slices which represent times when NCW

Table 9.1. Oxygen Isotope Values Used in the Composite $\delta^{18}O$ (Transects Shown in Fig. 9.6).

Site	Latitude (°N)	Species	Core	Section	Interval	$\delta^{18}O$
		17 Ma Time Slice				
667A	5	D. altispira	21	4	40	-1.14
667A	5	G. ruber	21	4	40	-1.51
667A	5	G. sacculifer	21	4	40	-1.34
366	6	D. altispira	20	2	130	-1.38
563	34	D. altispira	11	3	19	-0.75
408	63	G. praebulloides	34	6	75	0.52
		15 Ma Time Slice				
667A	5	D. altispira	19	5	36	-1.33
667A	5	G. sacculifer	19	5	36	-1.23
667A	5	G. sacculifer	19	5	36	-1.44
563	5	D. altispira	9	4	25	-0.88
563	5	D. altispira	9	4	25	-0.52
608	43	G. mayeri	34	4	105	-0.49
608	43	D. altispira	34	4	105	-0.01
408	63	G. praebulloides	32	2	142	0.57
		12.8 Ma Time Slice				
667A	5	D. altispira	18	2	36	-0.61
667A	5	G. mayeri	18	2	36	-0.54
667A	5	G. ruber	18	2	36	-0.67
667A	5	G. sacculifer	18	2	36	-0.68
366A	6	D. altispira	15	6	116	-0.58
563	34	D. altispira	8	1	34	0.68
608	43	G. bulloides	31	1	85	1.03
608	43	D. altispira	31	1	85	0.67
553	56	G. bulloides	7	6	80	0.98
553A	56	G. bulloides	7	6	80	0.69
553A	56	G. mayeri	7	6	80	0.99
553A	56	G. mayeri	7	6	80	0.69
553A	56	G. peripheroronda	7	6	80	0.21
408	63	G. bulloides	27	5	86	1.89
		11.8 Ma Time Slice				
366A	6	D. altispira	15	3	122	-0.39
563	34	D. altispira	5	4	110	0.26
563	34	G. nepenthes	5	4	110	0.55
563	34	G. nepenthes	5	4	110	0.60
608	43	D. altispira	27	7	11	1.04
608	43	D. altispira	27	7	11	1.10
608	43	G. bulloides	27	7	11	0.96
608	43	G. bulloides	27	7	11	1.06
608	43	G. mayeri	27	7	11	0.26
608	43	G. mayeri	27	7	11	0.12
407	63	G. bulloides	17	3	84	1.23
408	63	G. bulloides	24	6	0	1.55

was "on" and "off." Two comparisons were made, representing of NCW change before and after the middle Miocene $\delta^{18}O$ increase. Time slices for both comparisons were selected based on the similarity of benthic foraminiferal $\delta^{18}O$ values. This criterion minimizes the possibility that other factors may overprint the effects of NCW changes.

Transects for time slices at 17 Ma and 15 Ma were constructed to represent surface conditions prior to the middle Miocene $\delta^{18}O$ increase and were chosen as follows: the $\delta^{18}O$ minima before and after the Mi2 event, respectively (Miller et al., 1991). The absolute values for the 17 Ma and 15 Ma time slices are similar and are interpreted to reflect otherwise similar climatic conditions (Fig. 9.5) (Wright et al., 1992). The key difference between the two time slices is that NCW production decreased from peak production at ~17 Ma to undetectable levels by 15 Ma (Wright et al., 1992). Although based on only a few points, the $\delta^{18}O$ transects for the 17 Ma and 15 Ma time slices are similar. This indicates that changes in the flux of NCW had little effect on the $\delta^{18}O$ transect during the early middle Miocene, and hence, North Atlantic surface water hydrography (Fig. 9.6). If SST changes did occur, they would have been offset exactly by changes in $\delta^{18}O_{water}$ values, a less plausible scenario.

A similar comparison using the 12.8 Ma and 11.8 Ma transects is made to examine surface water changes associated with the re-initiation of NCW production during the late middle Miocene. Transects for 12.8 Ma and 11.8 Ma time slices postdate the middle Miocene $\delta^{18}O$ increase and represent times when NCW was beginning to re-establish (12.8 Ma) and when there was a significant flux (11.8 Ma) (Woodruff and Savin, 1989; Wright et al., 1992). These two time slices correspond to the Mi4 and Mi5 events of Miller et al., (1991). Benthic foraminiferal $\delta^{18}O$ values at 12.8 and 11.8 Ma were similar to the modern value indicating that deepwater temperatures were close to modern values during the late middle Miocene (Savin et al., 1975; Shackleton and Kennett, 1975; Miller et al., 1987). As in the case before the middle Miocene $\delta^{18}O$ increase, there is little difference between the 12.8 Ma and 11.8 Ma profiles, indicating no substantial change in surface water conditions associated with the initiation of and increase in NCW production during the late middle Miocene (Fig. 9.6).

Discussion

The two most prominent climate changes during the Neogene are the middle Miocene $\delta^{18}O$ increase and the development of large-scale North Hemisphere glaciation during the late Pliocene. Each climate event corresponded to a large-scale change in NCW production (Wright et al., 1992; Wright and Miller, 1996). Sediments from ODP Leg 151 show evidence for small-scale Northern Hemisphere glaciation at 14 and 13 Ma (Wolf-Welling et al., 1996), which corresponds to the Mi3 and Mi4 events recorded in the $\delta^{18}O$ record (Miller et al., 1991; Wright et al., 1992). It remains to be shown whether the initiation of the middle Miocene $\delta^{18}O$ increase at 15 Ma corresponded to a phase of Northern Hemisphere glaciation. If so, then a direct cause and effect can be made between the NCW decreases and the development of Northern Hemisphere ice sheets. A similar climate–deepwater relationship is observed for the Pliocene. Warm climates occurred when NCW fluxes were high during the early Pliocene (Raymo et al., 1992; Wright and Miller, 1996). By the late Pliocene (3.2 Ma), NCW production decreased and large-scale Northern Hemisphere glaciation commenced. In each case, it may be argued that reduced delivery of heat to the high northern latitudes precipitated the development of Northern Hemisphere ice sheets.

The coincidence between Northern Hemisphere glaciation and NCW decreases is intriguing; however, the timing and magnitude of change in this interval indicates that other changes in the ocean–atmosphere system must have accompanied the NCW changes to account for the long-term climate coolings. For example, the decrease in NCW flux occurred between 16 and 15 Ma and yet the $\delta^{18}O$ increase ended at 12.8 Ma, requiring too long a delay to account for the climate change (Wright and Miller, 1996). This conclusion is further supported by the similarity of the $\delta^{18}O$ transects at 17 Ma and 15 Ma and those at 12.8 Ma and 11.8 Ma; both compare surface water conditions at times of contrasting fluxes of NCW (Fig. 9.6). Despite large-scale deepwater changes, the surface waters recorded no change in the measured $\delta^{18}O$ gradients for either of the comparisons (Fig. 9.6). Furthermore, high latitude climates did not warm substantially after NCW resumed at 12 Ma, arguing against a primary role for NCW and the Greenland-Scotland

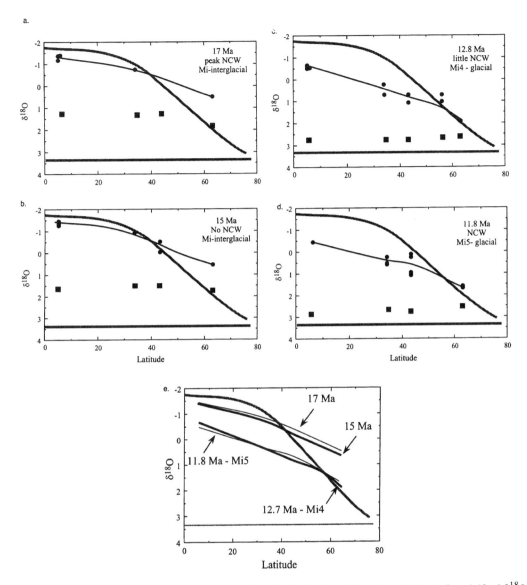

Fig. 9.6. Equator to pole surface water reconstructions from the North Atlantic based on planktonic foraminiferal $\delta^{18}O$ values for four time slices in the Miocene: (a) 17 Ma; (b) 15 Ma; (c) 12.8 Ma; (d) 11.8 Ma. A modern $\delta^{18}O$ transect from the North Atlantic (Wright and Fairbanks, unpublished data) and equilibrium benthic values (3.3 ‰) (Oppo and Fairbanks, 1987) are shown for reference as thick lines on each of the Miocene transects. (e) Comparison of the 17 Ma and 15 Ma $\delta^{18}O$ transects and 12.8 Ma and 11.8 Ma $\delta^{18}O$ transects which represent surface water conditions associated with the shutdown of NCW between 17 and 15 Ma and re-initiation of NCW between 12.8 and 11.8 Ma. Note that the 17 Ma and 15 Ma profiles and 12.8 Ma and 11.8 Ma profiles are similar.

Ridge in the long-term climate changes. A similar argument can be made for the initiation of large-scale glaciation in the Northern Hemisphere. If NCW was the primary control on the initiation of continental glaciation in the Northern Hemisphere, the absence of NCW during part of the middle

Miocene should have led to large continental ice sheets. It was during the late Pliocene and not during the middle Miocene shutdown of NCW that large-scale continental glaciation developed.

The evidence just presented indicates that the Greenland-Scotland Ridge appears to have been the primary control on deepwater circulation changes and may have been important in the development of Northern Hemisphere ice sheets (Vogt and Tucholke, 1989; Wright and Miller, 1996); however, this study did not consider the higher-frequency changes. Milankovitch cycles (40 kyr) have now been identified in much of the Miocene (Pisias et al., 1985; Flower and Kennett, 1995; Benson et al., 1995; Shackleton and Hall, in press; Flower et al., in press). The relationship between climate and NCW fluxes on orbital time scales has been traced into the Pliocene (Raymo et al., 1990, 1992). In particular, deep-sea $\delta^{18}O$ records have 40 kyr cycles during the interval between 15 and 12 Ma (Pisias et al., 1985) when the NCW flux was undetectable (Wright et al., 1992; Wright and Miller, 1996), indicating that high frequency climatic cycles do not require NCW production.

If the middle Miocene is an appropriate analog for Plio-Pleistocene climates, the higher-frequency NCW changes appear to be a consequence of climate cycles, although positive feedbacks associated with NCW production may have contributed to the climate warming. The large-scale cooling in the North Atlantic during the last glaciation that has been attributed to changes in the deepwater circulation, therefore, requires an alternative hypothesis. One explanation could be that the cold surface North Atlantic developed from the interaction between cold surface winds coming off the Laurentide Ice Sheet and the calving and transport of icebergs into the North Atlantic throughout this interval (e.g., Ruddiman and McIntyre, 1981).

Conclusions

There is ample evidence to suggest that the subsidence of the Greenland-Scotland Ridge has not followed the simple thermal subsidence patterns associated with normal oceanic crust. This is most likely related to changes in the flux of mantle plume material beneath Iceland. Consequently, the radial swell associated with the plume changes affected the sill depths along the Greenland-Scotland Ridge. The first-order NCW fluxes correspond well with these inferred changes along the ridge, implicating the mantle plume activity as the primary control on NCW fluxes during the Neogene. Coincidentally, the gross timing of changes in NCW flux and climate change appeared to make deepwater circulation

changes a viable mechanism to explain the middle Miocene $\delta^{18}O$ increase. However, changes along the Greenland-Scotland Ridge alone cannot explain the long-term trend towards a colder climate during the Neogene. Climate reconstructions show that NCW may have initiated and amplified the climate changes, but other changes in the ocean–atmosphere system are required to produce the middle Miocene $\delta^{18}O$ increase and the development of large-scale Northern Hemisphere glaciation during the Pliocene.

Acknowledgments

I thank K. Burke and T. Crowley for inviting this contribution and for their reviews. K. Miller, P. Vogt, and two anonymous reviewers provided constructive reviews which greatly improved the manuscript. I thank K.G. Miller and R.G. Fairbanks for their useful discussions regarding the relationship between deepwater circulation and climate changes and the interpretations of planktonic foraminiferal $\delta^{18}O$ values. J. Friez and D. Introne were invaluable for stable isotope analysis for the Miocene transects. This research was supported by OCE92-04143.

References

Aagaard, K., 1982, Inflow from the Atlantic Ocean to the Polar basin, in *The Arctic Ocean,* L. Rey (ed.), pp. 69–81, Wiley, New York.

Anderson, R.N., D.P. McKenzie, and J.G. Sclater, 1973, Gravity, bathymetry, and convection in the Earth, *Earth Planet. Sci. Lett., 18,* 391–407.

Barron, E.J., 1983, A warm, equable Cretaceous: The nature of the problem, *Earth Sci. Rev., 19,* 305–338.

Barron, E.J., 1987, Eocene equator-to-pole surface ocean temperatures: A significant climate problem? *Paleoceanography, 2,* 729–739.

Barron, J.A., et al., 1991, in *Proceedings of the Ocean Drilling Program, Scientific Results, vol. 119,* Ocean Drilling Program, College Station, Tex.

Bauldauf, J.G., 1987, Biostratigraphic and paleoceanographic interpretation of lower and middle Miocene sediments, Rockall Plateau region, North Atlantic Ocean, in *Initial Reports of the Deep Sea Drilling Project, vol. 94,* W.F.

Ruddiman, R.B. Kidd, E. Thomas et al. (eds.), pp. 1033–1043, U.S. Govt. Printing Off., Washington, D.C.

Belanger, P. E., W. B. Curry, and R. K. Matthews, 1981, Core-top evaluation of benthic foraminiferal isotopic ratios for paleo-oceanographic interpretations, *Paleogeogr. Palaeoclimatol. Palaeoecol., 33,* 205–220.

Benson, R.H., L.-A. Hayek, D.A. Hodell, and K. Rakic-El Bied, 1995, Extending the climatic precession curve back into the late Miocene by signature template comparison, *Paleoceanography, 10,* 5–20.

Berger, W.H., and E.L. Winterer, 1974, Plate stratigraphy and the fluctuation carbonate line, in Pelagic Sediments on Land Under the Sea, K.J. Hsu and H. C. Jenkins (eds.), *Spec. Publ. Int. Assoc. Sedimentol., 1,* 11–48.

Berggren, W.A., D.V. Kent, C.C. Swisher, III, and M.P. Aubry, 1995, A revised Cenozoic geochronology and chronostratigraphy, in *Geochronology, Time Scales, and Global Stratigraphic Correlatios: A Unified Temporal Framework for A Historical Geology,* W. A. Berggern, D.V. Kent, and J. Hardenbol (eds.), *Soc. Econ. Paleon. Mineral., Spec. Pub. No. 54,* 129–212.

Bott, M.H.P., 1983, Deep Structure and geodynamics of the Greenland-Scotland Ridge: An introductory review, in *Structure and Development of the Greenland-Scotland Ridge,* M.H.P. Bott, S. Saxov, M. Talwani, and J. Thiede (eds.), pp. 3–9, Plenum Press, New York.

Boyle, E.A., and L.D. Keigwin, Jr., 1987, North Atlantic thermohaline circulation during the past 20,000 years linked to high-latitude surface temperature, *Nature, 330,* 35–40.

Broecker, W.S., and T.H. Peng, 1982, *Tracers in the Sea,* 690 pp., ELDIGIO Press, New York.

Broecker, W.S., and G.H. Denton, 1990, The role of ocean-atmosphere reorganizations in glacial cycles, *Quaternary Science Reviews, 9,* 305–341.

Charles, C.D., and R.G. Fairbanks, 1992, Evidence from Southern Ocean sediments for the effect of North Atlantic deep-water flux on climate, *Nature, 355,* 416–419.

CLIMAP Project Members, and A. McIntyre, 1981, Seasonal reconstructions of the Earth's surface at the last glacial maximum, *Geol. Soc. America, Map and Chart Series, MC–36* (17 maps, 18 pp.

text), Geological Society of America, Boulder, Colo.

Covey, C., and E. Barron, 1988, The role of ocean heat transport in climatic change, *Earth Sci. Rev., 24,* 429–445.

Covey, C., and S.L. Thompson, 1989, Testing the effects of ocean heat transport on climate, *Global and Planetary Change, 1(4),* 331–341.

Craig, H., and L.I. Gordon, 1965, Deuterium and oxygen-18 variations in the oceans and marine atmosphere, in *Stable Isotopes in Oceanographic Studies and Paleotemperatures, Spoleto,* E. Tongiorgi (ed.), pp., 1–122, Consiglio Nazionale delle Ricerche, Laboratorio di Geologica Nucleare, Pisa, Italy.

Epstein, S., R. Buchsbaum, H. Lowenstam, H.C. Urey, 1953, Revised Carbonate-water temperature scale, *Bull. Geol. Soc. Am., 64,* 1315–1326.

Flower, B.P., and J.P. Kennett, 1995, Middle Miocene deepwater paleoceanography in the southwest Pacific: Relations with East Antarctic Ice Sheet development, *Paleoceanography, 10,* 1095–1112.

Flower, B.P., J.C. Zachos, and H. Paul, in press, Milankovitch-scal climate variability recorded near the Oligocene/Miocene boundary, in *Proceedings of the Ocean Drilling Program, Scientific Results, vol. 154,* N.J. Shackleton, W.B. Curry, and C. Richter (eds.), Ocean Drilling Program, College Station, Tex.

Gordon, A.L., 1981, Seasonality of Southern Ocean sea ice, *J. Geophys. Res., 86,* 4193–4197.

Gordon, A.L., 1986, Interocean exchange of thermocline water, *J. Geophys. Res., 91,* 5037–5046.

Graham, D. W., B. H. Corliss, M. L. Bender, and L. D. Keigwin, 1981, Carbon and oxygen isotopic disequilibria of Recent benthic foraminifera, *Mar. Micropaleontol., 6,* 483–497.

Greenlee, S.M., and T.C. Moore, 1988, Recognition and interpretation of depositional sequences and calculation of sea-level changes from stratigraphic data-offshore New Jersey and Alabama Tertiary, in Sea-Level Changes—An Integrated Approach, *SEPM Special Publication, 42,* 329–353.

Haq, B.U., J. Hardenbol, and P.R. Vail, 1987, Chronology of fluctuating sea levels since the Triassic, *Science, 235,* 1136–1167.

Hodell, D.A., and F. Woodruff, 1994, Variations in the strontium isotopic ratio of seawater during the Miocene: Stratigraphic and geochemical implications, *Paleoceanography, 9,* 405–426.

Jacobs, S.S., R.G. Fairbanks, and Y. Horibe, 1985, Origin and evolution of water masses near the Antarctic continental margin: Evidence from $H_2^{18}O/H_2^{16}O$ ratios in seawater, in Oceanology of the Antarctic Continental Shelf, S.S. Jacobs (ed.), pp. 59–85, *Antart. Res. Ser.,* vol. 43, AGU, Washington, D.C.

Jansen, E., and T. Veum, 1990, Evidence for two-step deglaciation and its impact on North Atlantic deep-water circulation, *Nature, 343,* 612–616.

Johansen, B., P.R. Vogt, and O. Eldholm, Reykjanes Ridge: further analysis of crustal subsidence and time-transgressive basement topography, *Earth Plant. Sci. Lett., 68,* 249-258.

Jones, E.J., M. Ewing, J.I. Ewing, and S.L. Ettreim, 1970, Influences of Norwegian Sea overflow water on sedimentation in the northern North Atlantic and Labrador Sea, *J. Geophys, Res., 75,* 1655–1680.

Keigwin, L.D., 1982, Isotopic paleoceanography of the Caribbean and east Pacific: Role of Panama uplift in late Neogene time, *Science, 217,* 350–353.

Keigwin, L.D., and G.A. Jones, 1994, Western North Atlantic evidence for millennial-scale changes in ocean circulation and climate, *J. Geophys. Res., 99,* 12,397–12,410.

Keigwin, L.D., and S.J. Lehman, 1994, Deep circulation change linked to Heinrich event 1 and Younger Dryas in a middepth North Atlantic core, *Paleoceanography, 9,* 185–194.

Keigwin, L.D., S.J., Lehman, and S. Johnsen, 1994, The role of the deep ocean in North Atlantic climate change between 70 and 130 kyr ago, *Nature, 371,* 323–326.

Keigwin, L.D., G.A. Jones, S.J. Lehman, and E.A. Boyle, 1991., Deglacial meltwater discharge, North Atlantic deep circulation, and abrupt climate change, *J. Geophys. Res., 96,* 16,811–16,826.

Keller, G., Depth stratification of planktonic foraminifers in the Miocene ocean, in *The Miocene Ocean: Paleoceanography and Biogeography*, J.P. Kennett (ed.), 177–195, GSA Mem. 163, 1985.

Kennett, J.P., 1977, Cenozoic evolution of Antarctic Glaciation, the Circum-Antarctic Ocean, and their impact on global paleoceanography, *J. Geophys. Res., 82,* 3843–3860.

Kroopnick, P., 1985, The distribution of ^{13}C of ΣCO_2 in the world oceans, *Deep Sea Res., 32,* 57–84.

Larsen, H.C., et al., 1994, *Proceedings of the Ocean Drilling Program, Initial Reports, vol. 152,* Ocean Drilling Program, College Station, Tex.

Lehman, S.J., and L.D. Keigwin, 1992, Sudden changes in North Atlantic circulation during the last deglaciation, *Nature, 356,* 757–762.

Luyendyk, B.P., J.R. Cann, et al., 1979, *Initial Reports of the Deep Sea Drilling Project, vol. 49,* U.S. Govt. Printing Off., Washington, D.C.

Maier-Reimer, E., U. Mikolajewicz, and T.J. Crowley, 1990, Ocean general circulation model sensitivity experiment with an open central American isthmus, *Paleoceanography, 5,* 349–366.

Manabe, S. and K. Bryan, Jr., 1985, CO_2-induced change in a coupled ocean-atmosphere model and its paleoclimatic implications, *J. Geophys. Res., 90,* 11689–11707.

Manabe, S., K. Bryan, and M.J. Spelman, 1990, Transient response of a global ocean-atmosphere model to a doubling of atmospheric carbon dioxide, *J. Phys. Ocean., 20,* 722–749.

McDougall, I., L.Kristjansson, K. Saemundsson, 1984, Magnetostratigraphy and geochronology of northwest Iceland, *J. Geophys. Res., 89,* 7029–7060.

Mikolajewicz, U., E. Maier-Reimer, T.J. Crowley, and K.-Y. Kim, 1993, Effect of Drake and Panamaniam gateways on the circulation of an ocean model, *Paleoceanography, 8,* 409–426.

Miller, K.G., 1992, Middle Eocene to Oligocene stable isotopes, climate, and deep-water history: The terminal Eocene Event? in *Eocene-Oligocene Climatic and Biotic Evolution*, D. Prothero and W.A. Berggren (eds.), pp. 160–177, Princeton University Press, Princeton, New Jersey.

Miller, K.G., and B.E. Tucholke, 1983, Development of Cenozoic abyssal circulation south of the Greenland-Scotland Ridge, in *Structure and Development of the Greenland-Scotland Ridge,* M. H. P. Bott, et al. (eds.), pp. 549–589, Plenum, New York.

Miller, K.G., and R.G. Fairbanks, 1985, Oligocene

to Miocene carbon isotope cycles and abyssal circulation changes, in *The Carbon Cycle and Atmospheric CO₂: Natural Variations Archean to Present. Geophys. Monogr. Ser.*, vol. 32, E.T. Sunquist and W.S. Broecker (eds.), pp. 469–486, AGU, Washington, D. C.

Miller, K.G., R.G. Fairbanks, and G.S. Mountain, 1987, Tertiary oxygen isotope synthesis, sea-level history, and continental margin erosion, *Paleoceanography, 2*, 1–19.

Miller, K.G., J.D. Wright, and R.G. Fairbanks, 1991, Unlocking the Icehouse: Oligocene–Miocene oxygen isotope, eustasy, and margin erosion, *J. Geophys. Res., 96*, 6829–6848.

Miller, K.G., et al., 1996, Drilling and dating New Jersey Oligocene–Miocene sequences: Ice volume, global sea level, and Exxon records, *Science, 271*, 1092–1095.

Mountain, G.S. and B.E. Tucholke, 1985, Mesozoic and Cenozoic Geology of the U.S. Atlantic Continental Slope and Rise, in *Geologic Evolution of the United States Atlantic Margin*, C.W. Poag (ed.), pp. 293–341, Van Nostrand Reinhold Company, New York.

Myhre, A.M., et al., 1995, *Proceedings of the Ocean Drilling Program, Initial Results, vol. 151*, Ocean Drilling Program, College Station, Tex.

Nunns, A.G., 1983, Plate tectonic evolution of the Greenland-Scotland Ridge and surrounding regions, in *Structure and Development of the Greenland-Scotland Ridge*, M. H. P. Bott, et al. (eds.), pp. 11-30, Plenum, New York.

Oppo, D.W., and R.G. Fairbanks, 1987, Variability in the deep and intermediate water circulation of the Atlantic Ocean during the past 25,000 years: Northern Hemisphere modulation of the Southern Ocean, *Earth Planet. Sci. Lett., 86*, 1–15.

Oppo, D.W., and S.J. Lehman, 1993, Mid-depth circulation of the subpolar North Atlantic during the last glacial maximum, *Science, 259*, 1148–1152.

Oppo, D.W., R.G. Fairbanks, and A.L. Gordon, 1990, Late Pleistocene Southern Ocean δ¹³C variability, *Paleoceanography, 5*, 43–54.

Pisias, N.G., N.J. Shackleton, and M.A. Hall, 1985, Stable isotope and calcium carbonate records from hydraulic piston cored Hole 574A: High-resolution records from the middle Miocene, in *Initial Reports of the Deep Sea Drilling Project, vol. 85*, L. Mayer and F. Theyer (eds.), pp. 735–748, U.S. Govt Printing Off., Washington, D.C.

Raymo, M.E., 1994, The Himalayas, organic carbon burial, and climate in the Miocene, *Paleoceanography, 9*, 399–404.

Raymo, M.E., and W.F. Ruddiman, 1992, Tectonic forcing of late Cenozoic climate, *Nature, 359*, 117–122.

Raymo, M.E., D. Hodell, and E.Jansen, 1992, Response of deep ocean circulation to initiation of Northern Hemisphere Glaciation (3-2 Ma), *Paleoceanography, 7*, 645–672.

Raymo, M.E., W.F. Ruddiman, N.J. Shackleton, and D.W. Oppo, 1990, Evolution of Atlantic-Pacific δ¹³C gradients over the last 2.5 m.y., *Earth Plant. Sci. Lett., 97*, 353–368.

Reid, J.L., 1979, On the contribution of the Mediterranean Sea outflow to the Norwegian-Greenland Sea, *Deep Sea Res. Part A, 26*, 1199–1223.

Rind, D., 1987, The doubled CO₂ climate: impact of the sea surface temperature gradient, *Am. J. Atm. Sci., 44(21)*, 3235–3268.

Rind, D., and M. Chandler, 1991, Increased ocean heat transports and warmer climate, *J. Geophys Res., 96*, 7437–7461.

Roberts, D.G., 1975, Tectonic and stratigraphic evolution of the Rockall Plateau and Trough, in *Petroleum and the Continental Shelf of Northwest Europe, vol. 1, Geology*, A.W. Woodland (ed.), pp. 77–89, Applied Science, Essex, England.

Ruddiman, W.F., 1972, Sediment redistribution on the Reykjanes Ridge: Seismic evidence, *Geol. Soc. Am. Bull., 83*, 2039–2062.

Ruddiman, W.F., and A. McIntyre, 1981, The North Atlantic Ocean during the last deglaciation, *Palaeogeogr., Palaeoclimatol., Palaeoecol., 35*, 145–214.

Saemundsson, K., 1986, Subaerial volcanism in the western North Atlantic, in *The Western North Atlantic Region, DNAG vol. M*, P.R. Vogt, and B.E Tucholke (eds.), pp. 69–86, Geological Society of America, Boulder, Colo.

Saemundsson, K., L. Kristjansson, I. McDougall, and N.D. Watkins, 1980, K-Ar dating, geological and paleomagnetic study of a 5-km lava succession in northern Iceland, *J. Geophys. Res., 85*, 3628–3646.

Savin, S.M., R.G. Douglas, and F.G. Stehli, 1975, Tertiary marine paleotemperatures, *Geol. Soc. Am. Bull., 86*, 1499–1510.

Savin, S.M., et al., 1985, The evolution of Miocene surface and near-surface marine temperatures: Oxygen isotopic evidence, in The Miocene Ocean: Paleoceanography and Biogeography, *GSA Memoir 163*, J.P. Kennett (ed.), pp. 49–82, Geological Society of America, Boulder, Colo.

Schilling, J.-G., 1986, Geochemical and isotopic variation along the Mid-Atlantic Ridge axis from 79°N to 0°N, in *The Western North Atlantic Region, DNAG vol. M*, P.R. Vogt, and B.E Tucholke (eds.), pp. 137–156 Geological Society of America, Boulder, Colo.

Schlich, R., et al., 1992, *Proc. Ocean Drill. Program, Sci. Results, vol. 120*, Ocean Drilling Program, College Station, Tex.

Schnitker, D., 1980, North Atlantic oceanography as possible cause of Antarctic glaciation and eutrophication, *Nature, 284*, 615–616.

Sclater, J.G., R.N. Anderson, and N.L. Bell, 1971, Elevation of ridges and evolution of the central eastern Pacific, *J. Geophys. Res., 76*, 7888–7915.

Shackleton, N.J., and A. Boersma, 1981, The climate of the Eocene ocean, *J. Geol. Soc. London, 138*, 153–157.

Shackleton, N.J., and M.A. Hall, in press, The late Miocene stable isotope record, Site 926, in *Proceedings of the Ocean Drilling Program, Scientific Results, vol. 154*, N.J. Shackleton, W.B. Curry, and C. Richter (eds.), Ocean Drilling Program, College Station, Tex.

Shackleton, N.J., and J.P. Kennett, 1975, Paleotemperature history of the Cenozoic and initiation of Antarctic glaciation: Oxygen and carbon isotopic analyses in DSDP Sites 277, 279, and 281, in *Initial Reports of the Deep Sea Drilling Project, vol. 29*, N.J. Shackleton and J. P. Kennett (eds.), pp. 743–755, U.S. Govt. Printing Off., Washington, D.C.

Shor, A. N., and R. Z. Poore, 1979, Bottom currents and ice rafting in the North Atlantic: Interpretation of Neogene depositional environment of Leg 49 cores, in *Initial Reports of the Deep Sea Drilling Project, vol. 49*, B.P. Luyendyk and J.R. Cann (eds.), pp. 859–872, U.S. Govt. Printing Off., Washington, D.C.

Sleep, N.H., 1990, Hotspots and Mantle Plumes: Some phenomenology, *J. Geophys. Res., 95*, 6715–6739.

Swift, J.H., 1984, The circulation of the Denmark Strait and Iceland-Scotland overflow waters in the North Atlantic, *Deep-Sea Research, 31* (11A), 1339–1355.

Swift, J.H., 1986, The Arctic Waters, in *The Nordic Seas*, B.G. Hurdle (ed.), pp. 129-153, Springer, New York.

Talwani, M., CC. Windisch, and M.G. Langseth, Jr., 1971, Reykjanes Ridge Crest: A detailed geophysical study, *J. Geophys. Res., 76*, 473–517.

Talwani, M., et al., 1976, *Initial Reports Deep Sea Drilling Project, vol 38*, U.S. Govt. Print. Off., Washington, D.C.

Thiede, J. and O. Eldholm, 1983, Speculations about the paleodepth of the Greenland-Scotland Ridge during late Mesozoic and Cenozoic times, in *Structure and Development of the Greenland-Scotland Ridge*, M.H.P. Bott, S. Saxov, M. Talwani, and J. Thiede (eds.), pp. 445–456, Plenum Press, New York.

Thiede, J., and A.M. Myhre, 1996, The paleoceanographic history of the North Atlantic –Arctic Gateways: synthesis of the Leg 151 drilling results, in *Proceedings of the Ocean Drilling Program, Scientific Results, vol. 151*, J. Thiede A. Myhre, J. Firth, G.L. Johnson, and W.F. Ruddiman (eds.), pp. 397–420, Ocean Drilling Program, College Station, Tex.

Thiede, J., A.M. Myhre, J.V. Firth, and Shipboard Scientific Party, 1995, Cenozoic northern hemisphere polar and subpolar ocean paleoenvironments (summary of ODP Leg 151 drilling results), in *Proceedings of the Ocean Drilling Program, Initial Results, vol. 151*, J. Thiede, A. Myhre, J. Firth (eds.), pp. 645–658, Ocean Drilling Program, College Station, Tex.

Tucholke, B.E., L.A. Raymond, P.R. Vogt, and N.C. Smoot, 1986, Bathymetry of the North Atlantic Ocean, Plate 2, in *The Western North Atlantic Region, DNAG vol. M*, P.R. Vogt, and B.E Tucholke (eds.), pp. 137–156 Geological Society of America, Boulder, Colo.

Vincent, E., J.S. Killingley, and W.H. Berger, 1985, Miocene oxygen and carbon isotope stratigraphy of the tropical Indian Ocean, in The Miocene Ocean: Paleoceanography and Biogeography, *GSA Memoir 163*, J.P. Kennett (ed.), pp. 103–130, Geological Society of America, Boulder, Colo.

Vogt, P.R., 1971, Asthenosphere motion recorded by the ocean floor south of Iceland, *Earth Plant. Sci. Lett., 13*, 153–160.

Vogt, P.R., 1972, The Faeroe-Iceland-Greenland Aseismic Ridge and the Western Boundary Undercurrent, *Nature, 239*, 79–81.

Vogt, P.R., 1983, The Iceland mantle plume: Status of the hypothesis after a decade of new work, in *Structure and Development of the Greenland-Scotland Ridge,* M.H.P. Bott, S. Saxov, M. Talwani, and J. Thiede (eds.), pp. 191–213, Plenum Press, New York.

Vogt, P.R., 1986, Seafloor topography, sediments, and paleoenvironments, in *The Nordic Seas,* B.G. Hurdle (ed.), pp. 237-410, Springer, New York.

Vogt, P.R., and O.E. Avery, 1974, Detailed magnetic surveys in the northeast Atlantic and Labrador Sea, *J. Geophys. Res. 79*, 363–389.

Vogt, P.R., and B.E. Tucholke, 1989, North Atlantic Ocean basin: Aspects of geologic structure and evolution, in *The Geology of North America,* Vol. A, *An Overview,* A.W. Bally and A.R. Palmer (eds.), pp. 5380, Geol. Soc. of Am., Boulder, Colo.

Vogt, P.R., R.K. Perry, R.H. Feden, H.S. Fleming, and N.Z. Cherkis, 1981, The Greenland-Norwegian Sea and Iceland environment: Geology and Geophysics, in *The Ocean Basins and Margins,* vol. 5, A.E.M. Nairn, et al. (eds.), pp. 493–598, Plenum Press, New York.

White, R., and D. McKenzie, 1989, Magmatism at rift zones: The generation of volcanic continental margins and flood basalts, *J. Geophys., Res. 94*, 7685–7729.

White, R.S., J.W. Brown, and J.R. Smallwood, 1995, The temperature of the Iceland plume and origin of outward-propagating V-shaped ridges, *J. Geol. Soc. London, 152,* 1039–1045.

White, R.S., et. al., 1996, Seismic images of crust beneath Iceland contribute to long-standing debate, *EOS Trans. AGU, 77,* 197–201.

Wold, C.N., 1994, Cenozoic sediment accumulation on drifts in the northern North Atlantic, *Paleoceanography, 9,* 917–941.

Wolf-Welling, T.C.W., M. Cremer, S. O'Connell, A. Winkler, and J. Thiede, 1996, Cenozoic Arctic gateway paleoclimate variability: Indications from changes in coarse-fraction composition, in *Proceedings of the Ocean Drilling Program, Scientific Results, vol. 151,* J. Thiede A. Myhre, J. Firth, G.L. Johnson, and W.F. Ruddiman (eds.), pp. 515–567, Ocean Drilling Program, College Station, Tex.

Woodruff, F., and S.M. Savin, 1989, Miocene deepwater oceanography, *Paleoceanography, 4,* 87–140.

Worthington, L.V., 1970, The Norwegian Sea as a mediterranean basin, *Deep-Sea Res., 17,* 77–84.

Wright, J.D, and K.G. Miller, 1993, Southern Ocean influences on late Eocene to Miocene deep-water circulation, in *The Antarctic Paleoenvironment: A Perspective on Global Change, 60 (Part 2),* J.P. Kennett and D. Warnke (eds.), pp. 1–25, American Geophysical Union, Washington D.C.

Wright, J.D., and K.G. Miller, 1996, Control of North Atlantic deep water circulation by the Greenland-Scotland Ridge, *Paleoceanography, 11,* 157–170.

Wright, J.D, K.G. Miller, and R.G. Fairbanks, 1991, Evolution of deep-water circulation: Evidence from the late Miocene Southern Ocean, *Paleoceanography, 6,* 275–290.

Wright, J.D, K.G. Miller, and R.G. Fairbanks, 1992, Miocene stable isotopes: Implications for deepwater circulation and climate, *Paleoceanography, 7,* 357–389.

Zachos, J.C., W.A. Berggren, M.-P. Aubrey, and A. Mackensen, 1993, Isotope and trace element geochemistry of Eocene and Oligocene foraminifers from Site 748, Kerguelen Plateau, in *Proceedings of the Ocean Drilling Program, Scientific Results, vol. 120,* S. Wise and R. Schlich (eds.), pp. 839-854, Ocean Drilling Program, College Station, Tex.

Zachos, J.C., L.D. Stott, and K.C. Lohmann, 1994, Evolution of early Cenozoic marine temperatures, *Paleoceanography, 9,* 353–387.

CHAPTER 10

Opening of Drake Passage and Its Impact on Cenozoic Ocean Circulation

Lawrence A. Lawver and Lisa M. Gahagan

In 1977, an article concerning the "Cenozoic evolution of Antarctic glaciation, the Circum-Antarctic Ocean and their impact on global paleoceanography" was published by Kennett. He correctly deduced many of the tectonic causes of the Cenozoic evolution of both Antarctic glaciation and global paleoceanography. Even without an absolute plate tectonic framework and direct knowledge of opening of Drake Passage between South America and Antarctica (Fig. 10.1), the conclusions in his paper are still valid. In the interim, compilation of a global database of marine magnetic anomalies and fracture zone lineations based on satellite derived

gravity has enabled us to produce an absolute plate tectonic framework for the Cenozoic (Müller et al., 1993). In addition, timing of the closure of the eastern Tethyan seaway has been worked out (Dercourt et al., 1986; Sengör et al., 1987) and constraints on the opening of a middle to deep water passageway between South America and Antarctica have been determined. Understanding of thermohaline circulation (Warren, 1983; Gordon, 1986; Gordon and Fine, 1996) has grown immensely since Kennett's article and it now appears that closure of equatorial seaways, in particular the Isthmus of Panama (Droxler et al., this volume, chapter 8;

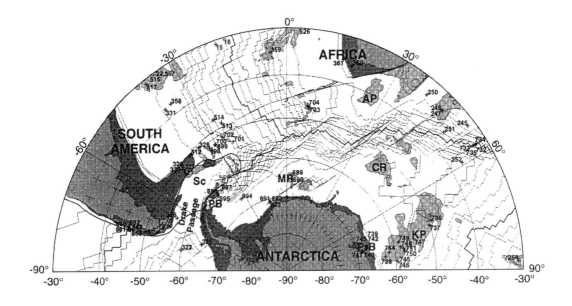

Fig. 10.1. Polar stereographic location map of southern ocean features. Continental margins are shaded out to 1000 m. Oceanic highs are shaded above 4000 m. Heavy lines delineate major plate boundaries. Light lines indicate magnetic isochrons. DSDP and ODP sites are indicated by Site number. AP=Agulhas Plateau, CR=Crozet Plateau, KP=Kerguelen Plateau, MR=Maud Rise, PB=Powell Basin, PrB=Prdyz Bay, Sc=Scotia Plate.

Mikolajewicz et al., 1993), may have had a more substantial long-term impact on Cenozoic and recent climate than the simple opening of Drake Passage. In short, while opening of Drake Passage and a passage south of Australia are necessary for the development of the Antarctic Circumpolar Current (ACC), simply opening these paleoseaways is not sufficient to produce vigorous circumpolar circulation.

Berggren and Hollister (1977) summarized plate tectonics and paleocirculation. They pointed out that oceanic circulation is dependent on the interaction of geography and climate. In their words, the Mesozoic ocean was tranquil under the influence of equable climates and oceanic thermohaline homogeneity. A reconstruction of Gondwana at 180 Ma indicates a single equatorial ocean with limited polar extent (Fig. 10.2). Circulation in such an ocean would be dominated by westward equatorial currents and limited, high latitude, middle or deep water circulation (Kennett, 1982). Berggren (1982) cites the presence of Cretaceous coals in high-latitude New Zealand and Alaska as evidence that the westward equatorial

current was directed poleward when it reached the western edge of the basin and transported significant heat to high latitudes along the western margin of the Cretaceous Pacific Ocean. It is thought that halothermal circulation with sinking of warm dense equatorial water prevailed and the long equatorial fetch (~290° of longitude) of the Cretaceous Pacific dominated world ocean circulation. As Gondwana broke apart, the single early Mesozoic ocean became separated into the three major oceans now present. According to Berggren and Hollister (1977), the Cenozoic climate deteriorated from the nearly constant Mesozoic climate to develop latitudinal thermal heterogeneity caused by high latitude cooling, which in turn led to accelerated surface and bottom water circulation in the oceans, especially along western margins of ocean basins. Accelerated bottom currents in the late Miocene led to erosion and redeposition as a major sedimentary process (Berggren and Hollister, 1977). They refer to the increased Cenozoic activity as "commotion in the ocean." Tucholke and Embley (1984) found evidence for regional Cenozoic erosion of the abyssal seafloor off South Africa. They

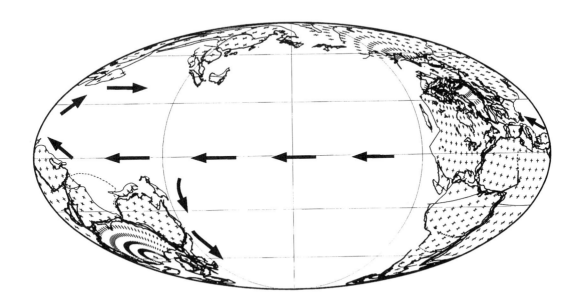

Fig. 10.2. Paleoreconstruction of the earth at 180 Ma shown as a Mollweide projection centered on 180°. Ocean circulation controlled by wind driven equatorial current dominated by halothermal circulation. Westward current probably extended from surface to ocean bottom (Berggren and Hollister, 1977). Coastlines are shown as heavy lines, medium lines are continental margins less than 200 m deep, and lightweight lines are oceanic plateaus.

Table 10.1. Major plate rotations interpolated for times indicated.

Time	SAM/AFR[a]			AFR/ANT[b]			SOB/ANT[c]		
	Lat.	Long.	Angle	Lat.	Long.	Angle	Lat.	Long.	Angle
60.0 Ma	62.1	32.3	23.34	2.7	-38.5-	10.93			
50.0 Ma	59.6	-31.6	20.40	9.3	-41.7	-9.24			
40.0 Ma	57.4	-32.5	16.46	17.0	-46.6	-7.19			
34.0 Ma	56.8	-33.7	13.78	12.8	-48.1	-5.66	64.8	134.9	-47.99
30.0 Ma	56.8	-34.4	11.94	11.8	-48.3	-4.80	90.0	0.0	0.00
20.0 Ma	57.9	-37.0	7.47	10.7	-47.9	-2.76			
10.0 Ma	59.9	-38.8	3.26	8.2	-49.4	-1.54			

[a] Shaw and Cande (1990)

[b] Royer et al. (1988), Royer and Sandwell (1989), and Royer and Chang (1991)

[c] Opening of Powell Basin, this paper

AFR = Africa; ANT = Antarctica; SAM = South America; SOB = South Orkney Block

feel the present erosion is a relict feature that originated in the late Miocene that was the result of increased glaciation of West Antarctica that produced large volumes of bottom water. Based on a deeper unconformity, they date the onset of significant abyssal circulation to the early Oligocene.

Methodology

The plate reconstructions shown in this chapter are based on a global database which consists of marine magnetic anomalies tied to the Cande and Kent (1995) timescale and fracture zone and transform fault lineations derived from satellite altimetry data (Gahagan et al., 1988; Sandwell and Smith, 1992). Plate motions (Table 10.1) between South America and Africa are taken from Shaw and Cande (1990) while Cenozoic motions between Africa and Antarctica are derived from Royer et al., (1988), Royer and Sandwell (1989), and Royer and Chang (1991). The revised Cande and Kent (1992) geomagnetic reversal timescale changed the date of the Eocene/Oligocene boundary from that of Berggren et al., (1985) at 36.6 Ma, to 33.7 Ma based on the bio- and magnetostratigraphic work of Odin et al. (1991). McIntosh et al. (1992) used $^{40}Ar/^{39}Ar$ to date ignimbrites from western North America and put the Eocene/Oligocene boundary at 33.4 Ma based on recalibrated Eocene-Oligocene geomagnetic polarity reversals. Most recently, Berggren and Aubry (1995) put the date of the Eocene/Oligocene boundary at 33.7 Ma based on their revised Cenozoic calcareous plankton magne-

tobiochronology and chronostratigraphy that used Cande and Kent (1995) as its geomagnetic timescale. Since most of the articles that discuss a major change in southern ocean water temperatures based on $\delta^{18}O$ and $\delta^{13}C$ proxies rely on the earlier Berggren et al. (1985) chronostratigraphy, those dates will be approximately reinterpreted to the Cande and Kent (1995) timescale, but the original unrevised age will be noted in brackets followed by [, B*]. For the most part, times will be cited only as definitive as early Oligocene or latest Eocene.

Early Cenozoic Antarctic Seaways

Early Cenozoic isostatic equilibrium maps of an ice-free Antarctica indicate that Drake Passage was not the first, high latitude, Cenozoic seaway in or around Antarctica. Whether or not Antarctica was totally ice-free in the period prior to Late Eocene as Miller et al., (1991) believe or Antarctica was "intermittently ice free" between 65 Ma and 50 Ma as Prentice and Matthew (1988) contend (although their Fig. 3 seems to indicate at most, only a very brief period of ice formation around 56 Ma), there was certainly a high southern latitude seaway connecting the Pacific Ocean with the Atlantic Ocean (see Fig. 8.9 of Wise et al., 1991) in the earliest Cenozoic. Whichever model is assumed, an ice-free Antarctica with glacial rebound considered, suggests that a significant shallow water seaway existed between East and West Antarctica in early Cenozoic. While such a seaway might have been limited by the presence of the Ellsworth Mountains and may have been no deeper than a few hundred meters in some places, Wise et al. (1991) show a mostly submerged west Antarctica on a deglacial adjusted geography of Antarctica. If one also considers the tectonic based higher Cretaceous and early Cenozoic sea level (Haq et al., 1988), then the early Cenozoic seaway between East and West Antarctica would have been deeper and more prominent (Fig. 10.3). It is clear though, that the present day Circum-Antarctic seaway was not the first, high southern latitude passageway.

Opening of Eocene/Oligocene Drake Passage

The earliest identified marine magnetic anomalies in the western Scotia Sea are dated at 28.7 Ma (Barker and Burrell, 1977; LaBrecque and Rabin-owitz, 1977). This evidence was thought to date opening of Drake Passage as Late Oligocene or later, even though major plate motions place the tip of the Antarctic Peninsula well to the east of the southern end of South America by 30 Ma (Lawver et al., 1985, 1992). King and Barker (1988) show a reconstruction of South America and Antarctica at 35 Ma that has any potential middle to deep water Drake Passage plugged up with an array of continental to semi-continental blocks. They dated opening of Powell Basin between the tip of the Antarctic Peninsula and the South Orkney block as 29–23 Ma or contemporoeous with the early opening of the western Scotia Sea. Evidence presented in Lawver et al., (1994) can be used to support a latest Eocene to early Oligocene opening of a middle to deep water "Drake Passage" through Powell Basin (Fig. 10.1). Heat flow measurements and age-versus-depth calcu-

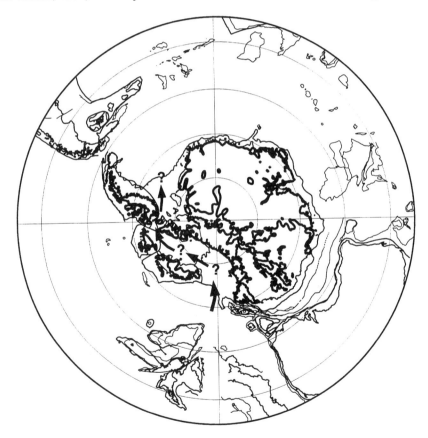

Fig. 10.3 Polar stereographic projection of reconstruction at 60 Ma. Table 10.1 lists rotations used. Heavy line is depth to bedrock from Drewry (1983). This is not isostatically compensated but sea level was almost 250 m higher than present. Arrows indicate where there may have been a 300 m to 500 m seaway between the Ellsworth Mountains (ELM) and the Antarctic Peninsula (AP).

lations for Powell Basin, a marginal basin to the east of the Antarctic Peninsula, give reasonable agreement and suggest an age for a fully open Powell Basin of 30.5 Ma. Such an age suggests Powell Basin began to open around the time of the Eocene-Oligocene boundary, about 5 million years earlier than the age of the oldest magnetic anomalies found in the western Scotia Sea (Barker and Burrell, 1977, LaBrecque and Rabinowitz, 1977). Although it is unclear what caused Powell Basin to open, it may have opened as a back-arc basin behind a northwestward directed subduction zone that resulted from subduction of Weddell Sea crust beneath the South Orkney block. We suggest that by 32 Ma, Powell Basin may have provided a middle to deep water passageway between Antarctica and South America. Figure 10.4 shows the 40 Ma paleoreconstruction with both Powell Basin closed and the South Tasman Rise blocking a deepwater passageway between Australia and Antarctica. Droxler et al., (this volume, chapter 8) indicate that some form of Caribbean passageway was open during Late Eocene.

Three polar stereographic reconstructions of Antarctica and the surrounding region (Fig. 10.5) show changes in possible seaways at 35 Ma, 32.5 Ma, and 30 Ma. Major plate movement (South America to Africa to Antarctica) indicates Drake Passage probably opened by 32.5 Ma even if Powell Basin did not function as a midwater passage. By 32.5 Ma, the South Tasman Rise had cleared North Victoria Land of East Antarctica (Fig. 10.5b), and the two events completed a high-latitude, circumpolar, midwater passageway around Antarctica at or near the time of the Eocene–Oligocene boundary. While it is clear that neither Drake Passage nor the passage south of Australia was open at 40 Ma (Fig. 10.4), by 32.5 Ma (Fig. 10.5b) both passageways were open to middle and deep water circulation Kennett (1977) was correct when he ascribed major oceanic changes to the time when the South Tasman Rise cleared North VictoriaLand. What he did not know was that Drake Passage opened at nearly the same time and may in fact have been open at a slightly earlier time if Powell Basin acted as an initial Drake Passage.

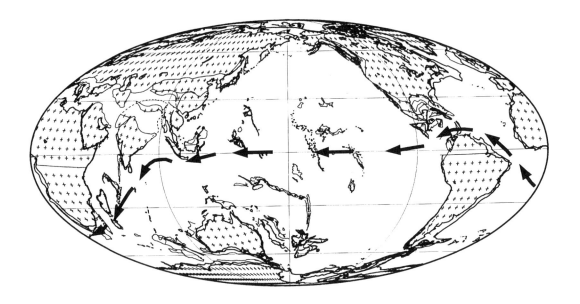

Fig. 10.4. Paleoreconstruction at 40 Ma shown as a Mollweide projection centered on 180°. Table 10.1 lists rotations used. The principal westward, surface-to-deep equatorial current was blocked by the collision of India with Eurasia. Primary ocean circulation was diverted from the 30°N Mediterranean circuit to south of Africa at 45°S. The South Tasman Rise still blocked deep water circulation to the south of Australia and there is no deepwater circulation between South America and Antarctica. Principal oceanic circulation is still predominantly equatorial.

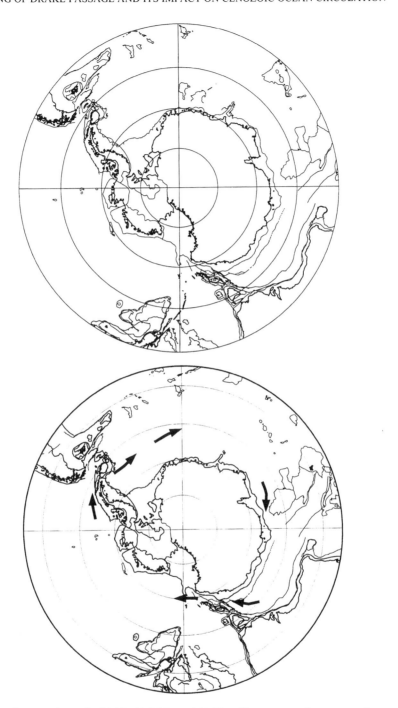

Fig. 10.5. Reconstructions are shown for 35 Ma, 32.5 Ma, and 30 Ma. The reconstructions use a polar stereographic projection, and the plate rotations are listed in Table 10.1. (a) 35 Ma reconstruction with only very shallow seaways between Australia and Antarctica and through Drake Passage region between South America and Antarctica. (b) 32.5 Ma reconstruction shows first indication that a mid-level or deep seaway could have existed between the South Tasman Rise and East Antarctica, (see arrows) and through Drake Passage. South Georgia is presently unconstrained due to a lack of identifiable magnetic anomalies in the central Scotia Sea but major plate motions between South America, Africa and Antarctica indicate a deep seaway through Drake Passage possibly through Powell Basin between the South Orkney block and the tip of the Antarctic Peninsula.

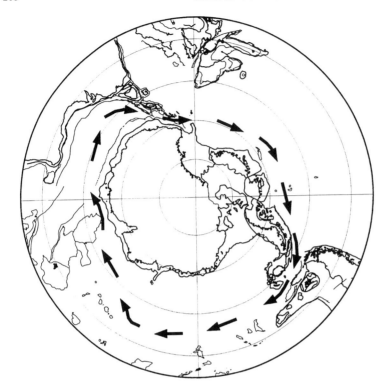

Fig. 10.5c. 30 Ma reconstruction with open Drake Passage. By this time, major plate motions indicate that there must have been a deep water passageway through Drake Passage since the assumed continental fragments found in the Scotia Sea could no longer have blocked a deepwater passageway. Heat flow and age versus depth calculations for Powell Basin (Lawver et al., 1994) indicate that it would have been fully open by this time.

Summary of Early Cenozoic Oceanic Circulation

Global reconstructions indicate that the 53 Ma $\delta^{18}O$ minimum, which is equated to a maximum in ocean water temperature, corresponds to the time of closure of the mostly equatorial eastern Neo-Tethyan seaway (Oberhänsli, 1992) as India collided with Eurasia. Lee and Lawver (1995) show a soft collision of India with Eurasia at 59 Ma, followed by a hard collision about 43 Ma. The soft collision is envisioned as the Indian block colliding with a Southern Tibet block and other continental fragments which slowed India from its northward rate of 170 mm/yr to a rate of 105 mm/yr between 58 Ma and 51 Ma as India's nearly northward direction changed to north-northwest and then slowed again to a rate of 90 mm/yr between 50 Ma and 44 Ma. The hard collision at 43 Ma finally slowed the northward motion of India to about 56 mm/yr. The dramatic slowing of the Indian block would also have a major effect on the total volume of mid-ocean ridge spreading centers and lead to a lower sea level (Kominz, 1984) which in turn would have ef-

fected any Early Cenozoic seaway between East and West Antarctica (Fig. 10.3).

As India moved northward during the Late Cretaceous and early Cenozoic, it forced the dominant westward flowing equatorial current into an ever narrower passage between India and Eurasia. The first response to closure between India and Eurasia may have been deflection of the mostly equatorial circulation, to the south of India and onward to the south of Africa (Fig. 10.4). Prior to the hard collision of India with Eurasia but after the initiation of a barrier to deepwater circulation between Western and Eastern Tethys, the equatorial current piled warm, saline water into the western reaches of the eastern Neo-Tethys where it intensified its salinity, sank and filled the deep basins of the world's oceans. This warm pulse of water may be what Stott et al., (1990) discuss as having occurred at the Paleocene/Eocene boundary. Although the Mediterranean was at 30°N during the early Cenozoic, it was still within the temperate zone with mean annual sea surface temperatures (SST) as high as 20°C during the Paleocene to late Eocene (Zachos et al., 1994). Zachos et al., (1994) conclude that while tropical SSTs stayed relatively constant during the late Paleocene to early Eocene, high-latitude SSTs increased from 9°C to 14°C. The increased constriction of the major "equatorial" current to an ever narrower passageway within the temperate zone assured that global seawater temperatures stayed warm through the Paleocene. When the Tethyan seaway closed completely, the main "equatorial" current diverted to the south of India and to the south of Africa. By 50 Ma, the southern tip of Africa was around 45° S. Simple closure of the Neo-Tethyan cul-de-sac and diversion of the primarily equatorial current from at most 30°N to at least 45°S caused a gradual cooling of the world's oceans. Zachos et al., (1994) estimate that SSTs at 45°S were as high as 15°C in the early Eocene but dropped to about 10°C by late Eocene as the "equatorial" heat was lost. While SSTs at 45°S stayed as high as 10°C into the early Oligocene, high latitude temperatures seem to have fallen to below 5°C south of 60°S from their high of 15°C in the early Eocene (Zachos et al., 1994).

Miller et al., (1987) smoothed their data to highlight an increase in $\delta^{18}O$ at the Eocene-Oligocene boundary but such a dramatic shift is not so apparent in the data presented by Zachos et al., (1994). While there is an apparent increase in $\delta^{18}O$ at about 32 Ma [35 Ma, B*] in the data Zachos et al., (1994) show, a more general increase in $\delta^{18}O$ can be inferred from 55 Ma until 25 Ma [28 Ma, B*], indicating continuous cooling. What is curious is there then appears an apparent decrease in $\delta^{18}O$ from 25 Ma [28 Ma, B*] until 22 or 23 Ma. Neither the flattening of the Miller et al., (1987) $\delta^{18}O$ line between 32 Ma [35 Ma, B*] and 15 Ma nor the apparent decrease at 25 Ma [28 Ma, B*] in the $\delta^{18}O$ of Zachos et al., (1994) correspond to an opening of Drake Passage at 30 Ma as based on the magnetic anomalies of Barker and Burrell (1977) or at 18 Ma as suggested by others. The apparent decrease in the $\delta^{18}O$ between 25 Ma and 22 Ma may have been a consequence of the continued northward motion of Africa.

Conclusion

It is very clear that a major change in water temperature occurred during the earliest Oligocene in the southern oceans (Kennett and Stott, 1990; Miller et al., 1991; Barrera and Huber, 1991, 1993; Mead et al., 1993; Wei, 1991). The dramatic change in mid-

water characteristics at the Eocene/Oligocene boundary is usually related to the opening of a high latitude, deepwater seaway (Kennett, 1982), and we now know that the high latitude seaway was the result of the nearly simultaneous opening of Drake Passage and a passage south of Australia. Initial opening of such a seaway or closure of an equatorial seaway may in fact have a reverse instantaneous effect compared to the expected final long term impact. For instance, initial opening of a deep water passage around Antarctica might first "pull" warm water southward before the longer term chilling of the world's oceans overwhelms the initial warm pulse. While opening of Drake Passage allowed development of the Antarctic Circumpolar Current, the ACC may not be the sole cause of massive Antarctic icesheets. In fact, Gordon (1986) argues that present-day transport of Subantarctic cold water from the Pacific to the Atlantic within Drake Passage is "of secondary importance," and accounts for perhaps only 25% of the total transport compared to the "warm water" route. Warren (1983) argues that the real cause of the major Pliocene icesheet formation in the world was the increase of warm, dense surface water in the North Atlantic that resulted from the closure of the midwater Panamanian passageway between South and North America. The higher temperature and greater salinity, perhaps influenced by outflow from the Mediterranean (Reid, 1981, 1994) led to greater evaporation and hence greater cooling. This led to higher production of cold, dense North Atlantic Deep Water and enhanced chilling of all the world's oceans.

Initiation of the ACC has been considered perhaps the most important event to affect Cenozoic climate (Kennett, 1982) and necessary to explain the development of the Miocene Antarctic icesheets. In fact, Cenozoic cooling of the world's oceans began at 53 Ma (Stott and Kennett, 1990; Stott et al., 1990), long before Drake Passage opened, and cooling of the world's oceans has continued from that time to present. Zachos et al., (1994) conclude on the basis of three independent lines of evidence that the early late Eocene was the first time in the Cenozoic that ephemeral ice-sheets appeared in Antarctica and it was not until the earliest Oligocene when larger and more extensive ice-sheets appeared. Although Birkenmajer (1990) reports fossiliferous glaciomarine clastic deposits with ice-rafted debris on King George Island,

Antarctic Peninsula that are capped with a early middle Eocene basaltic lava (K/Ar dated as 49.4 ± 5 Ma), he speculates that the deposits were produced by local glaciation restricted to the highest mountain tops of the region. While this may be evidence for initial Cenozoic Antarctic glaciation during the late Eocene, we conclude that evidence for pre-Oligocene Cenozoic glaciation in Antarctica is limited (Barron et al, 1991; Breza and Wise, 1992) and may only indicate the presence of isolated alpine glaciers or ephemeral ice-sheets.

We propose that while the presence of a middle to deep water passageway is necessary to the development of a strong ACC, a deep circumpolar passageway alone is not sufficient to produce a strong current and the subsequent permanent East Antarctic ice sheet. Without southward counterflow caused by closure of equatorial seaways, most of the midwater ocean circulation would remain in the temperature zone even if Drake Passage and the Australia–Antarctic Basin were open to permit circum-Antarctic midwater flow. Clearly, existence of a late Cretaceous to Cenozoic shallow passageway between East and West Antarctica did not disturb the tranquil homogeneous thermohaline ocean of Berggren and Hollister (1977). According to Miller et al., (1987), the permanent East Antarctic icesheet only developed in Miocene time, long after Drake Passage and the deepwater passage south of Australia opened. In fact, the closure of Australia/New Guinea with Southeast Asia during the middle Miocene which cut off the deepwater equatorial passage between the Pacific and Indian Oceans may have been the major influence on forcing development of a vigorous middle to deep water ACC. The final closure of the Isthmus of Panama in the Pliocene (Coates et al., 1992) blocked equatorial midwater and deepwater circulation between the Atlantic and Pacific (Droxler et al., this volume, chapter 8) and forced warm saline water into the North Atlantic which led to formation of North Atlantic Deep Water (Warren, 1983). Even though there is more fresh water input into the North Atlantic than the North Pacific, the warm equatorial water that formally exited to the Central Pacific via a low latitude Panamanian seaway, now gets swept into the North Atlantic and the dense warm water has a much higher evaporation rate than the cooler water of the North Pacific, so most saline deep water is produced in the North Atlantic (Warren, 1983).

Consequently, while opening of Drake Passage and a seaway between Australia and Antarctica in Early Oligocene led to an initiation of the Antarctic Circumpolar Current and a drop in the $\delta^{18}O$ value indicative of cooler temperatures, it was the combination of Cenozoic equatorial seaway closures, including closure of the Tethyan ocean in early Cenozoic, closure of Southeast Asia with Australia in Miocene and finally closure of the Isthmus of Panama in Pliocene time that had the greater impact on the world's climate and led to the Pliocene development of massive polar icesheets.

Acknowledgments

We thank Ken Miller, Jim Kennett, and the editors of this volume for substantial and critical reviews. We also thank Ed King who pointed out the error in our age versus heat flow value for the Powell Basin (Lawver et al., 1994). The age is not the uncorrected age of 38 Ma as discussed in that article but rather a corrected, fully opened Powell Basin by 30.5 Ma.

References

Barker, P. F., and J., Burrell, 1977, The opening of Drake Passage, *Mar. Geol., 25*, 15–34.

Barrera, E. and B. T. Huber, 1991, Paleogene and early Neogene oceanography of the southern Indian Ocean: Leg 119 foraminifer stable isotope results, in *Proc. ODP, Sci. Results, 119*, J. Barron, B. Larsen, et al. (eds.), pp. 693–717, Ocean Drilling Program, College Station, Tex.

Barrera, E. and B. T. Huber, 1993, Eocene to Oligocene oceanography and temperatures in the Antarctic Indian Ocean, in The Antarctic Paleoenvironment: A Perspective on Global Change, J. P. Kennett, and D. A. Warnke (eds.), pp. 49–65, *AGU Antarctic Research Series, 60.*

Barron, J., B. Larsen, and J. G. Baldauf, 1991, Evidence for late Eocene to early Oligocene Antarctic glaciation and observations on late Neogene glacial history of Antarctica: results from Leg 119, in *Proc. ODP, Sci. Results,* 119, J. Barron, B. Larsen, et al. (eds.), pp. 869–891, Ocean Drilling Program, College Station, Tex.

Berggren, W. A., 1982, Role of ocean gateways in climatic change, *Studies in Geophysics: Climate in Earth History*, pp. 118–125, National Academy Press, Washington, D. C.

Berggren, W. A., and C. D. Hollister, 1977, Plate tectonics and paleocirculation—commotion in the ocean, *Tectonophysics*, *11*, 11–48.

Berggren, W. A., and M. –P. Aubry, 1995, A revised Cenozoic calcareous plankton magneto-biochronology and chronostratigraphy, Spring AGU Meeting, *EOS*, abstract, 76, S97.

Berggren, W. A., D. V. Kent, J. J. Flynn, and J. A. Van Couvering, 1985, Cenozoic geochronology, *Geological Society of America Bulletin*, *96*, 1407–1418.

Birkenmajer, K., 1990, Tertiary glaciation in the South Shetland Islands, West Antarctica: Evaluation of data, in *Geological Evolution of Antarctica*, M. R. A. Thomson, J. A. Crame, and J. W. Thomson (eds.), pp. 629–632, Cambridge Univ. Press, Cambridge.

Breza, J. R., and S. W. Wise, Jr., 1992, Lower Oligocene ice–rafted debris on the Kerguelen Plateau: Evidence for East Antarctic continental glaciation, in *Proc. ODP, Sci. Results, 120*, W. S. Wise, Jr., R. Schlich, et al. (eds.), pp. 161–178, Ocean Drilling Program, College Station, Tex.

Cande, S. C., and D. V. Kent, 1992, A new geomagnetic polarity time scale for the Late Cretaceous and Cenozoic, *J. Geophys. Res., 97*, 13,917–13,951.

Cande, S. C., and D. V. Kent, 1995, Revised calibration of the geomagnetic polarity timescale for the Late Cretaceous and Cenozoic, *J. Geophys. Res., 100*, 6093–6095.

Coates, A. G., et al., 1992, Closure of the Isthmus of Panama: The near–shore marine record of Costa Rica and western Panama, *Geological Society of America Bulletin*, *104*, 814–829.

Dercourt, L. E., et al., 1986, Geological evolution of the Tethys belt from the Atlantic to the Pamirs since the Lias, *Tectonophysics*, *123,* 241–314.

Drewry, D. J., and S. R. Jordan, 1983, The Bedrock Surface of Antarctica, Sheet 3 of *Antarctica: Glaciological and Geophysical folio*, D. J. Drewry (ed.), Scott Polar Research Institute, Cambridge, England.

Gahagan L. M., et al., 1988, Tectonic fabric map of the ocean basins from satellite altimetry data. *Tectonophysics, 155*, 1–26.

Gordon, A., 1986, Interocean exchange of thermocline water. *J. Geophys. Res., 91*, 5037–5046.

Gordon, A. and R. A. Fine, 1996, Pathways of water

between the Pacific and Indian oceans in the Indonesian seas, *Nature, 379*, 146–149.

Haq, B. U., J. Hardenbol, P. R. Vail, and G. R. Baum, 1988. Mesozoic and Cenozoic chronostratigraphy and eustatic cycles, in *Sea–level Change: An Integrated Approach*, C. K. Wilgus, et al. (eds.), pp. 71–108, *Society of Economic Paleontologists and Mineralogists Special Publication, 42*.

Kennett, J. P., 1977, Cenozoic Evolution of Antarctic Glaciation, the Circum–Antarctic Ocean, and Their Impact on Global Paleoceanography, *J. Geophys. Res. 82*, 3843–3859.

Kennett, J. P., 1982, *Marine Geology*, Prentice-Hall, Inc., Englewood Cliffs, N. J.

Kennett, J. P., and L. D. Stott, 1990. Proteus and Proto–Oceanus: Ancestral Paleogene oceans as revealed from Antarctic stable isotopic results; ODP Leg 113, in *Proc. ODP, Sci. Results, 113*, P. R. Barker, J. P. Kennett, et al. (eds.), pp. 865-880, Ocean Drilling Program, College Station, Tex.

King, E. C., and P. F. Barker, 1988, The margins of the South Orkney microcontinent, *J. Geol. Soc., London, 145*, 317–331.

Kominz, M., 1984, Oceanic ridge volumes and sea–level change—An error analysis, in *Interregional Unconformities and Hydrocarbon Accumulation*, J. S. Schlee (ed.), pp. 108–123, Amer. Assoc. Petro. Geol. *Memoir 36*, Tulsa, Okla.

LaBrecque, J. L., and P. D. Rabinowitz, 1977, Magnetic anomalies bordering the continental margin of Argentina, *Map Ser. Cat. 826*, Amer. Assoc. Petrol. Geol., Tulsa, Okla.

Lawver, L. A., J. G. Sclater, and L. Meinke, 1985, Mesozoic and Cenozoic Reconstructions of the South Atlantic, *Tectonophysics, 114*, 233–254.

Lawver, L. A., L. M. Gahagan, and M. F. Coffin, 1992. The development of paleoseaways around Antarctica, in *The Antarctic Paleoenvironment: A Perspective on Global Change*, J. P. Kennett, and D. A. Warnke, (eds.), pp. 7–30, *A G U Antarctic Research Series, 56*.

Lawver, L. A., T. Williams, and B. J. Sloan, 1994, Seismic Stratigraphy and heat flow of Powell Basin, *Terra Antartica, 1*, 309–310.

Lee, Tung–yi, and L. A. Lawver, 1995, Cenozoic plate reconstruction of the Southeast Asia region, *Tectonophysics, 251*, 85–138.

McIntosh, W. C., J. W. Geismann, C. E. Chapin,

M. J. Kunk, and C. D. Henry, 1992, Calibration of the latest Eocene–Oligocene geomagnetic polarity time scale using ^{40}Ar/^{39}Ar dated ignimbrites, *Geology, 22*, 459–463.

Mead, G. A., D. A. Hodell, and P. F. Ciesielski, 1993, Late Eocene to Oligocene vertical oxygen isotopic gradients in the South Atlantic: Implications for Warm Saline Deep Water, in *The Antarctic Paleoenvironment: A Perspective on Global Change*, J. P. Kennett, and D. A. Warnke, (eds.), pp. 27–48, *AGU Antarctic Research Series, 60.*

Mikolajewicz, U., Maier–Reimer, E., Crowley, T. J., and Kim, K.–Y., 1993, Effect of Drake and Panamanian gateways on the circulation of an ocean model, *Paleoceanography, 8*, 409–426.

Miller, K. G., R. G. Fairbanks, and G. S. Mountain, 1987, Tertiary oxygen isotope synthesis, sea level history, and continental margin erosion, *Paleoceanography, 2*, 1–19.

Miller, K. G., M. D. Feigenson, J. D. Wright, and B. M. Clement, 1991, Miocene isotope reference section, Deep Sea Drilling Project Site 608: An evaluation of isotope and biostratigraphic resolution, *Paleoceanography, 6*, 33–52.

Müller, R. D., Royer, J. –Y., and Lawver, L. A., 1993, Revised plate motions relative to the hotspots from combined Atlantic and Indian Ocean hotspot tracks, *Geology, 21*, 275–278.

Oberhänsli, H., 1992, The influence of the Tethys on the bottom waters of the Early Tertiary Ocean, in *The Antarctic Paleoenvironment: A Perspective on Global Change*, J. P. Kennett, and D. A. Warnke, (eds.), pp. 167–184, *AGU Antarctic Research Series, 56.*

Odin, G. S., Montanari, A., Deino, A., Drake, R., Guise, P. G., Kreuzer, H., and D. C. Rex, 1991, Reliability of volcano-sedimentary biotite ages across the Eocene–Oligocene boundary (Apennines, Italy), *Chem. Geol. Isotope Geosci. Sect., 86*, 203–224.

Prentice, J. L., and R. K. Matthews, 1988, Cenozoic ice-volume history, development of a composite oxygen isotope record, *Geology, 17*, 963–966.

Reid, J., 1981, On the mid-depth circulation of the world ocean, in *Evolution of Physical Oceanography*, B. A. Warren and C. Wunsch (eds.), pp. 70–111, The MIT Press, Cambridge, Mass.

Reid, J., 1994, On the total geostrophic circulation of the North Atlantic Ocean: flow patterns, tracers, and transports, *Prog. Oceanog., 33*, 1–92.

Royer, J.–Y. and Sandwell, D. T., 1989, Evolution of the Eastern Indian Ocean since the Late Cretaceous: Constraints from Geosat altimetry, *J. Geophys. Res., 94*, 13,755–13,782.

Royer, J.–Y. and T. Chang, 1991, Evidence for relative motions between the Indian and Australian plates during the last 20 Myr from plate tectonic reconstructions: Implications for the deformation of the Indo–Australian plate, *J. Geophys. Res., 96*, 11779–11802.

Royer, J.–Y., P. Patriat, H. Bergh, and C. Scotese, 1988, Evolution of the southwest Indian Ridge from the Late Cretaceous (anomaly 34) to the Middle Eocene (anomaly 20), *Tectonophysics, 155*, 235–260.

Sandwell, D. T., and W. H. F. Smith, 1992, Global marine gravity from ERS–1, Geosat and Seasat reveals new tectonic fabric, *EOS Trans. AGU, 73*, 133.

Sengör, A. M. C., D. Altmer, A. Cin, T. Ustaömer, and K. J. Hsü, 1987, Origin and assembly of the Tethyside orogenic collage at the expense of Gondwana Land, in *Gondwana and Tethys*, M. G. Audley–Charles and A. Hallam (eds.), pp. 119–181, *Geological Society Special Pub. No. 37*, Geological Society London.

Shaw, P. R. and Cande, S. C., 1990, High–resolution inversion for South Atlantic plate kinematics using joint altimeter and magnetic anomaly data, *J. Geophys. Res., 95*, 2625–2644.

Stott, L., and J. P. Kennett, 1990. Cenozoic planktonic foraminifers and Antarctic Paleogene biostratigraphy: ODP Leg 113, Sites 689–690, in *Proc. ODP, Sci. Results, 113*, P. F. Barker, J. P. Kennett, et al. (eds.), pp. 549–569, Ocean Drilling Program, College Station, Tex.

Stott, L. D., J. P. Kennett, N. J. Shackleton, and R. M. Corfield, 1990. The evolution of Antarctic surface waters during the Paleogene, inferences from the stable isotopic composition of planktonic foraminifera, in *Proc. ODP, Sci. Results, 113*, P. F. Barker, J. P. Kennett, et al. (eds.), pp. 849–864, Ocean Drilling Program, College Station, Tex.

Tucholke, B. E., and R. W. Embley, 1984. Cenozoic regional erosion of the abyssal sea floor off South Africa, in *Interregional Unconformities and Hydrocarbon Accumulation*, J. S. Schlee (ed.), pp. 145–164, *Memoir 36*, Amer. Assoc. Petro. Geol., Tulsa, Okla.

Warren, B. A., 1983, Why is no deep water formed in the North Pacific? *J. Mar. Res., 41*, 327–347.

Wei, W., 1991, Evidence for an earliest Oligocene abrupt cooling in the surface waters of the Southern Ocean, *Geology, 19*, 780–783.

Wise, S. W., Jr., J. R. Breza, D. M. Harwood, and W. Wei, 1991, Paleogene glacial history of Antarctica, in *Controversies in Modern Geology*, J. A. McKenzie, D. W. Müller, and H. Weissert (eds.), pp. 133–171, Academic Press Ltd., London.

Zachos, J. C., L. D. Stott, and K. C. Lohmann, 1994, Evolution of early Cenozoic marine temperatures, *Paleoceanography, 9*, 353–387.

PART VI

ROLE OF BATHYMETRY

CHAPTER 11

Reconstruction of Realistic Early Eocene Paleobathymetry and Ocean GCM Sensitivity to Specified Basin Configuration

Karen L. Bice, Eric J. Barron, William H. Peterson

Oceanic circulation affects climate through the transport of heat, salt, and nutrients. Oceans are the second largest reservoir of carbon in the Earth's system and ocean circulation is a primary control on the addition of carbon to significant portions of the one of the largest carbon reservoirs: Earth's sedimentary deposits. The circulation of the oceans is controlled in part by the configuration of the ocean basins, including shelf morphology, seafloor topography, and the configuration of oceanic gateways. Considerable effort is being applied to the construction of atmosphere and ocean global general circulation models (GCMs), both coupled and uncoupled, which adequately incorporate the role of oceans in the climate system. Our attempts to use these models in the simulation of past climates are hindered by our ability to adequately reconstruct the boundary conditions to the modeled system. In the case of the ocean model, most attention has been given to the upper boundary conditions: surface wind stresses and the flux of heat and moisture across the atmosphere–ocean interface. Model simulations of the modern ocean employ observations of surface wind stress, temperature, and salinity as forcing to the model ocean. Lacking observations of these conditions for past intervals, paleo-ocean simulation studies generally employ surface forcing derived from atmospheric model simulations which allow for treatment of the effects of atmospheric composition, paleotopography, vegetation, snow and ice, and continental runoff. Less attention has concerned the lower ocean boundary condition, reconstruction of complex ocean basin configurations and seafloor topography. Recently, the horizontal and vertical resolution of ocean models applied to paleo-ocean studies has increased suffi-

ciently to prompt efforts to reconstruct high resolution (< 5°) paleobathymetry.

Oceanic gateways, shallow shelf seas and seafloor features such as seismic and aseismic ridges have long been recognized as controls on ocean circulation in modern oceanographic studies and numerical models of the present-day ocean. Current measurements in Drake Passage indicate that the speed of Antarctic circumpolar currents and the mass transport of deepwater out of the Weddell Sea are strongly influenced by the topography of local meridional seafloor ridges (Reid and Nowlin, 1971; Carmack and Foster, 1975). Early numerical model experiments suggested that the existence of zones of barrier-free flow at surface, intermediate and deep levels can affect the formation of intermediate and deep water masses and the meridional transport of heat. The volume of Antarctic Intermediate Water formed in the Southern Ocean is dependent on the depth of Drake Passage (Gill and Bryan, 1971; England, 1992). In numerous experiments, variation in the model shape of the Southern Ocean seafloor resulted in a change in the volume of transport by the Antarctic Circumpolar Current and the rate of production of Antarctic Bottom Water (Gill and Bryan, 1971; Bryan and Cox, 1972; Bye and Sag, 1972; Holland and Hirschman, 1972; Johnson and Hill, 1975; Mikolajewicz et al., 1993).

There is abundant evidence in the marine geologic record that such controls played a significant role in ancient ocean circulation patterns (Kennett 1977, 1986; Barker and Burrell, 1982; Berggren, 1982; Keller et al., 1987; Mountain and Miller, 1992; Oberhänsli, 1992; Pak and Miller, 1992; Robert and Chamley, 1992). In addition to having acted as controls on the distribution of organisms, an effect generally easily recognized in the sediment record, the presence or absence of barriers to flow in the ancient ocean is likely to have influenced water mass production and global oceanic nutrient and heat transport. The resulting global climatic effects are not as easily recognized in the sediment record because of the problem of multiple possible causes for observed climatic change. Various theoretical and numerical models have attempted to identify the important climatic influence of past ocean configurations (Brass et al., 1982a, b; Barron and Peterson, 1989, 1990, 1991; Maier-Reimer et al., 1990; Mikolajewicz et al., 1993; Bice, 1997). Maier-Reimer et al., (1990) investigated the effect of an open central American passageway, a condition appropriate for ~4 Ma, using the 3.5° x 3.5° grid, 11-layer Hamburg ocean model (Maier-Reimer and Hasselmann, 1987). They found that surface transport through the open gateway resulted in a surface salinity variation effective in significantly decreasing deepwater production in the North Atlantic basin. Subsequent sensitivity tests by the same group (Mikolajewicz et al., 1993) examined paleogeography appropriate for ~40 Ma and showed that a closed Drake Passage increased Antarctic Bottom Water production, which in turn decreased the rate of North Atlantic Deep Water production, in agreement with simulations of highly idealized modern oceans by Gill and Bryan (1971), Cox (1989), England (1992) and Toggweiler and Samuels (1993).

Barron and Peterson (1989, 1990, 1991) investigated the nature of middle Cretaceous and Cenozoic ocean circulation and the potential for formation of warm, saline deepwater given varying ocean configurations and atmospheric forcing. Their experiments used an early version of the model adapted from Semtner (1974) with coarse horizontal (5° x 5°) and vertical (4 layers) resolution. Barron and Peterson (1990) performed a series of simulations which was designed to examine the sensitivity of that model to the specification of two different mid-Cretaceous model ocean bathymetries, a bowl-like bathymetry and one modified to include several seafloor features. They observed that the model-simulated circulation patterns in the uppermost 50 m (model layer 1) were nearly indistinguishable and that the only differences between the two simulations was routing of deepwater around a Pacific seafloor plateau and differences in the

strength of two surface currents. They noted no differences in simulated temperatures and salinities. Based on their conclusion that the lack of a more detailed paleobathymetric reconstruction was not a critical limitation, Barron and Peterson (1991) subsequently used idealized bowl-like bathymetries in a series of Cenozoic ocean model simulations. However, the coarse resolution of the model may have severely limited the development and recognition of significant sensitivity in "realistic" ocean and "bowl" ocean experiments. With the increased horizontal (< 5°) and vertical (at least 20 layers) resolution of more recent ocean GCMs, it is necessary to reexamine the question of model sensitivity to bathymetry.

In the study described here, paleobathymetry for 55 Ma (approximately the early Eocene) has been reconstructed and used in simulations employing the Parallel Ocean Climate Model (POCM) (Semtner and Chervin, 1992). This approach uses digitized magnetic anomalies and the relationship between seafloor age and bathymetry derived from observations and idealized oceanic plate models. The approach has application to paleobathymetric reconstructions for the approximate time interval 70 – 0 Ma. An initial assessment of the sensitivity of early Eocene circulation to specified ocean basin configuration is made through comparison of Eocene simulations with realistic and simplified bathymetries using identical atmospheric forcing.

Reconstruction of Eocene Bathymetry for Ocean Modeling

It has long been noted that in much of the present ocean, water depth increases with increasing age of the underlying seafloor (Menard, 1969; Sclater et al., 1971). Oceanic plate material is created at spreading centers and cools, contracts, and increases in density with increasing age. The observed relationship between age and depth agree well with the results of thermal models of the oceanic crust, given assumptions of isostatic equilibrium, crustal properties and boundary conditions (Parsons and Sclater, 1977; Sclater et al., 1980; Jarvis and Peltier, 1982; Stein and Stein, 1992). With knowledge of the age of much of the present-day seafloor and the assumption of age-depth relationships based on observations and idealized crustal models, realistic bathymetry can be reconstructed for large portions of ancient ocean floors. (For a complete treatment of models of plate formation, cooling and the depth of the oceans, the reader is referred to Sclater et al., 1980, 1985; Peltier, 1989; Stein and Stein, 1992; and Nagihara et al., 1996; or the textbook of Fowler, 1990.)

Realistic early Eocene bathymetry was reconstructed using digitized seafloor magnetic lineations and the age-depth relationship of Parsons and Sclater (1977). This approach is one which can be used to reconstruct latest Cretaceous through Cenozoic bathymetry for a significant portion of the global seafloor. The technique was applied by Sclater and McKenzie (1973) and Sclater et al., (1977) in the reconstruction of Atlantic paleobathymetry. Cottereau (1992) has attempted to reconstruct high resolution Tethyan paleobathymetry using the seafloor age-depth relationship. However, in the Tethys Ocean, this application requires assumed spreading center positions as well as spreading rates. Thiede (1982) used this technique to reconstruct Atlantic Ocean paleobathymetry and estimated the error in such a reconstruction to be ±100–200 m in regions of average oceanic crust and ±300–500 m over aseismic rises. Thiede's maximum error estimate is comparable to the resolution at depth of most ocean general circulation models. (In the ocean model described in this chapter, the lower eight layers, representing ocean depths from 1335 – 5200 m, have an average layer thickness of 483 m.)

The model assumes the seafloor age-depth relationship (Parsons and Sclater, 1977)

$$d(t) = 2500 + 350(t)^{1/2} \qquad (11.1)$$

where $d(t)$ is seafloor depth in meters and t is oceanic crust age in millions of years. In this equation, the mean depth assumed for paleo-ocean spreading center ridges is 2500 m. Equation (11.1) describes most of the observed correspondence between seafloor age and bathymetry for seafloor younger than 70 million years. For oceanic crust older than 70 m. y., observed depths depart from a linear relationship and fit the curve

$$d(t) = 6400 - 3200e^{-t/62.8} \qquad (11.2)$$

The utility of this approach in the reconstruction of paleobathymetry depends on the preservation,

mapping, and dating of seafloor magnetic lineations. The term lineation here refers to the linear traces that demarcate the position and known extent of seafloor spreading anomalies, oceanic crust exhibiting an identified magnetic intensity and polarity (Jacobs, 1994). Seafloor anomalies are assigned absolute ages based on radiometric and paleontologic evidence and extrapolation from and interpolation between absolute age control points.

Because of its dependence on magnetic lineations, this approach will provide a better reconstruction of paleobathymetry in ocean basins surrounded by passive margins. Atlantic and Indian ocean paleobathymetry can be better constrained than that for much of the Pacific Ocean because of subduction along Pacific plate active margins. Because of the destruction of oceanic lithosphere through plate subduction, the magnetic lineation data available for a reliable reconstruction decrease rapidly with increasing age (Fig. 11.1). Figure 11.1 was constructed using the data of Sclater et al., (1980). The shaded bar graph shows seafloor area as a function of the age of oceanic crust. The solid curve indicates the cumulative percent of seafloor area older than X million years old. Approximately

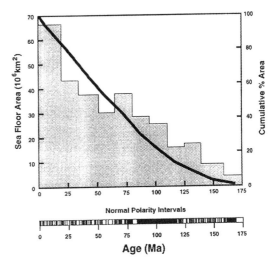

Fig. 11.1. Seafloor area as a function of age (Sclater et al., 1980) and normal polarity intervals for the time interval 175 – 0 Ma (Kent and Gradstein, 1986; Cande and Kent, 1992). Shaded bar graph shows seafloor area for each time interval indicated. The solid curve indicates cumulative percent of seafloor area older than X million years. Normal polarity intervals with duration of less than 200 kyr have been omitted for clarity.

20% of the present-day seafloor dates to the Cretaceous Quiet interval, a long-lived period of normal polarity lasting approximately 83–118 Ma (Cande and Kent, 1992). In the absence of extensive crustal sampling and dating, this seafloor can yield little information useful in the reconstruction of bathymetry using the method outlined here. For example, in reconstructing bathymetry for 40 Ma using the age-depth relationship of Parsons and Sclater (1977), about 42% (62% − 20%) of the present-day seafloor could potentially yield magnetic lineation information for the estimation of paleodepth. On the other hand, in reconstructing bathymetry for the Campanian ocean (~80 Ma), only about 12% (32% − 20%) of the present-day seafloor might yield useful data. One solution to this problem would be interpolation of seafloor ages across the Cretaceous Quiet zones using the assumption of a uniform seafloor spreading rate. At the horizontal and vertical resolution employed in the ocean model experiments described here, this assumption is unlikely to introduce error. Another improvement would incorporate published detailed reconstructions of significant portions of the oceans where available (e.g., Lawver et al., 1992).

In order to create a global reconstruction of early Eocene paleobathymetry, four sets of magnetic lineations were digitized: anomalies 25, 31, 34, and M0 on the map of Cande et al. (1989) (Fig. 11.2). Anomaly 25 was chosen to approximate the position of the seafloor spreading ridges at approximately 55 Ma. Anomalies 31, 34, and M0 have been identified in many ocean basins and therefore provide the best spatial and temporal coverage for reconstruction of a 55 million year-old seafloor. An approximation of the early Eocene shoreline was interpreted and digitized from the work of Vogt and Tucholke (1986) and Ronov et al. (1989). Each digitized line segment was assigned to the proper tectonic plate and was then rotated to its position relative to North America at 55 Ma. Plate rotation vectors were calculated from the stage poles of Barron (1987) with anomaly ages updated to the time scale of Cande and Kent (1992). A second rotation was performed in order to position all plates relative to the North American paleomagnetic pole at 55 Ma (Müller et al., 1993). The resulting distribution of magnetic lineations at 55 Ma is shown in figure 11.3. A bathymetric control value was then assigned for each lineation using the Parsons and Sclater (1977) relationship given in equation (11.1) (Table 11.1). The seafloor ages used in the calculation are approximately the present anomaly age

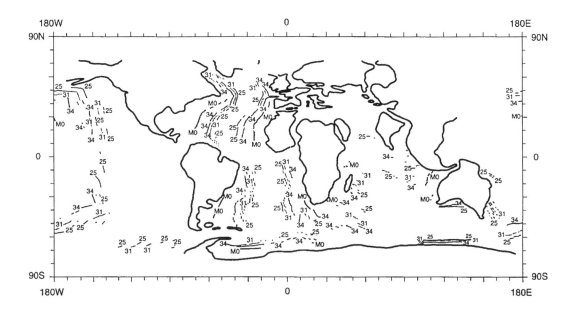

Fig. 11.2. Present positions of magnetic anomalies 25, 31, 34, and M0 (Cande et al., 1989). Bold black lines are the present-day position of the approximate early Eocene shoreline (Vogt and Tucholke, 1986; Ronov et al., 1989).

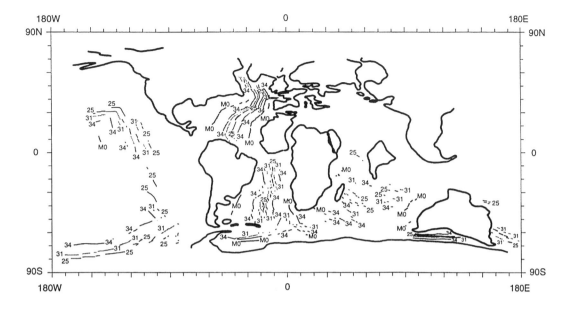

Fig. 11.3. Magnetic lineations and paleo-shoreline data (Fig. 11.2) rotated to their positions at 55 Ma.

(Cande et al., 1987) minus 55 m. y. The resulting data were interpolated to a 2° x 2° grid.

While much of the present seafloor is observed to fit the age-depth relationship of Parsons and Sclater (1977), regions of anomalous heat flow and flexural properties exhibit shallower water depths relative to cooler oceanic crust of the same age (Stein and Stein, 1993). The 2° bathymetry digital file was therefore edited to include seafloor topographic features such as the Emperor-Hawaiian seamount chain, the Pacific Superswell and Darwin Rise, the Kerguelen Plateau, and Ninetyeast Ridge. The assumption was made that the present mean elevation of these features approximates the Eocene value. As a first approximation (at 2° horizontal resolution and given the vertical resolution of the ocean model grid), the width and depth of the present shelf edge and base of the continental rise were assumed to be reasonable approximations of the positions of these features at 55 Ma. A uniform depth of 200 m was assumed for the shelf edge and 4000 m for the base of the continental rise. The early Eocene reconstruction is open to surface circulation in Drake Passage and the Australian/Antarctic Passage with a water depth of 500 m at their shallowest points. Tethyan bathymetry was interpreted from the work of Dercourt et al., (1986).

The subsidence and sedimentation histories of

Table 11.1. Age and depth for magnetic anomalies used in the reconstruction of early Eocene bathymetry.

Datum	Present age (m.y.)	Age (m.y.) at 55 Ma	Calculated depth (m) at 55 Ma (Eq. 11.1)	Assigned depth (m) for 2° reconstruction
Anomaly M0	118	63	5278	5200 (max. model depth)
Anomaly 34	83	28	4352	4350
Anomaly 31	69	14	3810	3800
Anomaly 25	56	1	2850	2500 (mid-ocean ridge)

Note: Present ages from Cande et al., (1989); calculated depths from equation of Parsons and Sclater (1977).

continental margins has not been treated in detail. In order to generate "realistic" global paleo-bathymetry for a number of different intervals using a consistent technique, it was necessary to use a fairly idealized treatment of bathymetry. For example, the effects of sediment deposition and compaction and depression of the oceanic crust through sediment loading were not taken into consideration. The reconstruction of paleobathymetry through back-stripping of sediments and the "rebound" of oceanic crust represents an alternate technique (Hay et al., 1989; Steckler et al., 1995; Wold, 1995), one which still involves considerable assumptions about, for example, sediment thicknesses in areas lacking direct or geophysical observations.

In experiments designed to examine the sensitivity of the ocean model to bathymetry, bowl oceans were created by specifying water depths of 185 m and 2750 m in the first two ocean grid points adjacent to the continents and 5200 m water depth in all other ocean grid points. One potentially important result of creating an idealized bowl-like bathymetry using this approach is widened and deepened ocean gateways. Figure 11.4 illustrates both realistic and bowl bathymetries specified in these experiments.

The Ocean Model and Atmospheric Forcing

The ocean GCM used in this study is the Parallel Ocean Climate Model (version 2.0) developed by A. Semtner and R. Chervin (Semtner and Chervin, 1992). POCM is a three-dimensional global ocean model derived from the Bryan and Cox model (Bryan, 1969; Cox, 1970) with specific modifications by Cox (1975, 1984) and Semtner (1974). The standard equations of motion are integrated over a rigid-lid ocean domain with a realistic basin configuration, allowing for complex shoreline and bottom geometry. POCM allows the specification of a maximum of 20 vertical layers and horizontal resolution as fine as 0.25° latitude by 0.25° longitude. In the experiments described in this chapter, 20 ocean layers and horizontal resolution of 2° x 2° were specified. Present-day simulations using POCM and climatological forcing (observations) are described by Semtner and Chervin (1992), McCann et al., (1994), and Washington et al., (1994). Following the approach of Semtner and

Chervin, no Arctic Ocean is included in these simulations. The maximum poleward extent of the model Eocene oceans is 72°N and 78°S.

Surface forcing to the ocean model is the wind stress, moisture flux, and 2-m air temperature simulated by the GENESIS (version 1.02) atmospheric GCM (Thompson and Pollard, 1995a, b). GENESIS is a three-dimensional atmospheric GCM consisting of a 12-layer atmosphere model coupled with soil, snow, sea-ice and 50-m mixed-layer slab ocean models. The model includes a land-surface transfer scheme which simulates the physical effects of vegetation. In the simulation employed in this study, horizontal resolution of 4.5° latitude by 7.5° longitude was used. Model output was then interpolated to POCM resolution. The GENESIS slab ocean model serves as a source and sink for heat and moisture and allows for the meridional transport of heat according to a specified heat transport configuration. In the Eocene simulation used to create ocean model forcing, zonally symmetric poleward ocean heat transport values of 0.3 times the present-day control values of Covey and Thompson (1989) were specified. Paleogeography was interpreted by L. Sloan (Sloan, 1994) from the paleogeographic reconstruction of Scotese and Golonka (1992). Sloan (1990, 1994) and Sloan and Barron (1992) have described several Eocene GCM experiments designed to investigate the roles of paleogeography, oceanic heat transport, and atmospheric composition in producing the warm climate and reduced (relative to the present-day) meridional sea surface temperature gradient inferred for the early Eocene. The GENESIS Eocene simulation used in ocean model experiments described here has a specified atmospheric carbon dioxide concentration of 340 ppm and globally uniform vegetation and soil characteristics. Mean annual values of temperature, winds and moisture flux are shown in figures 11.5 and 11.6. These data were obtained by averaging the last five years of a 35-year GENESIS simulation. The ocean surface moisture flux derived from GENESIS output includes simulated continental freshwater runoff. Idealized continental drainage basins were specified. Net positive continental precipitation was summed for each drainage basin and was added to specified adjacent coastal ocean model cells. Present-day ocean simulations forced by output from the GENESIS model are described by Bice (1997).

Two 1000-year integrations of POCM with early

A

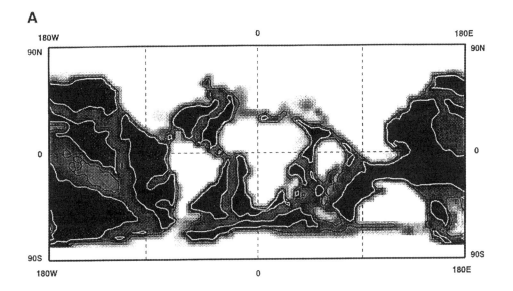

B

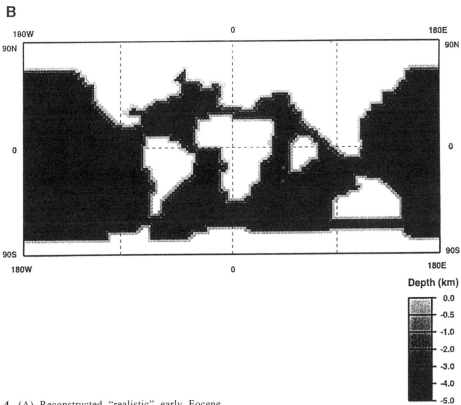

Depth (km)

0.0
-0.5
-1.0
-2.0
-3.0
-4.0
-5.0

Fig. 11.4. (A) Reconstructed "realistic" early Eocene bathymetry. The 4000 m depth is indicated by a white contour line. (B) Early Eocene "bowl" bathymetry. Horizontal resolution is 2° x 2°.

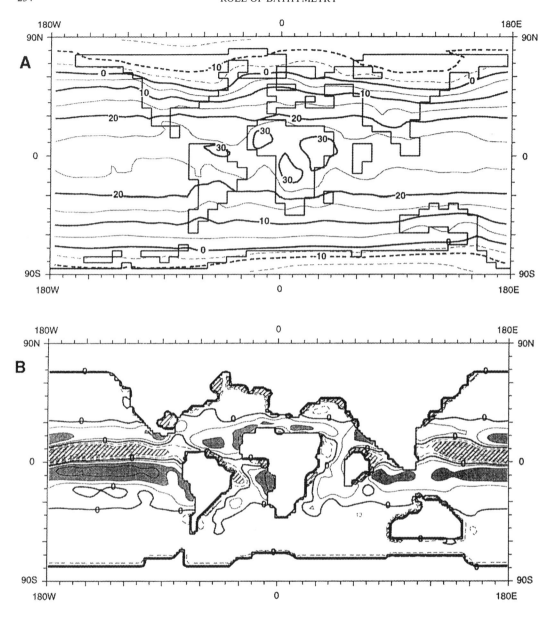

Fig. 11.5. Mean annual temperature (A) and moisture flux (B) from the GENESIS early Eocene simulation used to force the ocean model. Temperature contour interval is 5°C. Moisture flux (mm day^{-1}) is expressed as evaporation minus precipitation minus continental runoff. Moisture flux contour interval is 2 mm day^{-1}. Negative values (dashed lines) indicate regions of net precipitation; positive values (solid lines) are regions of net evaporation. Regions of > 4 mm day^{-1} (shaded) and less than -4 mm day^{-1} (hatched) are indicated.

Eocene atmospheric forcing were performed: a realistic ocean bathymetry case and a bowl bathymetry case. The model was forced with mean annual atmospheric conditions for model years 1 – 300 and full seasonal cycle conditions for model years 301 –

1000. An instantaneous "snapshot" of POCM results was archived every ten years. The global average layer temperatures shown in figure 11.7A are model December 31 values at the end of each 10-year integration. (The change from mean annual to

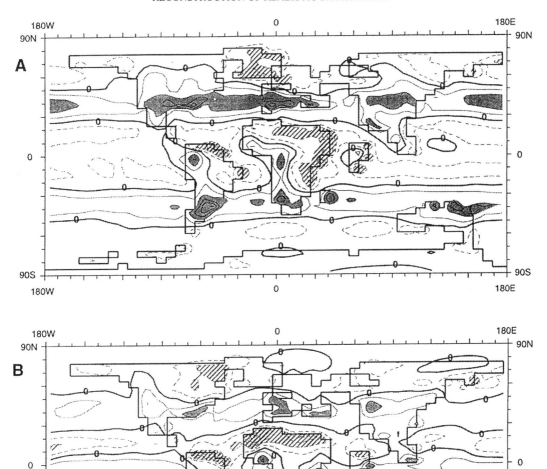

Fig. 11.6. Mean annual wind stress forcing (N m^{-2}) from the GENESIS simulation used to force the ocean model. (A) Eastward wind stress. (B) Northward wind stress. Contour interval is 0.05 N m^{-2}. Regions of greater than 0.1 N m^{-2} (shaded) and less than -0.1 N m^{-2} (hatched) are indicated.

full seasonal cycle forcing accounts for the jumps in December 31 temperature and temperature difference values at year 300. The effect is most pronounced in the uppermost 1000 m of the model, the layers which had nearly equilibrated after 300 years.) After 1000 model years, globally averaged model temperatures were at or very near steady-state in all model levels. The maximum rate of change of temperature at model year 1000 was 0.002°C yr^{-1} at 4125 m in the bowl ocean simulation. Each simulation was then run for another year with data archiving every 3 days. These data were averaged to produce mean annual model statistics.

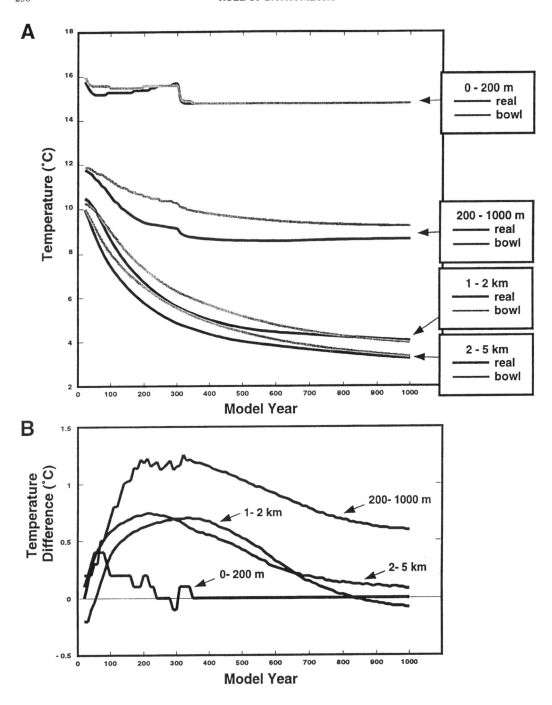

Fig. 11.7. Evolution of globally averaged ocean model layer temperatures during 1000-year simulations for realistic and bowl cases. (A) Layer temperatures averaged over four ocean intervals: 0–200 m, 200 m – 1000 m, 1– 2 km, and 2–5 km. Abrupt changes in shallow levels at model year 300 are caused by the change from mean annual forcing and mean annual data archiving to full seasonal forcing and model December 31 data archiving. (B) Layer temperature differences calculated as bowl case minus realistic case.

Model Sensitivity to Complex versus Simple Bathymetry

The evolution of globally averaged model layer temperature differences (bowl case minus realistic case) is shown in figure 11.7B. At steady-state, the global average December 31 temperatures differ by less than 1°C. The largest difference occurs in the depth interval 200–1000 m where the simulation with an idealized bowl bathymetry exhibits warming relative to the realistic bathymetry case. Vertical profiles of the mean annual temperatures (model year 1001) are given in figure 11.8. On a global average, mean annual basis, the bowl case is cooler than the realistic case in the upper 200 m and in the interval 1000–1750 m (Fig. 11.8) and is warmer at other depths by less than 1°C.

Temperature is the only model-predicted field which can presently be evaluated against quantitative estimates based on marine sediment analyses. Seawater paleotemperatures are calculated using oxygen stable isotope analyses ($\delta^{18}O$) of foraminiferal carbonate. In this calculation, error exists in the isotopic paleotemperature equation, the estimated $\delta^{18}O$ value of ambient seawater, and in the foraminiferal carbonate $\delta^{18}O$ measurements. Based on the isotopic paleotemperature equation error estimates of Erez and Luz (1983) and assuming a minimum error of ±0.1‰ in both seawater $\delta^{18}O$ and carbonate $\delta^{18}O$ measurements, the error in the isotopic paleotemperatures available for comparison against model-predicted temperatures is at least ±2°C. (See Bice, 1997, and Bice et al., 1998, for a more complete treatment of the comparison of model-predicted temperatures and foraminiferal isotopic paleotemperatures.)

If the estimated error in calculated isotopic paleotemperature is ±2°C, the sensitivity observed in the globally averaged temperatures (less than 1°C) can be considered insignificant. Likewise, the model layer meridional temperature differences (averaged across all longitudes) are insignificant with the exception of warming of over 2° in the bowl bathymetry case at 300 – 400 m water depth in the Northern Hemisphere mid-latitudes. However, local temperature differences within the model layers above 1000 m are greater than 2°C over large regions. Grid-point temperature differences between the two simulations (calculated as bowl case minus realistic case) vary from ±10°C in the uppermost

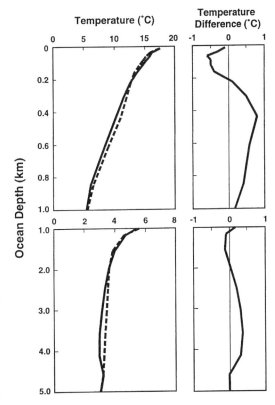

Fig. 11.8. (A) Vertical temperature profiles for realistic (solid line) and bowl case (dashed line) after 1000 model years. (B) Layer temperature differences calculated as bowl case minus realistic case. Values are global, mean annual averages. Note depth scale change between plots of 0–1 km and 1–5 km intervals.

1000 m to ±2°C at depths below 1000 m. The regions over which temperatures differ by 2°C or more in the uppermost 50 m (Fig. 11.9) highlight a systematic difference in upper ocean circulation between the two simulations. The bowl ocean exhibits strengthened mid-latitude gyre circulation relative to the realistic ocean configuration. In general, in the upper 100 m, this results in warming along western boundary currents and cooling along eastern boundary currents because of increased surface water transport poleward and equatorward, respectively. In the Southern Hemisphere Atlantic basin, for example, increased poleward transport by the Eocene equivalent of the Brazil Current results in warming on the western side of the South Atlantic while increased equatorward transport by a north-

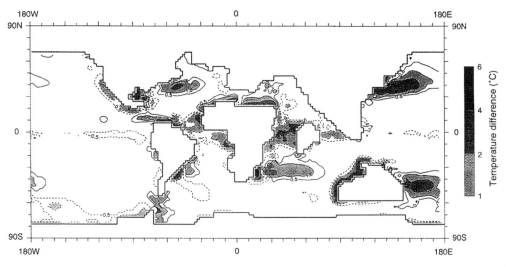

Fig. 11.9. Temperature difference (uppermost 50 m) calculated as bowl case minus realistic case. Positive values (solid lines) indicate that the bowl case is warmer than the realistic case. Dashed lines indicate regions of negative values. Zero temperature difference contours have been omitted for clarity. Contour lines shown are 0.5°, 1°, and 2° C. Shading indicates temperatures differences of between 1° and 6°C.

ward-flowing Benguela-like current causes cooling along the western subequatorial African coast. A large temperature difference contrast is shown for the region of divergence of the southward-flowing Brazil Current and the northward-flowing Falkland Current.

The increased intensity of gyre circulation in the bowl simulation relative to the realistic bathymetry case is reflected in an increase in the depth and strength of meridional overturning in the upper 600 m (Fig. 11.10). This is most pronounced in mid- and low-latitude cells of meridional overturning. However, the observed variation in overturning is not due to variation in the strength of equatorial easterly winds and mid-latitude westerly winds. The bowl configuration produced higher westward veloc-ities in the equatorial current resulting in increased upwelling of cool water and slightly cooler surface temperatures in the region of equatorial divergence. This effect is most pronounced in the narrower equa-torial seas (Atlantic and equatorial Tethyan re-gions).

Meridional volume transport in the Southern Ocean, averaged across all longitudes, is substantially reduced from a maximum of 65 Sv in the realistic bathymetry case to a maximum of 37 Sv in the bowl case (Fig. 11.10). The bowl basin configuration with a wider, deeper Drake Passage allows greater circumpolar flow and reduced Antarctic Bottom Water formation. This result is consistent with other model studies of the role of gateways in Southern Hemisphere circulation (Cox, 1989; England, 1992; Mikolajewicz et al., 1993; Toggweiler and Samuels, 1993, 1995).

The change from a realistic to a simplified bowl bathymetry results in no change in the overall structure of the surface circulation. All the major surface currents simulated in the realistic bathymetry case are apparent in the bowl bathymetry case. However, the surface circulation and deeper through-flow in ocean passageways and in the narrower Tethyan and proto-Indian Ocean re-gions are sensitive to the change. In the northern Tethys Ocean, removal of a significant seafloor ridge to produce a flat-bottom ocean apparently re-sults in better organized anticyclonic flow in the eastern Tethys (Fig. 11.11A). Where flow was pre-dominately away from the northern African coast-line in the real bathymetry case, regional flow is along the coastline in the bowl bathymetry case. The effect is decreased upwelling along the northern African coast, resulting in temperatures as much as 9°C warmer in the upper 150 m of the bowl case. Globally, the largest surface salinity difference be-tween the two simulations occurs in the northern Tethys Ocean where bowl case salinities are nearly

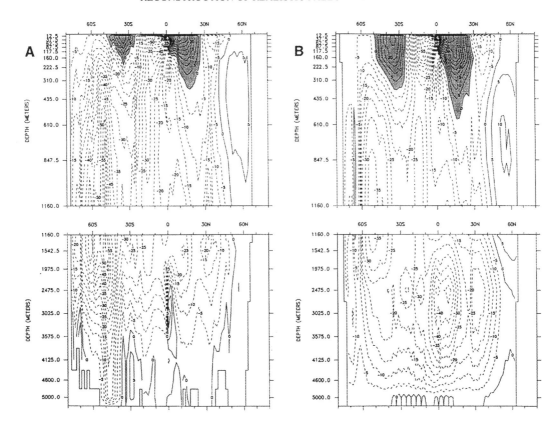

Fig. 11.10. Model-simulated globally averaged meridional volume transport (Sv) for (A) realistic bathymetry case and (B) bowl bathymetry case. Contour interval is 5 Sv (1 Sv = 10^6 m^3 s^{-1}). Positive values (solid lines) indicate regions of clockwise flow around local maxima. Cells of meridional overturning beneath the northern equatorial easterly winds and Southern Hemisphere mid-latitude westerly winds have been shaded to highlight deepening and intensification of meridional transport in the bowl simulation.

1‰ higher than those simulated for the realistic bathymetry case (Fig. 11.12).

Construction of a bowl ocean configuration results in an artificially widened and deepened Australian/Antarctic Passage. This allows the beginning development of a surface cyclonic gyre with a westward-flowing coastal current (like the present-day East Wind Drift) in the seaway and reduced eastward surface velocities in the bowl case (Fig. 11.11B). The opposite effect is observed in the widened and deepened bowl Drake Passage: smoothing of the ocean bottom in the passageway in the bowl case allows better organized eastward flow and through-flow at deeper levels (Fig. 11.11C).

The maximum difference in grid-point salinities predicted in the two experiments occurs in the up-

permost 200 m where salinities in the bowl bathymetry case are generally higher than those in the realistic bathymetry case. Maximum local salinity differences are +3‰ in the upper 50 m (Fig. 11.12), +2.4‰ in the interval 50–200 m and +0.5‰ in the interval 200–100 m. Below 1000 m, grid-point salinities differ by 0.2‰ or less.

It is difficult to define the significance of differences in the model-predicted salinities because there presently exists no quantitative paleosalinity proxy. However, one possible measure of the significance of surface salinity differences involves use of the model-predicted salinity in the calculation of the oxygen isotopic composition of surface seawater. Empirical isotopic paleotemperature equations such as that of Erez and Luz (1983) require knowledge of the ambient water $\delta^{18}O$ value (δ_{sw}), a

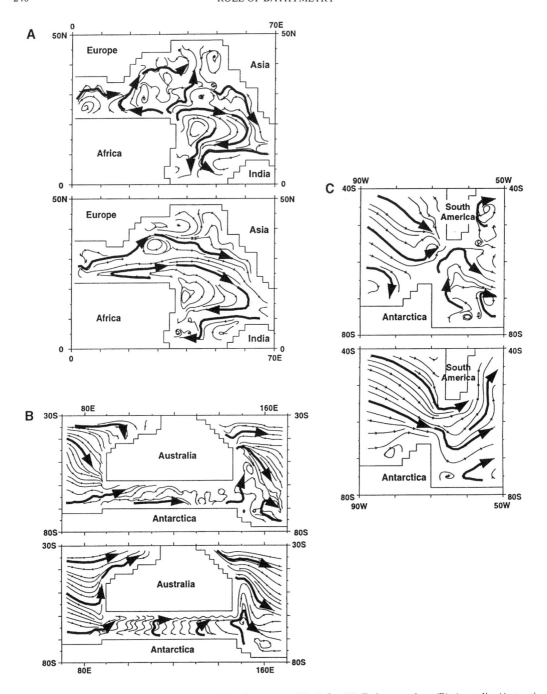

Fig. 11.11. (A) Model-simulated surface circulation (uppermost 50 m) for (A) Tethyan region, (B) Australian/Antarctic Passage, and (C) Drake Passageway. Upper maps show surface circulation for the realistic bathymetry case and lower maps show the bowl bathymetry case.

parameter for which there also exists no proxy. The use of a latitude-dependent δ_{sw} relationship (Zachos et al., 1994) allows for no longitudinal variability in δ_{sw}. One possible solution to the problem is the

use of model-simulated salinity and empirical salinity-δ_{sw} relationships in order to estimate seawater $\delta^{18}O$.

Open ocean surface water $\delta^{18}O$ and salinity are

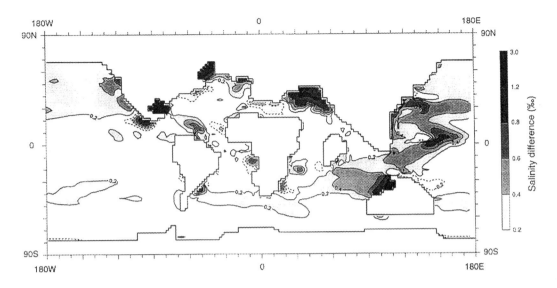

Fig. 11.12. Salinity difference (uppermost 50 m) calculated as bowl case minus realistic case. Contour interval is 0.2 ppt. Positive values (solid lines) indicate higher salinity in the bowl case. Negative values (dashed lines) indicate lower salinity in the bowl case. Zero salinity difference contours have been omitted for clarity.

both strongly controlled by evaporation and precipitation. There is therefore a strong linear correlation between open ocean surface water $\delta^{18}O$ and salinity (Broecker, 1989; Fairbanks et al., 1992; Zachos et al., 1994). Bice (1997) examined the use of model-predicted salinities and various salinity-δ_{sw} relationships in the calculation of early Eocene surface seawater $\delta^{18}O$. She showed that the model-simulated salinity is a viable tool for estimating the isotopic composition of surface seawater for past time periods, given the best possible reconstruction of boundary conditions. However, careful consideration must be given to the salinity-δ_{sw} relationship used. The Atlantic and Pacific open ocean salinity-δ_{sw} equation of Broecker (1989) allowed calculation of temperatures from carbonate $\delta^{18}O$ measurements on the surface-dwelling planktonic foraminifer *Morozovella* that were within 2°C of the model-predicted surface temperatures at 16 of 18 drilling program sites. Model-predicted and isotopic paleotemperatures differed by more than 2°C at Sites 94 (Gulf of Mexico, paleolatitude 27°N) and 152 (Central American Passageway, paleolatitude 14°N). Use of either of the two equatorial Atlantic salinity-δ_{sw} equations of Fairbanks et al., (1992) produced an improved fit between isotope and model

temperatures for these low latitude sites while the isotopic paleotemperatures at sites equatorward of 27° (drilling program Sites 364 and 865) remained within or very nearly within 2°C of the model-predicted temperature. (All calculated seawater $\delta^{18}O$ values were adjusted by -1.0‰ to correct for ice-free conditions (Shackleton and Kennett, 1975).

In this approach, the sensitivity of the calculated δ_{sw} value to the specified salinity is a function of the slope of the salinity-δ_{sw} equation used. Calculations made using the average Atlantic and Pacific equation of Broecker (at sites poleward of 27° latitude) are more sensitive to salinity than those made for sites equatorward of 27° for which the equatorial Atlantic equations of Fairbanks et al., (1992) were used. In the equation of Broecker, a difference in salinity of 0.4‰ leads to a difference in calculated δ_{sw} of 0.2‰ (Standard Mean Ocean Water) and an isotopic paleotemperature difference of approximately 1°C (using the temperature equation of Erez and Luz, 1983). If an accuracy of only 2°C is expected for the isotopic paleotemperatures, then surface salinity may vary by as much as 0.8‰. Therefore, for middle and high latitude regions, salinity differences of greater than 0.8‰ might be considered to be significant. Salinity differences of this magnitude between the realistic and bowl ocean experiments are observed along the northwestern coast of Australia and in the Gulf of Mexico, North Atlantic, and northern Tethyan Seaway.

Seawater $\delta^{18}O$ values calculated using the equatorial Atlantic equations of Fairbanks et al., (1992) are less sensitive to specified salinity. Using the western equatorial Atlantic Ocean salinity-δ_{sw} equation, a difference in salinity as great as 2‰ is required to produce a difference in δ_{sw} of 0.4‰ (SMOW) and an isotopic paleotemperature difference of approximately 2°C. Equatorward of 27° latitude, differences in surface salinity of this magnitude occur only in the western equatorial Pacific and northwestern margin of the Central American Passageway.

Discussion and Conclusion

This study suggests that smoothed basin topography and changes in the width and depth of ocean gateways causes a change in gyre transport, equatorial currents, and mid- and low-latitude meridional overturning. This sensitivity results in local temperature and salinity variations which may be significant in the uppermost 200 m of the model ocean. The globally averaged layer temperatures differ by less than 1°C, but grid-point differences as great as 10°C are observed in the upper 1000 m. Grid-point salinity differences of as much as 3‰ occur at depths above 200 m. The globally averaged meridional volume transport exhibits deepened and strengthened overturning in equatorial and mid-latitude cells in the upper 600 m which is not the result of a change in surface wind forcing. The greatest sensitivity of surface circulation to specified bathymetry is observed in narrow ocean gateways and in the narrower (Atlantic, Tethyan, and proto-Indian) ocean basins. It should be noted that Haq (1984) proposed a "Tethys Current" which carried water from the low latitude Southern Hemisphere region south of Indonesia northwestward through the Tethys Ocean and into the North Atlantic. Although the details of surface flow through narrow gateways is sensitive to specified bathymetry and atmospheric forcing (Bice, 1997), this feature is not predicted by any early Eocene experiment performed using POCM.

Gyre intensification might result from reduced friction due to smoothing of the continental

margins in the bowl case. However, it is difficult to be certain what causes the relative cooling and warming in the bowl case because a flat-bottom bathymetry was specified in the entire global ocean, rather than varying bottom topography in one region at a time or modifying the configuration of one gateway at a time. The removal of significant seafloor topographic features and dramatic changes in the gradient of the continental slope can act to alter the horizontal pressure gradient force and therefore the local geostrophic balance (Pickard and Emery, 1990). Previous model studies have noted the importance of this effect in regions of barriers to flow (Gill and Bryan, 1971; Maier-Reimer et al., 1990; England, 1992; Mikolajewicz et al., 1993; Toggweiler and Samuels, 1995).

In the surface ocean, the combined effect of increased equatorial upwelling and increased boundary current transport in the bowl simulation is a reduced meridional sea surface temperature gradient. This reduction is more pronounced in the relatively narrow ocean basins than for the Pacific Ocean or global ocean average. In a coupled ocean-atmosphere model experiment, such a reduction in gradient could lead to changes in evaporation, atmospheric circulation, and water vapor transport. To the extent that the meridional sea surface temperature gradient controls the position of the zonal jets in the overlying atmosphere, one ocean-atmosphere feedback resulting from a change in the temperature gradient could be a change in the position of the Antarctic Polar Frontal Zone and a change in the flux of Antarctic Intermediate Water (Emery, 1977; Pickard and Emery, 1990).

It has been suggested that increased gyre circulation is a potential mechanism for increased poleward heat transport by the oceans during periods of high latitude warmth (although the mechanism for such an increase is problematic during a period of decreased atmospheric winds such as the early Eocene). We have observed increased gyre transport in response to a change in the specified ocean bathymetry. The observed differences between these two simulations may help to identify increased gyre transport if it has occurred in past time periods. Figure 11.9 suggests that increased gyre circulation might be manifested in marine sediments as simultaneous warming and cooling at equivalent paleolatitudes on either side of the same ocean basin. In the transition into a period of net global warming in which increased gyre heat transport occurs, this cross-basinal warming/cooling might be observed as greater warming along western ocean boundaries than the coincident warming along eastern boundaries.

Also relevant to the understanding of the nature of intervals of high latitude warmth is the following: although the increase in gyre circulation results in a net increase in poleward heat transport in the uppermost 100 m of the model ocean (Fig. 11.13), no net high latitude warming is observed. Little equatorial cooling is observed. This suggests that either increased gyre heat transport alone is insufficient to explain high latitude warmth or a much larger increase in surface heat transport is needed in order to achieve high latitude warming if the cause is increased gyre transport.

Results from the present study indicate sensitivity to specified ocean bathymetry at local and regional scales. On a global average basis, however, model layer temperature and salinity differences are insignificant. Therefore, if the purpose of a model experiment is to simulate only global average conditions, it is not necessary to reconstruct realistic paleobathymetry. However, local circulation, temperature and salinity differences between the realistic and bowl bathymetry cases indicate that, in ocean modeling experiments where the purpose is to reconstruct local or meridional conditions, it is necessary to reconstruct realistic paleobathymetry. This

conclusion is supported by an additional sensitivity test (results not shown here) which was performed using the present-day continental configurations. Atmospheric forcing from a GENESIS present-day simulation was applied to a POCM simulation in which a modern bowl bathymetry was created using the same technique described above. The results of comparison of the realistic and bowl cases for the present-day control are consistent with those described here for the early Eocene.

This conclusion contrasts with that reached by Barron and Peterson (1990) using the earlier, coarser resolution ocean model and mid-Cretaceous geography. Given the coarse horizontal (5° x 5°) and vertical resolution (4 layers, with depths of 50 m, 500 m, 2000 m, and 4000 m) of that precursor model, it is not surprising that Barron and Peterson (1990) noted little sensitivity to bathymetry. With the exception of the Pacific basin, the mid-Cretaceous oceans were relatively narrow, especially when reconstructed on a 5° x 5° grid. With only four model depths, it would be difficult to construct two significantly different basin geometries in the narrower model ocean basins, the region where changes in the width and depth of numerous gateways would have played an important role in global ocean circulation.

More recently, Seidov and Maslin (1995) showed that the results of modern and last glacial maximum North Atlantic basin simulations, using the Geophysical Fluid Dynamics Laboratory community primitive equation model, were qualitatively insensitive to the specification of flat bottom versus real bottom topography. Their sensitivity tests used 2° x 2° horizontal resolution and both 6-layer and 12-layer model configurations. The lack of sensitivity noted by Seidov and Maslin (1995) may result from the fact that the model basin configuration (limited to the North Atlantic between 10° and 80°N) was varied, but the southern boundary forcings (the conditions at a vertical "wall" at 10°N) were not. Because of the demonstrated effect of varying the configuration of Southern Ocean gateways, some variation in the southern boundary condition to such a basin-scale model would be expected if the assumption of a flat bottom global ocean is applied on a global basis. In further North Atlantic basin model experiments, Seidov and Prien (1996) indicated that, although bathymetry does affect circulation significantly,

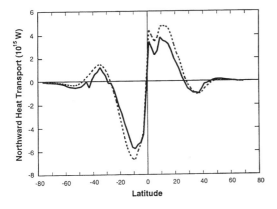

Fig. 11.13. Meridional profile of the mean annual northward ocean heat transport for the realistic bathymetry (solid curve) and bowl bathymetry (dashed line) cases. Heat transport is integrated over the uppermost 100 m and across all ocean basins.

differences arising from this boundary condition might be less important than those caused by a dramatic change in atmospheric forcing to the ocean model, as in the case of glacial-interglacial model studies.

Acknowledgments

The Parallel Ocean Climate Model was developed by Albert Semtner of the Naval Postgraduate School and Robert Chervin of the National Center for Atmospheric Research. The GENESIS Earth systems model was developed by Starley Thompson and Dave Pollard of the Interdisciplinary Climate Systems Section at the National Center for Atmospheric Research. We thank Lisa Cirbus Sloan (University of California Santa Cruz) for use of her early Eocene GENESIS model paleogeography and James L. Sloan (Penn State) for technical assistance with plate rotation programs. The manuscript was improved by the comments of Kevin Burke, Tom Crowley, Lisa Sloan, and an anonymous reviewer. This work was supported by NSF grant ATM-9113944 to Eric J. Barron.

References

Barker, P. F., and J. Burrell, 1982, The influence upon Southern Ocean circulation, sedimentation, and climate of the opening of Drake Passage, in *Antarctic Geoscience*, C. Craddock (ed.), pp. 377–385, University of Wisconsin Press, Madison, Wisc.

Barron, E. J., 1987, Cretaceous plate tectonic reconstructions, *Palaeogeogr., Palaeoclimatol., Palaeoecol., 59*, 3–29.

Barron, E. J., and W. H. Peterson, 1989, Model simulation of the Cretaceous ocean circulation, *Science, 244*, 684–686.

Barron, E. J., and W. H. Peterson, 1990, Mid-Cretaceous ocean circulation: Results from model sensitivity studies, *Paleoceanography, 5*, 319–337.

Barron, E. J., and W. H. Peterson, 1991, The Cenozoic ocean circulation based on ocean General Circulation Model results, *Palaeogr., Palaeoclimatol., Palaeoecol., 83*, 1–28.

Berggren, W. A., 1982, The role of ocean gateways in climatic change, in *Climate in Earth History*, W. H. Berger and J. C. Crowell (eds.), pp. 118–125, National Academy, Washington, D.C.

Bice, K. L., 1997, An investigation of early Eocene deep water warmth using uncoupled atmosphere and ocean general circulation models: Model sensitivity to geography, initial temperatures, atmospheric forcing and continental runoff, Ph.D. dissertation, 363 pp., Pennsylvania State University, Pa.

Bice, K. L., L. C. Sloan, and E. J. Barron, 1998, High resolution global ocean circulation simulation of the early Eocene: Comparison with foraminiferal isotope paleotemperatures, in *Warm Climates in Earth History*, B. T. Huber, et al. (eds.), in review.

Brass, G. W., E. Saltzmann, J. L. Sloan, II, J. R. Southam, W. W. Hay, W. T. Holser, and W. H. Peterson, 1982a, Ocean circulation, plate tectonics, and climate, in *Climate in Earth History*, W. H. Berger and J. C. Crowell (eds.), pp. 83–89, National Academy, Washington, D.C.

Brass, G. W., J. R. Southam, and W. H. Peterson, 1982b, Warm saline bottom water in the ancient ocean, *Nature, 296*, 620–623.

Broecker, W. S., 1989, The salinity contrast between the Atlantic and Pacific Oceans during glacial time, *Paleoceanography, 4*, 207–212.

Bryan, K., 1969, A numerical model for the study of the world ocean, *J. Comp. Phys., 4*, 347–376.

Bryan, K., and M. D. Cox, 1972, The circulation of the world ocean: A numerical study. Part I, A homogeneous model, *J. Phys. Oceanogr., 2*, 319–335.

Bye, J. A. T., and T. W. Sag, 1972, A numerical model for circulation in a homogeneous world ocean, *J. Phys. Oceanogr., 2*, 305–318.

Cande, S. C., and D. V. Kent, 1992, A new geomagnetic polarity time scale for the Late Cretaceous and Cenozoic, *J. Geophys. Res., 97*, 13,917–13,951.

Cande, S. C., J. L. LaBrecque, R. L. Larson, W. C. Pitman, III, X. Golovchenko, and W. F. Haxby, 1989, *Magnetic Lineations of the World's Ocean Basins, Lamont-Doherty Geological Observatory Contribution No. 4367*, American Association of Petroleum Geologists, 1 sheet, Tulsa, Okla.

Carmack, E. C., and T. D. Foster, 1975, On the flow of water out of the Weddell Sea, *Deep-Sea Res., 22*, 711–724.

Cottereau, N., 1992, Paleobathymetric reconstruction of the Tethys at the end of the Jurassic, *Ab-*

stracts, 29th International Geological Congress, p. 374, Kyoto, Japan.

Covey, C., and S. T. Thompson, 1989, Testing the effects of ocean heat transport on climate, *Palaeogr., Palaeoclimatol., Palaeoecol., 75,* 331–341.

Cox, M. D., 1970, A mathematical model of the Indian Ocean, *Deep-Sea Res., 17,* 47–75.

Cox, M. D., 1975, A baroclinic numerical model of the world ocean: Preliminary results, in *Numerical Models of Ocean Circulation,* pp. 107–118, National Academy, Washington, D.C.

Cox, M. D., 1984, A primitive equation three-dimensional model of the ocean, *Technical Report 1, NOAA Geophysical Fluid Dynamics Laboratory,* Princeton University, Princeton, N.J.

Cox, M. D., 1989, An idealized model of the world ocean. Part I: The global-scale water masses, *J. Phys. Oceanogr., 19,* 1730–1752.

Dercourt, J., et al., 1986, Geologic evolution of the Tethys belt from the Atlantic to the Pamirs since the Lias, *Tectonophysics, 123,* 241–315.

Emery, W. J., 1977, Antarctic Polar Frontal Zone from Australia to the Drake Passage, *J. Phys. Oceanogr., 7,* 811–822.

England, M. H., 1992, On the formation of Antarctic intermediate and bottom water in ocean general circulation models, *J. Phys. Oceanogr., 22,* 918–926.

Erez, B., and J. Luz, 1983, Experimental paleotemperature equation for planktonic foraminifera, *Geochim. Cosmochim. Acta, 47,* 1025–1031.

Fairbanks, R. G., C. D. Charles, and J. D. Wright, 1992, Origin of global meltwater pulses, in *Radiocarbon After Four Decades: An Interdisciplinary Perspective,* R. E. Taylor, et al. (eds.), pp. 473–500, Springer-Verlag, New York.

Fowler, C. M. R., 1990, *The Solid Earth: An Introduction to Global Geophysics,* Cambridge University Press, Cambridge.

Gill, A. E., and K. Bryan, 1971, Effects of geometry on the circulation of a three-dimensional southern-hemisphere ocean model, *Deep-Sea Res., 18,* 685–721.

Haq, B. U., 1984, Paleoceanography: A synoptic overview of 200 million years of ocean history, in *Marine Geology and Oceanography of Arabian Sea and Coastal Pakistan,* B. U. Haq and J. D. Milliman (eds.), pp. 201–231, Van Nostrand Reinhold, New York.

Hay, W. W., C. A. Shaw, and C. N. Wold, 1989, Mass-balanced paleogeographic reconstructions, *Geologische Rundschau, 78,* 207–242.

Holland, W. R., and A. D. Hirschman, 1972, A numerical calculation of the circulation in the North Atlantic Ocean, *J. Phys. Oceanogr., 2,* 336–354.

Jacobs, J. A., 1994, *Reversals of the Earth's Magnetic Field,* Cambridge University Press, Cambridge.

Jarvis, G. T., and W. R. Peltier, 1982, Mantle convection as a boundary layer phenomenon, *Geophys. J. Royal Astronom. Soc., 68,* 389–427.

Johnson, J. A., and R. B. Hill, 1975, A three-dimensional model of the Southern Ocean with bottom topography, *Deep-Sea Res., 22,* 745–751.

Keller, G., T. Herbert, R. Dorsey, S. D'Hondt, M. Johnsson, and W. R. Chi, 1987, Global distribution of late Paleogene hiatuses, *Geology, 15,* 199–203.

Kennett, J. P., 1977, Cenozoic evolution of Antarctic glaciation, the circum-Antarctic ocean, and their impact on global paleoceanography, *J. Geophys. Res., 82,* 3843–3860.

Kennett, J. P., 1986, Miocene paleoceanography and plankton evolution, in *Mesozoic and Cenozoic Oceans, Geodyn. Ser., vol. 15,* K. L. Hsü (ed.), pp. 119–122, AGU, Washington, D.C.

Kent, D. V., and F. M. Gradstein, 1986, A Jurassic to Recent chronography, in *The Western North Atlantic Region, The Geology of North America, vol. M,* P. R. Vogt and B. E. Tucholke (eds.), pp. 45–50, Geological Society of America, Boulder, Colo.

Lawver, L. A., L. M. Gahagan, and M. F. Coffin, 1992, The development of paleoseaways around Antarctica, in *The Antarctic Paleoenvironment: A Perspective on Global Change, Ant. Res. Ser., vol. 56,* J. P. Kennett and D. A. Warnke (eds.), pp. 7–30, AGU, Washington, D.C.

Maier-Reimer, E., and K. Hasselmann, 1987, Transport and storage of CO2 in the ocean: An inorganic ocean-circulation carbon cycle model, *Clim. Dynamics, 2,* 63–90.

Maier-Reimer, E., U. Mikolajewicz, and T. J. Crowley, 1990, Ocean general circulation model sensitivity experiment with an open central American isthmus, *Paleoceanography, 5,* 349–366.

McCann, M. P., A. J. Semtner, and R. M. Chervin,

1994, Transports and budgets of volume, heat, and salt from a global eddy-resolving ocean model, *Clim. Dynamics, 10*, 59–80.

Menard, H. W., 1969, Elevation and subsidence of oceanic crust, *Earth Planet. Sci. Lett., 6*, 275–284.

Mikolajewicz, U., E. Maier-Reimer, T. J. Crowley, and K.-Y. Kim, 1993, Effect of Drake and Panamanian gateways on the circulation of an ocean model, *Paleoceanography, 8*, 409–426.

Mountain, G. S., and K. G. Miller, 1992, Seismic and geologic evidence for Early Paleogene deepwater circulation in the western North Atlantic, *Paleoceanography, 7*, 423–439.

Müller, R. D., J.-Y. Royer, and L. A. Lawver, 1993, Revised plate motions relative to the hotspots from combined Atlantic and Indian Ocean hotspot tracks, *Geology, 21*, 275–278.

Nagihara, S., C. R. B. Lister, and J. G. Sclater, 1996, Reheating of old oceanic lithosphere: Deductions from observations, *Earth Planet. Sci. Lett., 139*, 91–104.

Oberhänsli, H., 1992, The influence of the Tethys on the bottom waters of the early Tertiary ocean, in *The Antarctic Paleoenvironment: A Perspective on Global Change, Ant. Res. Ser., vol. 56*, J. P. Kennett and D. A. Warnke (eds.), pp. 167–184, AGU, Washington, D.C.

Pak, D. K., and K. G. Miller, 1992, Paleocene to Eocene benthic foraminiferal isotopes and assemblages: Implications for deepwater circulation, *Paleoceanography, 7*, 405–422.

Parsons, B., and J. G. Sclater, 1977, An analysis of the variation of ocean floor bathymetry and heat flow with age, *J. Geophys. Res., 82*, 803–827.

Peltier, W. R., 1989, Mantle convection and plate tectonics: The emergence of paradigm in global geodynamics, in *Mantle Convection*, W. R. Peltier (ed.), pp. 1–22, Gordon and Breach, New York.

Pickard, G. L., and W. J. Emery, 1990, *Descriptive Physical Oceanography*, Pergamon, Elmsford, N.Y.

Reid, J. L., and W. D. Nowlin, Jr., 1971, Transport of water through the Drake Passage, *Deep-Sea Res., 18*, 51–64.

Robert, C., and H. Chamley, 1992, Late Eocene-Early Oligocene evolution of climate and marine circulation: Deep-sea clay mineral evidence, in *The Antarctic Paleoenvironment: A Perspective*

on *Global Change, Ant. Res. Ser., vol. 56*, J. P. Kennett and D. A. Warnke (eds.), pp. 97–117, AGU, Washington, D.C.

Ronov, A. B., V. Khain, and A. Balukhovsky, 1989, *Atlas of Lithological-paleogeographical Maps of the World, Mesozoic and Cenozoic of Continents and Oceans*, USSR Academy of Sciences, Leningrad.

Sclater, J. G., and D. P. McKenzie, 1973, Paleobathymetry of the South Atlantic, *Geol. Soc. Am. Bull., 84*, 3203–3216.

Sclater, J. G., R. N. Anderson, and M. L. Bell, 1971, Elevation of ridges and the evolution of the central eastern Pacific, *J. Geophys. Res., 76*, 7888–7915.

Sclater, J. G., S. Hellinger, and C. Tapscott, 1977, The paleobathymetry of the Atlantic Ocean from the Jurassic to the present, *J. Geol., 85*, 509–552.

Sclater, J. G., C. Jaupart, and D. Galson, 1980, The heat flow through oceanic and continental crust and the heat loss of the Earth, *Rev. Geophys. Space Phys., 18*, 269–311.

Sclater, J. G., L. Meinke, A. Bennett, and C. Murphy, 1985, The depth of the ocean through the Neogene, in *The Miocene Ocean: Paleogeography and Biogeography*, J. P. Kennett (ed.), pp. 1–19, *Mem. Geological Society of America, 163*, Boulder, Colo.

Scotese, C. R., and J. Golonka, 1992, *PALEOMAP Paleogeographic Atlas*, PALEOMAP Progress Report No. 20, Department of Geology, University of Texas at Arlington, 43 pp.

Seidov, D., and M. Maslin, 1995, Glacial to interglacial changes of the North Atlantic thermohaline conveyor, *ICP V Program and Abstracts, Fifth International Conference on Paleoceanography*, p. 149, University of New Brunswick, Fredericton, New Brunswick.

Seidov, D., and M. Prien, 1996, A coarse resolution North Atlantic ocean circulation model: An intercomparison study with a paleoceanographic example, *Annales Geophysicae, 14*, 246-257.

Semtner, A. J., 1974, *An oceanic general circulation model with bottom topography*, Technical Report 9, Dept. of Meteorology, University of California, Los Angeles, 99 pp.

Semtner, A. J., and R. M. Chervin, 1992, General circulation from a global eddy-resolving model, *J. Geophys. Res., 97*, 5493–5550.

Shackleton, N. J., and J. P. Kennett, 1975, Pale-otemperature history of the Cenozoic and the initiation of Antarctic glaciation: Oxygen and carbon isotope analyses in DSDP Sites 277, 279, and 281, J. P. Kennett and R. E. Houtz (eds.), pp. 743–755, *Initial Rep. Deep Sea Drill. Proj., 29.*

Sloan, L. C., 1990, Determination of critical factors in the simulation of Eocene global climate, with special reference to North America, Ph.D. dissertation, 282 pp., Pennsylvania State University, Pa.

Sloan, L. C., 1994, Equable climates during the early Eocene: Significance of regional paleogeography for North American climate, *Geology, 22,* 881–884.

Sloan, L. C., and E. J. Barron, 1992, A comparison of Eocene climate model results to quantified paleoclimatic interpretations, *Palaeogr., Palaeoclimatol., Palaeoecol., 93,* 183–202.

Steckler, M. S., L. Lavier, and M. Séranne, 1995, From carbonate ramps to clastic progradation: Morphology and stratigraphy of continental margins during Tertiary global change, *ICP V Program and Abstracts, Fifth International Conference on Paleoceanography,* p. 83, University of New Brunswick, Fredericton, New Brunswick.

Stein, C. A., and S. Stein, 1992, A model for the global variation in oceanic depth and heat flow with lithospheric age, *Nature, 359,* 123–129.

Stein, C. A., and S. Stein, 1993, Constraints on Pacific midplate swells from global depth-age and heat flow-age models, in *The Mesozoic Pacific: Geology, Tectonics, and Volcanism, Geophys. Monogr. Ser. vol. 77,* M. S. Pringle, et al. (eds.), pp. 53–76, AGU, Washington, D.C.

Thiede, J., 1982, Paleogeography and paleobathymetry: Quantitative reconstructions of ocean basins, in *Tidal Friction and the Earth's Rotation II,* P. Brosche, and J. Sündermann (eds.), pp. 229–239, Springer-Verlag, N. Y.

Thompson, S. L., and D. Pollard, 1995a, A global climate model (GENESIS) with a land-surface transfer scheme (LSX), Part I: Present climate simulation, *J. Clim., 8,* 732–761.

Thompson, S. L., and D. Pollard, 1995b, A global climate model (GENESIS) with a land-surface transfer scheme (LSX), Part II: CO_2 sensitivity, *J. Clim., 8,* 1104–1121.

Toggweiler, J. R., and B. Samuels, 1993, Is the magnitude of the deep outflow from the Atlantic Ocean actually governed by southern hemisphere winds? in *The Global Carbon Cycle,* M. Heimann (ed.), pp. 303–331, Springer-Verlag, New York.

Toggweiler, J. R., and B. Samuels, 1995, Effect of Drake Passage on the global thermohaline circulation, *Deep-Sea Res., 42,* 477–500.

Vogt, P. R., and B. E. Tucholke, 1986, Paleogeography: Late Cretaceous to Holocene, in *The Western North Atlantic Region,* P. R. Vogt and B. E. Tucholke (eds.), Plate 10, Geological Society of America, Boulder, Colo.

Washington, W. M., G. A. Meehl, L. VerPlank, and T. W. Bettge, 1994, A world ocean model for greenhouse sensitivity studies: Resolution intercomparison and the role of diagnostic forcing, *Clim. Dynamics, 9,* 321–344.

Wold, C. N., 1995, Palaeobathymetric reconstruction on a gridded database: The northern North Atlantic and southern Greenland-Iceland-Norwegian sea, in *The Tectonics, Sedimentation and Palaeoceanography of the North Atlantic Region, Geological Society Special Publication No. 90,* R. A. Scrutton, et al. (eds.), pp. 271–302, The Geological Society, London.

Zachos, J. C., L. D. Stott, and K. C. Lohmann, 1994, Evolution of early Cenozoic marine temperatures, *Paleoceanography, 9,* 353–387.

PART VII

TECTONICS AND CO₂

CHAPTER 12

Sensitivity of Phanerozoic Atmospheric CO_2 to Paleogeographically Induced Changes in Land Temperature and Runoff

Robert A. Berner

In a discussion of the role of tectonic boundary conditions in climate reconstruction, it is important to examine not only how changes in paleogeography directly affect climate (e.g., Fawcett and Barron, this volume, Chapter 2), but also how such changes indirectly affect climate via changes in the level of atmospheric CO_2. Variations in the size and elevation of continents and migration of the continents through different climate zones will affect relief, land temperature, and river runoff which are major factors that control the rate of uptake of CO_2 via rock weathering. By means of the atmospheric greenhouse effect, changes in atmospheric CO_2 in turn affect land temperature and runoff. In this way there is a feedback between tectonics and climate by way of atmospheric CO_2.

Over millions of years the level of atmospheric CO_2 is controlled predominantly by the exchange of carbon between rocks and the combined atmosphere/biosphere/ocean system. On this time scale, exchange of CO_2 between each of the atmospheric, biospheric and oceanic reservoirs, which dominates the cycle of carbon on shorter time scales, plays a secondary role to the formation and breakdown of rocks. Considerable discussion and modeling of various aspects of the multimillion year or "long-term carbon cycle" have been presented since the late 1970's (e.g., Holland, 1978; Walker et al., 1981; Berner et al., 1983; Lasaga et al., 1985; Volk, 1987; 1989, Kump, 1989; Berner, 1991, 1994; Francois and Walker, 1992; Caldeira and Kasting, 1992). These models help to illuminate which factors are potentially important in affecting atmospheric CO_2 content and they help to organize thinking about long term climate and CO_2 changes. Also, by employing self-consistent mass conservation, the models avoid fundamental problems that plague other approaches to global climate change that are not based on carbon mass conservation (e.g., Edmond et al., 1995).

As part of this work the GEOCARB model (Berner, 1991, 1994) has been constructed to calculate the level of atmospheric CO_2 over Phanerozoic time and to show how results depend on different estimates of the important input parameters of the model. Processes, represented schematically in Figure 12.1, that are treated quantitatively by the GEO-CARB model include: uptake of atmospheric CO_2 by the weathering of Ca–Mg silicate minerals on the continents, the release of CO_2 to the atmosphere by the oxidative weathering of organic matter in shales, the removal of carbon from the surficial reservoir by deposition of carbonate minerals and organic matter in oceanic sediments, and the diagenetic/metamorphic/volcanic thermal breakdown of carbonates and organic matter, with consequent degassing of CO_2, upon deep burial of the sediments (including burial into the mantle).

The fundamental basis of the GEOCARB and similar models rests on the reactions:

$$CO_2 + CaSiO_3 \leftrightarrow CaCO_3 + SiO_2 \quad \text{(i)}$$

$$CO_2 + MgSiO_3 \leftrightarrow MgCO_3 + SiO_2 \quad \text{(ii)}$$

$$CO_2 + H_2O \leftrightarrow CH_2O + O_2 \quad \text{(iii)}$$

Written from left-to-right, reactions (i) and (ii) represent the weathering of calcium and magnesium silicate minerals with the transfer of carbon from the atmosphere to carbonate minerals buried in sediments. (These two weathering reactions summa-

Long Term Carbon Cycle

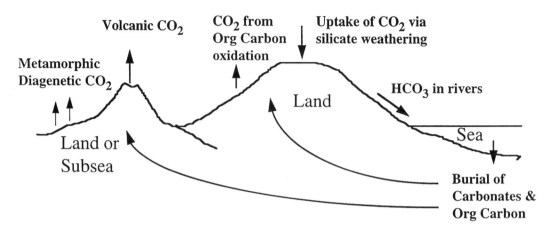

Fig. 12.1. Diagrammatic representation of the long term (rock) carbon cycle

rize many intermediate steps including photosynthetic fixation of CO_2, root respiration, organic litter decomposition in soils, the reaction of carbonic and organic acids with primary silicate minerals, the conversion of CO_2 to HCO_3^- in soil and ground water, the flow of riverine HCO_3^- to the sea, and the precipitation of oceanic HCO_3^- as carbonates in bottom sediments. For a detailed discussion consult Berner, 1995). Reaction (iii) summarizes the burial of organic matter in sediments representing an excess of global photosynthesis over respiration. Written from right-to-left, reactions (i) and (ii) represent the thermal decomposition of carbonates upon deep burial with the consequent degassing of CO_2 back to the atmosphere. Reaction (iii), written from right-to-left, represents either the oxidative weathering of ancient organic matter on the continents and/or the thermal decomposition of buried organic matter to reduced carbon-containing gases with the subsequent oxidation of these gases to CO_2 by atmospheric O_2.

In the previous publications (Berner, 1991, 1994) the GEOCARB model was used to illustrate the sensitivity of atmospheric CO_2 mainly to changes in continental relief, the evolution of land plants, and solar evolution as they affect chemical weathering, and variations in seafloor spreading rate and carbonate content as they affect degassing. Results show that during the early and middle Paleozoic the rise of vascular land plants between 380 and 350 Ma was the dominant influence on bringing about a change in atmospheric CO_2 level. Population of the continents by plants brought about a large slow drop in CO_2 due both to the acceleration of silicate rock weathering and to the production of bacterially resistant organic matter (e.g., lignin) which resulted in the enhanced burial of organic matter in sediments. Once the land biota were established other factors became more important including the uplift of mountains as they affect the rate of silicate weathering and changes in the rate of global degassing of CO_2. Throughout the Phanerozoic the slow increase in solar radiation also had an effect on atmospheric CO_2 because the rate of CO_2 uptake by rock weathering is positively correlated with temperature. Thus, past changes in atmospheric CO_2 have been a function of a complex interaction of solar, biological, and geological processes.

In this chapter we look at two additional factors. They are changes in land temperature and runoff that result from changes in paleogeography and the effect that these changes may have had on the weathering rate of Ca–Mg silicates and the level of atmospheric CO_2 over Phanerozoic time. Also investigated is the range of CO$_2$ values that would result from a reasonable range of estimates of land temperature and runoff for specific time periods.

GEOCARB modeling examines a number of individual processes as they affect the carbon cycle and atmospheric CO$_2$. It is realized that interactions between some of the processes, as well as additional processes, may lead to non-linearities that could affect results. An example is change in continental paleolatitude or paleotopography and how this might affect the distribution of vegetation which might in turn affect weathering rate. Another is change in continental position to the point where much land is permanently covered by ice resulting in the inhibition of rock weathering. Such complexities are beyond the scope of this chapter and must await further modeling studies.

Method

A detailed description of the GEOCARB model is given in earlier articles (Berner, 1991, 1994) and will not be repeated here. Instead, modifications of this model will be discussed. Here a new dimensionless parameter $f_H(t)$ is introduced to express the effects of changes in paleogeography on global mean land temperature as it affects the rate of Ca–Mg silicate weathering. This is added to a number of dimensionless parameters expressing the effects on silicate weathering of changes in relief $f_R(t)$, continental runoff $f_D(t)$, the evolution of land plants $f_E(t)$ and negative feedback from changes in global mean temperature and plant growth due to changes in atmospheric CO$_2$ plus solar radiation $f_B(t)$. Thus, the amended expression for the weathering of Ca and Mg silicates is given by

$$F_{wsi(t)} = f_R(t)\, f_D(t)^{0.65}\, f_E(t)\, f_H(t)\, f_B(t)\, F_{wsi}(0) \quad (12.1)$$

where F_{wsi} refers to the uptake of CO$_2$ from the atmosphere, by the weathering of Ca and Mg silicates with the carbon eventually transferred to buried limestones and dolostones, and t and 0 refer to some past time and the present, respectively.

The expression for $f_H(t)$ is based on laboratory and field studies of the effect of temperature on the rate of silicate weathering (Berner, 1994, 1995). It is

$$f_H(t) = \exp\,[0.09(T_L(t)-12.4)] \quad (12.2)$$

where $T_L(t)$ = global mean land temperature at time t and 12.4 is that for the present earth. Values for $T_L(t)$ are obtained by interpolation from the data of Otto-Bliesner (1995) for the continents at specific times in the geologic past (514, 458, 433, 425, 390, 342, 306, 255, 237, 195, 130, 69, 50, and 0 Ma), calculated by her from a low-resolution general circulation model (GCM) model for idealized flat sealevel continents with no prescribed ice sheets. Also, Otto-Bliesner's results are for present day atmospheric CO$_2$ level and solar luminosity. This is appropriate for the GEOCARB modeling in that the effects of changes in luminosity and CO$_2$ on temperature are included separately in the feedback parameter $f_B(t)$. In other words, $f_H(t)$ is for the effect on weathering of changes in global mean land temperature due to changes in continental position and size, at constant CO$_2$, constant solar radiation, and constant continental relief. By contrast, $f_B(t)$ expresses the effect on weathering of changes in solar luminosity and CO$_2$, but at constant relief, size, and position, and $f_R(t)$ expresses the effect on weathering of changes in elevation/relief at constant CO$_2$, luminosity, size, and position.

The effect of changes in runoff due to changes in paleogeography are expressed by the parameter $f_D(t)$ which is defined as:

$$f_D(t) = \text{global runoff per unit area } (t)/\text{global runoff per unit area } (0)$$

Note that $f_D(t)$ is raised to the 0.65 power in equation (12.1) to express the fact that silicate mineral dissolution cannot keep pace with water flow resulting in dilution at higher flow rates (Berner, 1994). Theoretical treatment of the dependence of $f_D(t)$ on paleogeography is similar to that applied to the land temperature parameter $f_H(t)$. Values of $f_D(t)$ used here and in previous GEOCARB modeling (Berner, 1994) are also taken from the work of Otto-Bliesner (1995) and are based on flat sea level continents at present levels of CO$_2$ and solar radiation. Thus, $f_D(t)$, like $f_H(t)$, expresses only the effects of changing continental size and position on weathering rate. The effects of changing global temperature, forced by CO$_2$ and solar radiation, on runoff as part of the global hydrological cycle, are contained within the expression $f_B(t)$.

Note also that $f_D(t)$ is not multiplied by changes

in total global land area to get total water flux to the sea. This is because (Berner, 1994) most weathering of Ca and Mg silicates occurs in areas of high relief such as mountains (Stallard, 1992). Changes in total land area over geological time are brought about by the flooding or emergence of only coastal lowlands. Such areas generally are underlain by sedimentary rocks which are relatively poor in primary Ca and Mg silicates and rich in less-reactive clay weathering products. Thus, changes in total continental land area should not affect Ca and Mg silicate weathering nearly as much as changes in continental relief. With high relief, protective mantles of clay weathering residues can be readily stripped by physical erosion thereby exposing the underlying primary, highly weatherable Ca and Mg silicates to soil solutions. High relief also allows for enhanced runoff due to both rapid flushing of the rocks and to orographically-induced rainfall (e.g., Otto-Bliesner, Chapter 6).

Even though the effect of changes in total land area are not directly incorporated in the expression for runoff, it is still necessary to include land area somehow in the weathering expression (eq. 12.1) so that total water flux (runoff x area) is expressed in the correct units. This is done in the GEOCARB model by including the effects of changes in the land area of *mountainous regions* in the term $f_R(t)$, albeit indirectly, through the use of strontium isotopes as a proxy for all effects of mountain uplift on silicate weathering (enhanced erosion, higher relief, more mountain area, enhanced runoff, etc.) (e.g., see Raymo, 1991; Richter et al., 1992; Berner, 1994). Nevertheless, there is some controversy concerning the effects of mountain uplift on weathering and the use of strontium isotopes to express this (e.g., Lasaga et al, 1994), and for the sake of completeness a curve is plotted for an additional parameter $f_{AD}(t)$ which is dimensionally correct and which is defined as

$$f_{AD}(t) = f_A(t) f_D(t) \qquad (12.3)$$

where $f_A(t)$ is defined as total land area(t)/total land area (0). (See Berner, 1991 for the actual values used for $f_A(t)$.) When using $f_{AD}(t)$ in place of $f_D(t)$ in equation (12.1) it is also raised to the power 0.65 to express riverine dilution. This results in $f_A(t)$ also being raised to 0.65 which is not unreasonable in that increases in land area normally involve large

arid tracts that do not affect runoff. (What has been said here about Ca and Mg silicates is not true for Ca and Mg carbonates because limestones are highly reactive and dissolve rapidly during weathering regardless of elevation or relief, for example, Holland, 1978; Berner and Berner, 1996. Thus, because there is little dilution of dissolved carbonate and because carbonate dissolution occurs everywhere, $f_A(t)$ is included, and to the first power, in the GEOCARB expression for carbonate weathering.)

For the sake of completeness some minor changes in the GEOCARB II model used in this chapter are mentioned here. They are that the period of the rise and spread of vascular land plants, based on the study of Algeo et al (1995), is put between 380 and 350 Ma instead of 350 and 300 Ma, as used in the original model, and a straightforward Newton-Raphson technique (3 iterations) is now used to solve for RCO$_2$ from values of $f_B(t)$. (For details consult Berner, 1994.)

Results and Discussion

Plots of the two parameters of major interest $f_H(t)$ (land temperature parameter) and $f_D(t)$ (runoff parameter) versus time are shown in Figure 12.2. Comparison of the effect on atmospheric CO$_2$, of changes in global mean land temperature due to changes in paleogeography, is shown in Figure 12.3 for variable $f_H(t)$ (from Figure 12.2) and for $f_H(t)=1$. The parameter RCO$_2$ represents the ratio of the mass of atmospheric CO$_2$ at some past time (t) to that at present. As can be seen, the inclusion of variable land temperature does not produce dramatic changes in overall results. For most of the Phanerozoic the effects of land plant evolution, mountain uplift, solar evolution and volcanic/metamorphic degassing overshadow the effects of drifting continents on atmospheric CO$_2$. (Note, however that the effects of prescribed polar ice on the mean temperature of land undergoing weathering is not included in the Otto-Bliesner data used to construct Fig. 12.2.) The most notable paleogeographic effect is during the early Paleozoic (520–420 Ma) where fluctuations in land temperature bring about considerable differences from the plot for constant $f_H(t)$. The local minimum in the RCO$_2$ curve during the late Ordovician may, in fact, be indicative of cooling which could have

Paleogeographic Parameters

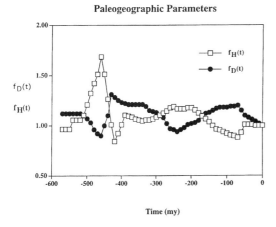

Fig. 12.2. Plots of $f_H(t)$ and $f_D(t)$ vs time. $f_H(t)$ and $f_D(t)$ represent the effects of changing mean land temperature and runoff, respectively, on the rate of weathering (see text). Values based on the temperature and runoff calculations of Otto-Bliesner (1995).

contributed to the short lived glaciation at the end of the Ordovician (Crowley and Baum, 1995). Recent work (Gibbs et al., 1997) indicates the high sensitivity of late Ordovician glaciation to the co-existing level of atmospheric CO$_2$.

The effect of changes in global runoff are shown in Figure 12.4. Here curves for variable $f_D(t)$ (based on the data of Figure 12.2) and for $f_D(t)=1$ are compared. As one can see, there is very little effect on RCO$_2$ of variations in $f_D(t)$. Also shown in

Figure 12.4 is the effect of combining runoff with total land area in terms of the parameter $f_{AD}(t)$. This does cause considerable increase in RCO$_2$ for the early Paleozoic because of the small area of total land at that time (Ronov, 1976; Scotese et al., 1979). However, as pointed out above, it is considered unreasonable to let silicate weathering respond to changes in total land area because most area changes are due to the exposure or inundation of coastal lowlands that are underlain mainly by sediments that are relatively impoverished in readily weatherable Ca and Mg silicates.

Since estimates of runoff and land temperature over Phanerozoic time are only approximations, it is instructive to investigate the sensitivity of the level of atmospheric CO$_2$ to a reasonable range of values of $f_D(t)$ and $f_H(t)$. To do this two fixed times in the Paleozoic were chosen which exhibit very different RCO$_2$ values according to the GEOCARB model. They are 440 Ma (Ordovician) and 300 Ma (Upper Carboniferous) and results are shown in Figures 12.5 and 12.6. (It should be noted that consideration of land temperature and runoff as independent variables is somewhat unrealistic because of the inverse correlation shown in Figure 12.2). For comparison, the situation where runoff is multiplied by total land area, in terms of the parameter $f_{AD}(t)$, is also shown in the figures. The strong response to temperature emphasizes the importance of continental drift as it affects the paleolatitude of the continents and consequently

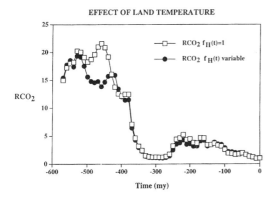

Fig. 12.3. Plot of RCO$_2$ vs time for constant global mean land temperature $(f_H(t)=1)$ and for variable land temperature $(f_H(t)$ data from Figure 12.2). RCO$_2$ is the ratio of the mass of atmospheric CO$_2$ at some past time to that at present.

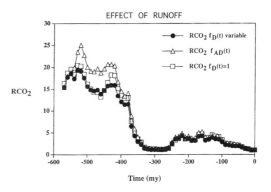

Fig. 12.4. Plot of RCO$_2$ vs time for different formulations of runoff. $f_{AD}(t)$ is the ratio for some past time to that at present of runoff per unit area $f_D(t)$ multiplied by total land area $f_A(t)$.

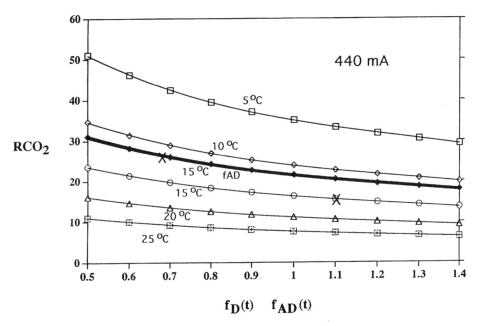

Fig. 12.5. Plot of RCO₂ vs $f_D(t)$ or $f_{AD}(t)$ for different global mean land temperatures (plotted as contours) for the Upper Ordovician (440 Ma). The heavier curve marked f_{AD} is for $f_{AD}(t)$; all other curves are for $f_D(t)$. Values denoted by an X represent best estimates based on Figure 12.2. (The X to the left is for $f_{AD}(t)$.)

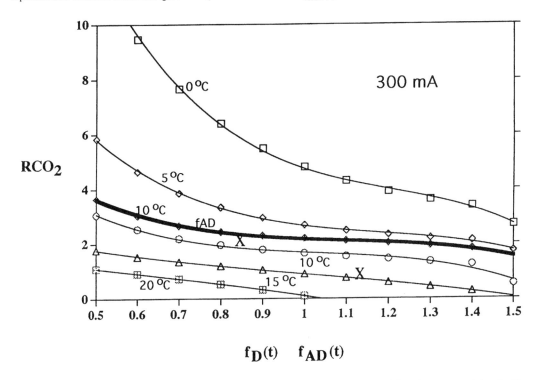

Fig. 12.6. Plot of RCO₂ vs $f_D(t)$ or $f_{AD}(t)$ for different global mean land temperatures (plotted as contours) for the Upper Carboniferous (300 Ma). The heavier curve marked f_{AD} is for $f_{AD}(t)$; all other curves are for $f_D(t)$. Values denoted by an X represent best estimates based on Figure 12.2. (The X to the left is for $f_{AD}(t)$.) Note much smaller vertical scale than for Figure 12.5.

mean continental temperature (Worsley et al., 1994). The weaker dependence on runoff can be explained by the use of $f_D(t)^{.65}$ (or $f_{AD}(t)^{.65}$) in the weathering expression (12.1). This sublinear dependency, as pointed out earlier, is due to the effects of dilution during the weathering of silicate minerals.

Figure 12.5 shows that for reasonable values of $f_D(t)$ (between 0.5 and 1.4) and mean land temperature (between 5° and 25°C), calculated values for RCO$_2$ at 440 Ma are very high, possibly as high as 30. By comparison, for 300 Ma (Fig. 12.6), RCO$_2$ values are very low, approaching those at present (RCO$_2 \approx 1$). This major difference can be accounted for mainly by the rise and evolution, between these times, of vascular plants and their inducement of both enhanced silicate weathering and enhanced organic carbon burial (Berner, 1991; 1994). Best estimates, based on the $f_H(t)$ and $f_D(t)$ values of Figure 12.2, are shown in the figures by the symbol X. (Plotted X's on the left in each figure represent the inclusion of land area in the runoff parameter as $f_{AD}(t)$.) From Figure 12.6 it can also be seen that there is a limit to how high a land temperature, combined with high runoff, can be tolerated before RCO$_2$ goes to zero. This means, that if the modeling is at all correct, at runoff values within ±20% of that at present (a reasonable range — T. Crowley, personal communication, 1996), a global mean land temperature of >20°C is highly unlikely for the Upper Carboniferous. This is in agreement with the calculations of Crowley et al. (1996) of much lower mean land temperatures for this time.

So far discussion has been confined only to the effect of continental area and position on land temperature and runoff as they affect weathering rate and atmospheric CO$_2$. What about the effect of changes in CO$_2$ and solar luminosity on land temperature and runoff? In the GEOCARB model forcing by CO$_2$ and solar radiation are included in the feedback parameter $f_B(t)$. The $f_B(t)$ function is derived from a combination of expressions for mineral dissolution rate and global runoff as a function of global mean surface temperature with an equation, based on the results of GCM calculations, for global mean surface temperature as a function of atmospheric CO$_2$ level and solar luminosity. For this the GCM modeling results of Marshall et al (1994) for temperature and Manabe and Stouffer

(1980) for runoff have been used. This results in a simple expression for global mean temperature fitted to GCM results (Berner, 1994)

$$T(t) - T(0) = \Gamma \ln RCO_2 - Ws(t/570) \qquad (12.4)$$

where T is global mean surface temperature (land plus sea) in °C, t is past time in Ma, and the terms Γ and Ws represent curve fit parameters. (The second term on the right represents the slow linear increase of solar radiation over Phanerozoic time).

Since GCM results vary from one group to another, the sensitivity of RCO$_2$ to different values of the key parameters Γ and Ws of equation (12.4) has been presented elsewhere (Berner, 1995) and is reproduced here in Figure 12.7. As one can see, the amplitude of CO$_2$ variations depends strongly on which set of parameters are used in the feedback expression $f_B(t)$. Furthermore, there is an additional problem. The question is: do the values of Γ and Ws, as well as the functional dependence of global mean temperature and global runoff on CO$_2$ and solar radiation, change appreciably with changing paleogeography? What about changes in land temperature with changes in CO$_2$, solar radiation, and the presence or absence of large ice sheets? These questions have not been investigated here, but they need to be addressed in future work.

Conclusions

Calculations in this chapter show that changes in land runoff and land temperature during Phanerozoic time, due to changing continental size and position, have had only a minor effect on global weathering rate as it affects atmospheric CO$_2$. The effects on weathering of solar evolution, the rise and evolution of land plants, and changes in the relief of the continents due to mountain uplift dominate over continental size and position. However, this conclusion may be revised in future work when the effects of ice sheets on both runoff and the mean temperature of land undergoing weathering (not total land which includes the ice sheets themselves) are considered. For the contribution of the CO$_2$ greenhouse effect to global climate, I tentatively conclude that continental size and position are less important than solar evolution, plant evolution and continental relief.

The greatest effect of paleogeography occurred

Variation of Γ and Ws

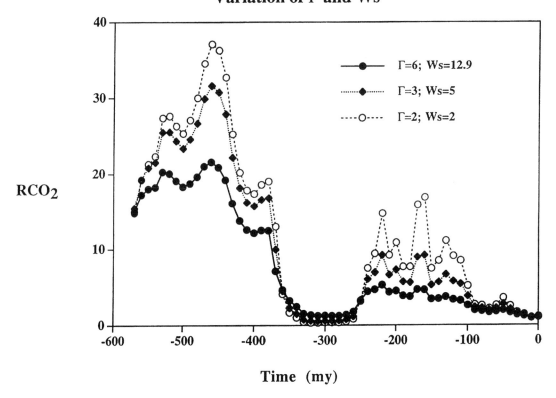

Fig. 12.7. Plot of RCO_2 vs time showing sensitivity to the climate feedback parameters Γ and W_s. The parameters Γ and W_s represent the sensitivity of global mean temperature to atmospheric CO_2 level and solar radiation respectively based on an empirical equation fitted to the results of general circulation modeling (see equation 12.4). The highest values are based on the work of Marshall et al (1994), intermediate values on that of Manabe and Bryan (1985) and Manabe (personal communication), and the lowest values are minima allowable for physically possible values of RCO_2. (After Berner, 1995)

during the early Paleozoic. Variation of runoff and land temperature resulted in a moderate drop in CO_2 during the late Ordovician which may help to explain the glaciation that occurred at that time. This reduces one of the major problems of the earlier modeling which showed very high calculated CO_2 levels at the same time as the Ordovician glaciation.

The calculated amplitude of CO_2 variation over Phanerozoic time depends strongly on the GCM parameters employed in constructing the climate feedback function for CO_2. Future work needs to consider what effect changing paleogeography has on GCM's as they affect the climate feedback function in carbon cycle modeling. This illustrates the necessity for further collaboration between geochemists and paleoclimatologists.

Calculated values of atmospheric CO_2 from GEOCARB modeling generally agree with independent estimates based on measurements of the carbon isotopic composition of paleosols and sedimentary rocks (Ehleringer and Cerling, 1995; Mora et al, 1996; for a recent summary consult Berner, 1997). This conclusion is not altered by the results of the present paper. Results also generally agree with independently deduced paleoclimates, with low CO_2 occurring at the same time as major glaciations and high CO_2 at times of intense global warming (Crowley and Baum, 1992). At any rate the results of the present paper reaffirm the earlier conclusion that the atmospheric greenhouse effect has been a major factor in controlling global climate over Phanerozoic time.

Acknowledgments

I am indebted to Tom Crowley and Bette Otto-Bliesner for discussions and for providing data. Also, reviews by K. Caldeira and G. Bluth were very helpful. The research was supported by NSF Grant EAR-9417325 and DOE Grant DE-FG02-95ER-14522.

References

Algeo, T. J., R. A. Berner, J. B. Maynard, and S. E. Scheckler, 1995. Late Devonian oceanic anoxic events and biotic crises: "rooted" in the evolution of vascular land plants?, *GSA Today, 5,* 45, 64–66.

Berner, E. K. and Berner, R. A., 1996. *Global environment: water, air and geochemical cycles,* Prentice-Hall, Upper Saddle River, NJ.

Berner, R. A., 1991. A model for atmospheric CO$_2$ over Phanerozoic time, *Am. J. Sci., 291,* 339–376.

Berner, R. A., 1992. Weathering, plants, and the long-term carbon cycle, Geochim. *Cosmochim. Acta, 56,* 3225–3231.

Berner, R. A., 1994. GEOCARB II: A revised model for atmospheric CO$_2$ over Phanerozoic time, *Am. J. Sci., 294,* 56–91.

Berner, R. A., 1995. Chemical weathering and its effect on atmospheric CO$_2$ and climate, in *Chemical Weathering of Silicate Minerals,* White, A. F. and S. L. Brantley (eds.), *Reviews in Mineralogy, 31,* pp. 565–583.

Berner, R. A., 1997. The rise of plants and their effect on weathering and atmospheric CO$_2$, *Science, 276,* 544–546.

Berner, R. A., A. C. Lasaga, and R. M. Garrels, 1983. The carbonate-silicate geochemical cycle and its effect on atmospheric carbon dioxide over the past 100 million years, *Am. J. Sci., 283,* 641–683.

Caldeira, K. and J. F. Kasting, 1992. The life span of the biosphere revisited, *Nature 360,* 721–723.

Crowley, T. J. and S. K. Baum, 1992, Modeling late Paleozoic glaciation, *Geology, 20,* 507–510.

Crowley, T. J. and S. K. Baum, 1995. Reconciling late Ordovician (440 Ma) glaciation with very high (14X) CO$_2$ levels. *J. Geophys. Res., 100,* 1093–1101.

Crowley, T. J., K- J. J. Yip, S. K. Baum and S. B. Moore, 1996. Modeling Carboniferous coal formation, *Paleoclimates, 2,* 159–177.

Edmond, J. M., M. R. Palmer, C. I. Measures, B. Grant, and R. F. Stallard, 1995. The fluvial geochemistry and denudation rate of the Guayana Shield in Venezuela, Colombia and Brazil, *Geochim. Cosmochim. Acta, 59,* 3301–3325.

Ehleringer, J. R. and T. E. Cerling, 1995. Atmospheric CO$_2$ and the ratio of intercellular to ambient CO$_2$ concentrations in plants, *Tree Physiology, 15,* 105–111.

Francois, L. M. and J. C. G. Walker, 1992. Modeling the Phanerozoic carbon cycle and climate: constraints from the ^{87}Sr/^{86}Sr isotopic ratio of seawater, *Am. J. Sci., 292,* 81–135.

Gibbs, M. T., E. J. Barron, and L. R. Kump, 1997. An atmospheric pCO$_2$ threshold for glaciation in the late Ordovician, *Geology, 25,* 447–450.

Holland, H. D., 1978. *The Chemistry of the Atmosphere and Oceans,* Wiley, New York.

Kump, L. R., 1989. Alternative modeling approaches to the geochemical cycles of carbon, sulfur and strontium isotopes, *Am. J. Sci., 289,* 390–410.

Lasaga, A. C., R. A. Berner, and R. M. Garrels, 1985. An improved geochemical model of atmospheric CO$_2$ fluctuations over the past 100 million years, in *The Carbon Cycle and Atmospheric CO$_2$: Natural Variations Archean to Present,* Sundquist, E. T. and W. S. Broecker (eds.), pp. 397–411, *Am. Geophys. Union Geophysical Monograph, 32.*

Lasaga, A. C., J. M. Soler, J. Ganor, T. E. Burch, and K. Nagy, 1994. Chemical weathering rate laws and global geochemical cycles, *Geochim Cosmochim Acta, 58,* 2361–2386.

Manabe, S. and R. J. Stouffer, 1980. Sensitivity of a global climate model to an increase of CO$_2$ concentration in the atmosphere, *J. Geophys. Res. 85,* 5529–5554.

Marshall, S., R. J. Oglesby, J. W. Larson, and B. Saltzman, 1994. A comparison of GCM sensitivity to changes in CO$_2$ and solar luminosity. *Geophys. Res. Lett., 21,* 2487–2490.

Mora, C. I., S. G. Driese, and L. A. Colarusso, 1996. Middle to late Paleozoic atmospheric CO$_2$ levels from soil carbonate and organic matter, *Science, 271,* 1105–1107.

Otto-Bliesner, B., 1995, Continental drift, runoff and weathering feedbacks: Implications from climate model experiments. *J. Geophys. Res.,* *100,* 11,537–11548.

Raymo, M. E., 1991. Geochemical evidence supporting T. C. Chamberlin's theory of glaciation, *Geology, 19,* 344–347.

Richter, F. M., D. B. Rowley, and D. J. DePaolo, 1992. Sr isotope evolution of seawater: the role of tectonics, *Earth. Planet. Sci. Lett., 109,* 11–23.

Ronov, A. B., 1976. Global carbon geochemistry, volcanism, carbonate accumulation, and life, *Geochemistry International (translation of Geokhimiya), 13,* 172–195.

Scotese, C. R., B. K. Bambach, C. Barton, R. Van der Voo, and A. Ziegler, 1979. Paleozoic base maps, *J. Geology, 87,* 217–277.

Stallard, R. F., 1992. Tectonic processes, continental freeboard, and the rate-controlling step for continental denudation, in *Global Biogeochemical Cycles,* Butcher, S. S., R. J.

Charlson, G. H. Orians, and G. V. Wolfe (eds.), pp 93–121, Academic Press, New York.

Volk, T., 1987. Feedbacks between weathering and atmospheric CO$_2$ over the last 100 million years, *Am. J. Sci., 287,* 763–779.

Volk, T., 1989. Rise of angiosperms as a factor in long-term climatic cooling, *Geology, 17,* 107–110.

Walker, J. C. G, P. B. Hays, and J. F. Kasting, 1981. A negative feedback mechanism for the long-term stabilization of Earth's surface temperature, *J. Geophys. Res., 86,* 9776–9782.

Worsley, T. R., T. L. Moore, C. M. Fraticelli, and C. P. Scotese, 1994. Phanerozoic CO$_2$ levels and global temperatures inferred from changing paleogeography, in *Paleoclimate, Tectonics, and Sedimentation during Accretion, Zenith and Breakup of a Supercontinent,* in Klein, G. D. (ed.), pp. 57–73, Spec. Paper 288, Geological Society of America, Boulder, Colo.

CHAPTER 13

The Transition From Arc Volcanism to Exhumation, Weathering of Young Ca, Mg, Sr Silicates, and CO$_2$ Drawdown

Douglas N. Reusch and Kirk A. Maasch

Paleoclimate modelers need estimates of the partial pressure of atmospheric carbon dioxide (pCO$_2$) because CO$_2$ influences the mean global temperature through the greenhouse effect (Arrhenius, 1896). Carbon dioxide enters the atmosphere as a result of high temperature decarbonation reactions and low temperature oxidation of organic C. Carbon dioxide leaves the atmosphere either through Ca, Mg silicate weathering and subsequent burial as carbonate or through photosynthesis and subsequent burial as organic C (Ebelmen, 1845). Changes in these C fluxes, brought about by various biological, oceanographic, and tectonic factors, affect atmospheric pCO$_2$ and, consequently, Earth's climate (Plass, 1956; Berner, 1990).

We here consider the impact on atmospheric pCO$_2$ of a particular tectonic scenario — the transformation of an aggrading arc system into an eroding orogen. Such a transformation may take place after a buoyant passive continental margin enters an active subduction zone. Miocene exhumation of the northern New Guinea arc terrane, following subduction of northern Australia, is a good example of this process.

In regard to the atmospheric CO$_2$ balance, the transformation of an aggrading volcanic arc into a site of exhumation warrants attention because it suggests that a large net CO$_2$ source is replaced by a large net CO$_2$ sink. Consumption of atmospheric CO$_2$ occurs both through weathering of highly soluble Ca, Mg silicates and through organic C burial resulting from higher sedimentation rates (France-Lanord et al., 1993) and possibly nutrient-enhanced productivity (Derry et al., 1992; Raymo, 1994).

We emphasize the combined importance of lithology and exhumation, in addition to climate (Berner, this volume, chapter 12) and vegetation, as rate-limiting factors for silicate weathering. Exhumation equals rock uplift less surface uplift (England and Molnar, 1990). Rapid exhumation rate is irrelevant to the C flux due to silicate weathering unless the material being exhumed includes Ca, Mg silicates. Also, the terrestrial exposure of Ca, Mg silicates does not have a great impact on the C flux due to silicate weathering unless the material is being actively exhumed (and other factors are not limiting).

For C cycle models relying on Sr and Os isotopic mass balance, a major point for consideration is the nonradiogenic character of the young, highly soluble Ca, Mg silicates in arc systems and related ophiolites (oceanic crust and upper mantle). These materials contrast sharply with old, highly radiogenic sial (continental crust) that does not contain such a high abundance of soluble Ca, Mg silicate. The idea that collisions between arcs and passive continental margins might affect the isotopic composition of seawater is not new (e.g., Berner and Rye, 1992; Turekian et al., 1995).

Following a review of the climate, relevant geochemical, and tectonic systems, we argue that the transformation of a volcanic arc into an eroding orogen is likely to decrease atmospheric pCO$_2$. An unconformity carved deeply into the northern New Guinea arc terrane during the Miocene affords the basis for estimating silicate weathering fluxes. We then consider the documented cooling events of late Precambrian through Phanerozoic time, noting changes in seawater composition and tectonic evidence that suggest arc exhumation and contemporaneous CO$_2$ drawdown. Finally, we address the implications of this idea for C cycle models and estimates of paleo-atmospheric pCO_2.

Background

The following section provides background for the reader unfamiliar with the climate, relevant geochemical, and tectonic systems. The Sr and Os systems bear on the C system because temporal variations in seawater $^{87}Sr/^{86}Sr$ and $^{187}Os/^{186}Os$ ratios constrain the C fluxes due to silicate, carbonate, and organic C weathering.

Review of Climate System

Solar output, distance from the sun, albedo, distribution of continents, ocean circulation, and atmospheric composition all influence Earth's climate. The globally averaged surface temperature of Earth is approximately 15°C, in part due to the greenhouse effect. Solar radiation entering Earth's atmosphere is predominantly short wave, while radiation leaving Earth is long wave. Water vapor, CO_2, CH_4 (methane), and other trace gases in Earth's atmosphere absorb long wave radiation. While incoming and outgoing radiation tend to remain nearly in balance, the net effect of energy absorption by greenhouse gases is to warm the surface of the planet. Without an atmosphere, Earth's surface temperature would be less than -20°C, well below the freezing point of water. The absorption of longwave radiation responds logarithmically to atmospheric pCO_2. Surface temperature therefore responds more to a decrease in pCO_2 than to an equal increase in pCO_2. Decreased atmospheric water vapor and increased areal extent of sea ice amplify the cooling effect of lowered pCO_2.

Review of Carbon System

The relationships between C in the atmosphere and C in rocks (Fig. 13.1) were, remarkably, established in the first half of the 19th century (Ebelmen, 1845; Berner and Maasch, 1996). The principal C reservoirs are carbonate minerals in limestones and dolostones (5000×10^{18} moles), reduced organic C in sedimentary kerogen (1250×10^{18} moles), and inorganic C in the oceans and atmosphere (less than 3×10^{18} moles). The atmospheric reservoir of CO_2 is especially small (0.02–0.05×10^{18} moles),

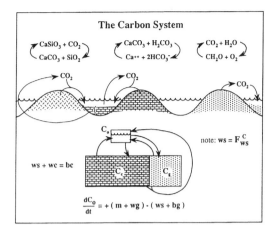

Fig. 13.1. The carbon system: reservoirs, reactions, fluxes, and mass balance. Principal carbon reservoirs are carbonate minerals (C_c), organic matter (C_g), and the atmosphere–ocean (C_o). Reactions include silicate dissolution–decarbonation, carbonate dissolution–precipitation, and photosynthesis–respiration. Weathering fluxes (w–) diminish the rock reservoirs from the top; metamorphic degassing (m–) diminishes them from beneath. Burial fluxes (b–) augment the rock reservoirs. Note that total carbonate burial (bc) equals the sum of carbon fluxes from weathering of carbonates (wc) and weathering of silicates (ws). The long-term carbon mass balance in the atmosphere–ocean reservoir (C_o) is determined by two sources (m– and wg) and two sinks (ws and bg).

approximately $^1/_{60}$ of the ocean reservoir. The amount of C in the atmosphere–ocean reservoir is sensitive to changes in the source or sink fluxes because the magnitude of this reservoir is much smaller than these fluxes. Yet, due to strong negative (i.e., stabilizing) feedbacks, atmospheric pCO_2 apparently has occupied a limited range for most of Earth's history (Walker et al., 1981).

The three important reactions involving carbon, shown schematically, are as follows:

$$CaSiO_3 + CO_2 = CaCO_3 + SiO_2 \tag{1}$$

$$H_2O + CO_2 = CH_2O + O_2 \tag{2}$$

$$CaCO_3 + H_2CO_3 = Ca^{++} + 2HCO_3^- \tag{3}$$

Reaction (1) describes the weathering of Ca, Mg silicates and, in the opposite direction, the thermal decarbonation of carbonates. Reaction (1) dominates the long-term carbon cycle simply because carbonates constitute the single largest reservoir of C. Reaction (2) describes the organic C subcycle. Burial of organic C represents net planetary photosynthesis and the weathering of organic C represents net planetary respiration. Carbonate dissolution and precipitation (reaction 3) do not affect the long-term C cycle because CO$_2$ sequestered from the atmosphere during carbonate weathering quickly returns when the carbonate precipitates. In contrast, the much slower carbonate-silicate cycle proceeds at a pace determined by tectonic processes.

Four input fluxes (sources) and two output fluxes (sinks) determine the mass balance of the atmosphere–ocean C reservoir (Fig. 13.1). By noting that the total carbonate burial flux equals the sum of the carbonate weathering and silicate weathering fluxes and by lumping the metamorphic fluxes, mass balance simplifies to the difference between the two sources, metamorphic degassing and organic C weathering, and two sinks, silicate weathering and organic C burial.

Inputs of CO$_2$ to the atmosphere result from the thermal degassing of inorganic and organic C and the oxidative weathering of organic C (Fig. 13.2). Degassing takes place in subduction systems, along hot oceanic ridges, and above hot mantle plumes. The fraction of C returned quickly through the accre-

tionary prism and volcanic conduits has been debated (Caldeira, 1992; Selverstone and Gutzler, 1993). Today, thermal degassing supplies 9×10^{18} moles of CO$_2$ myr^{-1} and oxidative weathering supplies 6×10^{18} moles of CO$_2$ myr^{-1} (Goddéris and François, 1995). For the past, seafloor-spreading rate serves as a proxy for the thermal degassing flux.

Terrestrial silicate weathering and organic C burial constitute the principal sinks for atmospheric CO$_2$ (Fig. 13.2). Submarine silicate weathering (François and Walker, 1992) is probably minor (Caldeira, 1995). Factors influencing the organic C burial flux include productivity, which is limited by pCO$_2$, nutrients, and light; sedimentation rate; and water column anoxia (e.g., Walker, 1991). A host of rate-limiting factors influence the terrestrial silicate weathering flux. Rainfall affects weathering (e.g., Dunne, 1978) because a continuous flux of water through the soil promotes continuous reaction (the solution is always far from equilibrium). Runoff matters less than total precipitation (White and Blum, 1995). Vegetation promotes weathering because it elevates soil pCO$_2$ (Berner, 1992). Relief permits the continuous exhumation of fresh material that is otherwise sealed by weathering products as in the Amazon basin (Stallard and

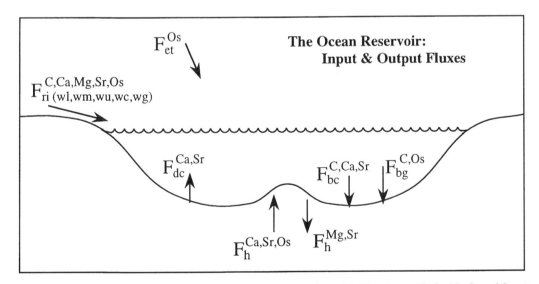

Fig. 13.2. The ocean reservoir: input and output fluxes relevant to C cycle models. The elements C, Ca, Mg, Sr, and Os enter and exit the ocean by a variety of processes. The dissolved load of rivers (ri) includes solutes from various sources: weathering of sialic materials (wl), weathering of mafic materials (wm), weathering of ultramafic materials (wu), weathering of carbonates (wc), and weathering of organic carbon (wg). Other sources include extraterrestrial materials (et), dissolution of carbonates (dc), and hydrothermal discharge (h). Sinks include hydrothermal recharge (h), burial of carbonate (bc), and burial of organic matter (bg).

Edmond, 1983; Edmond et al., 1996). Surface uplift, for example of the Tibetan Plateau, is significant because of the peripheral exhumation and monsoonal precipitation. The rocks undergoing weathering must contain Ca, Mg silicate to remove CO$_2$ from the atmosphere because other reactions do not form carbonate minerals in abundance. The silicate minerals formed at high temperature in mafic rocks are more soluble than the silicate minerals formed at low temperature in granitic rocks. Grain size affects the surface area exposed to weathering solutions, and glassy materials are especially soluble (Stallard, 1995). Studies of dislocations have shown that weathering is concentrated along crystal defects (Berner and Holdren, 1977). In this regard, Stallard (1995) has noted lower dissolution rates in older orogens, perhaps due to annealing of dislocations over time.

An important negative feedback results from silicate weathering. This feedback has been postulated to be the primary stabilizer of Earth's temperature (Walker et al., 1981; Volk, 1987). It is actually a quadruple negative feedback, because increased pCO$_2$ leads to increased temperature, rainfall, and vegetation, all four of which accelerate the removal of CO$_2$ from the atmosphere by silicate weathering. Conversely, low pCO$_2$ decelerates silicate weathering.

The marine bulk carbonate $\delta^{13}C$ record has been used to constrain C fluxes and sizes of the sedimentary reservoirs. Isotopic fractionation of C results from the preferential incorporation of light ^{12}C into plant matter during photosynthesis. Changes in the $\delta^{13}C$ of dissolved inorganic C, recorded in carbonate tests, reflect imbalances between the mean $\delta^{13}C$ of total inputs to the ocean and that of outputs. These imbalances arise from variable source composition (Beck et al., 1995), changes in the fractional burial of organic C (Shackleton, 1987), and variable fractionation, i.e. changes in the isotopic composition of organic C removed (Derry and France-Lanord, 1996).

Review of Sr System

Strontium bears on the C system because it behaves geochemically like Ca, which is typically 50–200 times more abundant than Sr in common rocks (Faure, 1986). Strontium enters the ocean via rivers from the weathering of carbonate and silicate mate-

rials and from the seafloor by dissolution of carbonates (Fig. 13.2). Along oceanic ridges, Sr in seawater exchanges with Sr in basalts during hydrothermal circulation. Strontium exits the ocean through the precipitation of carbonate minerals.

Old sialic rocks contain abundant radiogenic ^{87}Sr due to their abundant parent ^{87}Rb and their age. Conversely, young basalts contain large amounts of Sr but small amounts of radiogenic ^{87}Sr (due to sparse parent ^{87}Rb and their young age). The abundance of the parent Rb in sialic crust is due to its large ionic radius, which causes it to be expelled from the mantle.

The isotopic composition of the Sr in seawater is determined by the relative fluxes of Sr from weathering and dissolution of carbonates with intermediate $^{87}Sr/^{86}Sr$ ratios, weathering of old sialic rocks with high $^{87}Sr/^{86}Sr$ ratios, and interaction with young igneous rocks with low $^{87}Sr/^{86}Sr$ ratios. Interaction with nonradiogenic basalts at hydrothermal sites is important and, as we emphasize below, terrestrial weathering of nonradiogenic materials also affects the $^{87}Sr/^{86}Sr$ ratio of seawater.

Review of Os System

The Os system is less well understood than the Sr system but is currently under investigation (e.g., Peucker-Ehrenbrink et al., 1995). Scavenging of Os by organic matter links the Os system to the C system. Osmium also accumulates in metalliferous sediments due to scavenging by iron oxyhydroxides that originate at hydrothermal vents. Seawater Os is primarily a mixture of Os released from sediments or metasedimentary rocks, ultramafic rocks in the seafloor and in terrestrially exposed ophiolites, and cosmic debris (Fig. 13.2).

Rhenium and Os are preferentially incorporated in metallic and sulfide phases and therefore are depleted in the silicate crust. As with the Sr system, continental crust is characterized by high parent/daughter ratios, and consequently high $^{187}Os/^{186}Os$ ratios, but for the opposite reason — during partial melting of the mantle to form crust, little Re but almost no Os enters the melt. Ultramafic rocks are an important reservoir of Os with low $^{187}Os/^{186}Os$ ratios. In sedimentary environments, organic matter scavenges Re in addition to Os, resulting in sediments with high concentrations of ^{187}Os. The Os from sialic crust is locally extremely radiogenic (Peucker-Ehrenbrink and Ravizza, 1996), perhaps because of a metasedimentary influence.

The marine Os isotope record, to a first approximation, is controlled by the relative fluxes of radiogenic Os derived from shales (high average concentration) and nonradiogenic Os derived from ultramafic rocks (average concentration of 5300 ppt compared with 50 ppt in the crust). Metalliferous sediments are thought to be an ideal material for reconstructing the marine $^{187}Os/^{186}Os$ record (Ravizza, 1993). In these sediments, non-seawater sources of Os, including terrigenous Os in mineral detritus and cosmic Os in extraterrestrial dust, constitute a negligible fraction of the total Os (Ravizza and McMurtry, 1993; Peucker-Ehrenbrink, 1996).

Review of Tectonic System

Important surface manifestations of Earth's tectonic system include mantle plume volcanism and various plate tectonic phenomena. Mantle plumes may generate large volumes of Ca, Mg silicate materials. Plate motions lead to cycles of seafloor growth and destruction (Wilson cycles) involving continental rifting, seafloor-spreading, subduction, and collision (Fig. 13.3). Characteristic petrotectonic assemblages accumulate in a limited number of tectonic settings through the stages of a Wilson cycle. Thick sequences of arkose, evaporites, and bimodal volcanic suites fill rift valleys. Ophiolitic assemblages form at spreading centers while sequences of quartzites, carbonates, and shales build slowly at passive margins. Subduction of oceanic crust leads to the growth of magmatic arcs, flanking volcanogenic sediments, and accretionary prisms (Hamilton, 1988). Collisions involving continental crust create metamorphic-granitic complexes and thick accumulations of orogenic sediment.

Tectonic System – Carbon Cycle Relationships

Tectonic processes, including mantle plume activity and plate motions, rearrange the global pattern of CO₂ sources and sinks through time. Episodic mantle plumes trigger the release of CO₂; conversely, weathering of terrestrial flood basalts con-

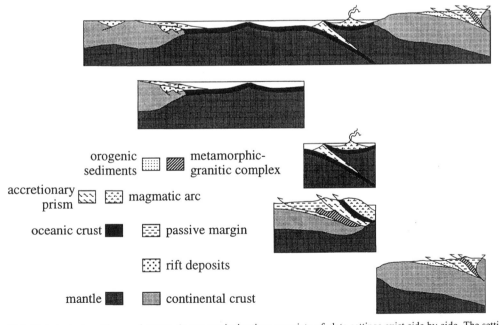

Fig. 13.3. Plate tectonic settings and stages. At any particular time, a variety of plate settings exist side by side. The settings are temporally related through the growth and destruction of ocean basins (Wilson cycle), including stages of continental rifting, seafloor spreading, subduction, and collisions. Petrotectonic assemblages accumulate in a variety of settings: rift deposits, oceanic crust and passive margins, magmatic arcs and accretionary prisms, metamorphic–granitic complexes and orogenic sediments. The arc–continent collision stage occurs when a passive continental margin enters an arc system, which generally leads to exhumation of the volcanic arc.

sumes CO$_2$. Based on general considerations of the Wilson cycle (Fig. 13.3) and of the four C fluxes affecting atmospheric pCO$_2$ (Fig. 13.1), we argue that the most likely plate tectonic mechanism for abrupt CO$_2$ drawdown is the exhumation of magmatic arcs following their collision with continental margins. Seafloor spreading and subduction create net CO$_2$ sources; if other factors remain constant, a decrease in spreading rate should bring about a corresponding decrease in atmospheric pCO$_2$. Such decreases, however, are generally gradual rather than abrupt. Compared to seafloor spreading and subduction, rifting events and collisions are relatively episodic. Continental crust, ruptured during rifting events, contains an abundance of K, Na silicates but a paucity of Ca, Mg silicates; furthermore, rift sediments accumulate so rapidly in grabens under dry climatic conditions that little silicate weathering takes place. Thus, rifting is not likely to enhance silicate weathering significantly. Collisions, on the other hand, have the potential to perturb the C cycle in an episodic fashion.

Collisions perturb the C cycle in a variety of ways, some tending to elevate but most tending to lower atmospheric pCO$_2$. Thermal degassing of tectonically buried sediments (Caldeira, 1992) and weathering of organic C in exhumed sediments (Beck et al., 1995) both add CO$_2$ to the atmosphere, although Selverstone and Gutzler (1993) argued that the first of these two processes is not particularly effective. Conversely, silicate weathering (Chamberlin, 1899; Raymo et al., 1988) and burial of organic C lower the inventory of C in the atmosphere–ocean reservoir. Burial of organic C may be enhanced by trapping of organic matter in orogenic sediments (France-Lanord et al., 1993) and possibly by eutrophication of the ocean triggered by increased chemical weathering (Derry et al., 1992; Raymo, 1994). In addition, during the transition from subduction to collision, decreased volcanic activity and deceleration of spreading centers should decrease the supply of CO$_2$ to the atmosphere.

During collisions between arcs and passive continental margins, three of the four fluxes affecting atmospheric pCO$_2$ change in the direction of CO$_2$ drawdown. First, sedimentary C is no longer conveyed to depth by subduction. Second, exhumation of the arc creates fertile conditions for silicate weathering. Third, organic C burial increases as discussed above.

The lithologic and isotopic compositions of materials exposed in eroding arcs (e.g., Paleogene Himalaya and Neogene New Guinea) differ from those exposed in intracontinental thrust zones (e.g., Neogene Himalaya). The rock types associated with a New Guinea-type arc include an ophiolitic basement comprising ultramafic and mafic igneous rocks, a thick arc sequence with a high proportion of basalt and andesite, and volcanogenic sediments deposited adjacent to the magmatic arc. In sharp contrast, the rock types exhumed during intracontinental thrusting (Neogene Himalaya) are primarily quartzo-feldspathic gneisses, pelitic schists, and minor calc-silicate marbles. France-Lanord et al., (1993) suggested that, since early Miocene time, 80% of the Bengal Fan has been derived from a narrow strip of crystalline rocks exhumed between the Main Central Thrust and South Tibet Detachment. The remaining sediment has been derived from passive margin and rift deposits, including some rift sediments with very high ^{87}Sr / ^{86}Sr ratios (>0.8). It has been suggested that large positive excursions of marine ^{87}Sr / ^{86}Sr correspond with the exhumation of sialic crust brought about by Himalayan-style intracontinental thrusting (Edmond, 1992; Richter et al., 1992). In regard to the C system, however, God(é)ris and François (1995) imply that this style of exhumation does not affect significantly the global C flux due to silicate weathering.

New Guinea Exhumation and Weathering Fluxes

Weathering fluxes related to the exhumation of magmatic arcs in New Guinea during the Neogene and along the southern margin of Tibet during the Paleogene can be estimated. An erosional surface cut into the plutonic roots of the arc documents the removal of the volcanic carapace and affords the basis for the mass balance calculations that follow. The sum of material dissolved from the compositionally distinct basement and CO$_2$ removed from the atmosphere must equal the material precipitated in the ocean.

The collision of New Guinea with Australia must have increased the area of Ca, Mg silicates exposed to the weathering environment. The New Guinea arc, 2500 km long (Fig. 13.4) and at least 100 km wide, was transformed from an aggrading volcanic

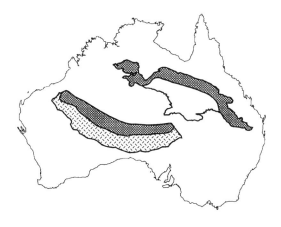

Fig. 13.4. Comparison of the relative sizes of the northern New Guinea and Trans-Himalayan (Gangdese) arcs. Both arcs are approximately 2500 km long, or about half the width of the Australian continent (shown for scale). The composite arc terrane of northern New Guinea (after Pigram and Davies, 1987), consisting of igneous complexes and derived sediments, is depicted by the dark shaded area (upper right). The Trans-Himalayan arc, exhumed during the Paleogene, and the Neogene Himalayan intracontinental thrust belt (after Searle et al., 1987) are depicted, respectively, by dark and light shaded areas (lower left).

pile to a site of exhumation during the middle Miocene (Hamilton, 1979). At this time, New Guinea lay 15 degrees south of the equator and, as today, was surrounded by ocean. Annual rainfall, orographically enhanced, may have been on the order of 5 m. Under such circumstances, the chemical denudation rate would have been nearly 80×10^3 kg $km^{-2} yr^{-1}$ (Dunne, 1978), which equates to 10^{17} moles of CO_2 removed from the atmosphere myr^{-1}. This is a conservative estimate because the watersheds in the Dunne study are underlain by a variety of lithologies and exhumation rate is not especially high.

Exhumation rate can also be used to estimate the C flux from the atmosphere due to silicate weathering. Upper Miocene sedimentary cover, up to 5 km thick in the Sepik Basin, overlies lower Miocene amphibolite grade basement (Hamilton, 1979). This unconformable relationship implies an exhumation rate of approximately 3 km myr^{-1}. Silicate weathering therefore consumed CO_2, assuming that 25% of the arc materials dissolved, at the rate of approximately 10^{18} moles myr^{-1}. In the absence of information on bulk composition of the

detritus, 25% dissolution seems reasonable considering that virtually all of the rate-limiting factors were optimized. The arc materials are highly soluble (Stallard, 1995). They were subjected to a high mean annual temperature and to high precipitation. Extreme relief in the northern arc terrane would have been a significant factor (Stallard and Edmond, 1983); today, resistant nonvolcanic lithologies in central New Guinea rise to over 5000 m elevation. Tropical vegetation, which flourishes on the nutrient-rich arc rocks, also would have contributed to a maximal weathering rate.

A crude estimate of chemical weathering of the Papuan ophiolite suggests that its exhumation could have increased the global river flux of Os by 25% and led to a detectable decrease of approximately 1 $^{187}Os/^{186}Os$ unit in the marine record. The calculation assumes 10% chemical weathering of Os-bearing phases (Peucker-Ehrenbrink et al., 1995) in ultramafic rocks exhumed at the rate of 1 km myr^{-1} in a belt 1000 km long and 10 km wide. (This calculation is highly dependent on the solubility of the Os phase, which could either be a soluble sulfide or a highly insoluble phase.)

At the same time that consumption of CO_2 was increasing, waning arc volcanism is calculated to have reduced the supply of CO_2 to the atmosphere by approximately 5%, assuming that CO_2 production is approximately proportional to length of plate boundary.

In relation to the global CO_2 budget, the effect of dissolving Ca, Mg silicates in New Guinea is thought to have been significant. A flux of 10^{17} to 10^{18} moles of CO_2 myr^{-1} from the atmosphere represents between 1.4% and 14% of the modern total silicate weathering rate (7×10^{18} moles myr^{-1}) estimated by Berner (1990).

Comparison of Climate, Seawater Composition, and Tectonic Records

During the past billion years, Earth's climate has fluctuated between ice-free warm modes and cold modes during which glaciers scoured the continents (Fischer, 1984; Frakes et al., 1992). Prominent global cooling events, documented by oxygen isotopic records, sedimentary facies, and fossils, occurred during Plio-Pleistocene, middle Miocene, late Eocene-early Oligocene, Permo-Carboniferous, late Ordovician, and late Precambrian times (Fig. 13.5).

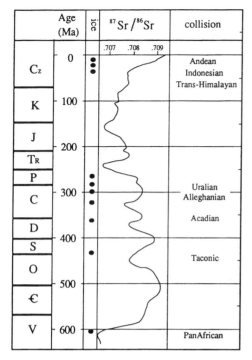

Fig. 13.5. Climate, marine $^{87}Sr/^{86}Sr$, and tectonic records, 630 − 0 Ma. Black dots denote times of widespread glaciation (after Frakes et al., 1992). Composite marine $^{87}Sr/^{86}Sr$ record constructed from several published sources (Burke et al., 1982; Kaufman et al., 1993; Hodell and Woodruff, 1994; Mead and Hodell, 1995). Times of selected arc exhumation events are indicated on the right.

Various geologic records show that the timing of changes in seawater composition and tectonic events is consistent with the hypothesis that arc exhumation contributed to global cooling. Following chemical weathering, arc materials impart their unique geochemical and isotopic signatures upon coeval marine precipitates. Low $^{87}Sr/^{86}Sr$ and $^{187}Os/^{186}Os$ ratios characterize many of the materials in arc systems. These isotopic signatures contrast with those of the old sialic materials in continental crust.

Pre-Cenozoic Cooling Events

The late Precambrian Varanger glaciations occurred near the beginning of a steep and prolonged rise in seawater $^{87}Sr/^{86}Sr$ (Kaufman et al., 1993) attributed to exhumation of sialic crust during the Pan-African collision (Hoffman, 1991). Thus, global cooling occurred at a time when collision systems were evolving from subduction systems.

Carbon isotopic records show a pronounced decrease at this time, attributed to input of sedimentary organic C (Kaufman et al., 1993).

Late Ordovician glaciation followed a period of extensive black shale deposition (Leggett, 1978). Neodymium isotopic studies indicate an increase in the weathering of mafic rocks in the Iapetus Ocean during this interval (Keto and Jacobsen, 1987). The Taconic orogeny is interpreted as an arc-continent collision, which took place in the tropics. In the orogenic sediments, detrital chromite but sparse olivine, pyroxene, and serpentine (Hiscott, 1978; Nelson and Casey, 1979; Rowley and Kidd, 1981) suggests significant chemical dissolution of ultramafic rocks. A widespread unconformity was sculpted into materials of arc affinity, including marine volcanic and related sedimentary rocks. The Silurian cover sequence consists of shallow marine to terrestrial sediments that are generally more compositionally mature than the earlier sediments.

Other cooling events occurred during the Devonian and Permo-Carboniferous. Poorly dated tillites suggest two glacial maxima, during the Pennsylvanian and the Permian (Frakes et al., 1992). Marine C isotopic records show a prolonged increase through this interval that has been related to increased organic C burial and land plant evolution (e.g., Berner, 1990). Peaks in the seawater $^{87}Sr/^{86}Sr$ record (Burke et al., 1982) correspond roughly with maxima in the extent of tillites. These peaks also correspond roughly with the Alleghanian and the Uralian continental collisions.

Paleogene Cooling

Abundant isotopic, paleontologic, and lithologic evidence exists for global cooling beginning at approximately 50 Ma (Fig. 13.6). A gradual increase in benthic foraminiferal $\delta^{18}O$ during Eocene time was followed by an abrupt increase in early Oligocene time (Mead and Hodell, 1995). The marine record shows a decrease in $^{187}Os/^{186}Os$ ratios near the Eocene–Oligocene boundary (Turekian et al., 1995). Calcite tests, apparently undisturbed by diagenesis, from the basal Oligocene at ODP Site 748 record a decrease in Sr/Ca ratios and an increase in Mg/Ca ratios just before the prominent $\delta^{18}O$ shift (Zachos et al., 1992). (We note that ultramafic rocks have anomalously low Sr/Ca and $^{187}Os/^{186}Os$ ratios and anomalously high Mg/Ca ratios.) Marine $^{87}Sr/^{86}Sr$ ratios began to rise steeply in the late Eocene at the rate of 40×10^{-6} per million years, somewhat before the time of the abrupt $\delta^{18}O$ shift in the early Oligocene (Mead and Hodell, 1995).

Cooling ages of the calc-alkaline Gangdese batholith and the onset of molasse deposition (Searle et al., 1987; Gansser, 1995) indicate rapid exhumation beginning in the late Eocene, approximately 50–40 Ma. Of note, the convergence rate between India and northern Asia slowed dramatically from 20 cm yr^{-1} to less than 5 cm yr^{-1} when India first made contact with Tibet. This deceleration would have substantially decreased the supplies of CO₂ and nonradiogenic Sr from the Carlsberg Ridge.

Neogene Cooling

Benthic foraminiferal $\delta^{18}O$ records (Fig. 13.6) indicate gradual warming through the early Miocene followed by an abrupt cooling and ice growth event (1 ‰ increase in $\delta^{18}O$) during the middle Miocene. The widespread belief that this oxygen isotopic shift might have been caused entirely by ice growth in Antarctica has been challenged recently based on lack of evidence for significant ice growth and evidence for constant polar desert conditions through this interval (Marchant et al., 1993).

After increasing steeply (80×10^{-6} per year) between 20 and 16 Ma, marine $^{87}Sr/^{86}Sr$ ratios became almost constant approximately a million years before the $\delta^{18}O$ shift and cooling event (Hodell and Woodruff, 1994). Ravizza (1993) shows the beginning of the Neogene rise in marine $^{187}Os/^{186}Os$ ratios at approximately 15 Ma. Recent work has verified this Neogene rise but has not ruled out the possibility of a decrease in $^{187}Os/^{186}Os$ ratios, related to ophiolite weathering, before the cooling event (Reusch et al., 1996). In this regard, a marine $^{187}Os/^{186}Os$ record from Mn nodules (Burton et al., 1995) suggests a decrease at approximately 16–15 Ma following an increase between 20 and 16 Ma.

The Monterey Formation of Miocene age contains abundant carbonate, silica, and organic C (Vincent and Berger, 1985). Raymo (1994) suggested that high rates of organic C burial might be

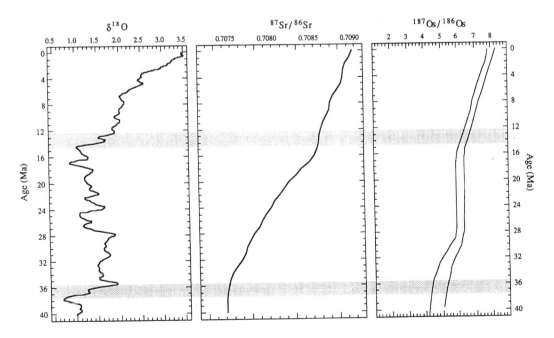

Fig. 13.6. Cenozoic marine isotopic records from 40–0 Ma: (a) smoothed $\delta^{18}O$ curve based on compilation of benthic foraminiferal oxygen isotope measurements (after Wright and Miller, 1993; Mix et al., 1995), (b) smoothed record of the $^{87}Sr/^{86}Sr$ composition of seawater (after Hodell and Woodruff, 1994; Mead and Hodell, 1995; Farrell et al., 1995), and (c) estimated envelope of the $^{187}Os/^{186}Os$ composition of seawater (after Peucker-Ehrenbrink et al., 1995). Gray shaded bars mark the Plio-Pleistocene, middle Miocene, and early Oligocene global cooling events.

linked to high rates of chemical weathering and delivery of nutrients to the ocean. She pointed to the
Himalayas as a likely source of solutes (as discussed
below, the New Guinea region should also have
been supplying nutrients at this time). In the
Himalayas, the Main Central Thrust became active
during the early Miocene as indicated by isotopic
cooling ages in the range 21–17 Ma (Raymo,
1994). The steep rise in seawater $^{87}Sr/^{86}Sr$ ratios
between 20 and 16 Ma has been attributed to this activity (e.g., Richter et al., 1992). The delay of more
than a million years between intensified weathering
of radiogenic materials during the early Miocene
and global cooling during the middle Miocene presents a problem for the Himalayan hypothesis as
does the likelihood that the radiogenic materials
were not particularly Ca, Mg silicate-rich. The occurrence of abundant organic C in the lower part of
the Monterey Formation, deposited during the
Miocene thermal optimum, is consistent with the
inference that pCO_2 was high before the middle
Miocene cooling event; increasing marine
$^{87}Sr/^{86}Sr$ ratios from 20–16 Ma are also consistent
with this inference.

The Australia–New Guinea collision began during Oligocene time and proceeded diachronously
from west to east culminating in the middle
Miocene. Before the collision, a composite terrane
of amalgamated arcs (Pigram and Davies, 1987) lay
largely below sea level. The passive margin of
Australia descended beneath the accretionary prism
until the collision halted due to buoyancy of the
continental material. The descending oceanic slab
presumably detached, a process that has considerable ramifications for surface uplift and exhumation
of the overlying plate. Possibly, the continental
plate plowed beneath the arc system, exhuming in
order the accretionary prism, forearc, and magmatic
arc. Australia moved north at the constant rate of 8
cm yr^{-1}; at this rate, it would have plowed under the
forearc in 2–3 million years. After the arc was accreted, a new subduction zone with opposite polarity formed off the north coast of New Guinea. This
new subduction zone contains the arc in its hanging
wall, which has since been elevated by underplating
at the base of the new accretionary prism. The timing of the arc erosion is constrained by an unconformity separating upper Miocene sediments from
pre-middle Miocene arc rocks that are generally
greenschist but locally amphibolite-grade

(Hamilton, 1979). While at least the southern and
northern fringes of the forearc basin were lifted up
and exhumed, portions of the central forearc may
have remained basins through the collision, receiv-
ing up to 5000 m of sediment. Significantly, the
largest ophiolite in the world rests on middle
Miocene sediment in Papua (Rangin et al., 1990).
Other parts of Indonesia, notably Borneo, which
includes an extensive arc–ophiolite terrane, were
also exhumed during the Miocene.

The most recent long-term global cooling began
during the late Pliocene following a warm early
Pliocene. A short-lived plateau in the marine
$^{87}Sr/^{86}Sr$ record during the late Pliocene (4 Ma to
2.5 Ma) precedes a steep increase during the
Pleistocene (Farrell et al., 1995). We cannot offer a
passive margin–arc collision mechanism for this
cooling event, but note that in the northern, tropical Andes, uplift was rapid (0.7 mm yr^{-1}) by 3 Ma
(Benjamin et al., 1987).

Discussion of Stratigraphic Analysis

Seawater composition and tectonic records are consistent with the notion that CO_2 drawdown via arc
exhumation was a contributing factor to a number of
documented late Precambrian and Phanerozoic cooling events.

An objection to the hypothesis that global cooling events were triggered by arc exhumation is
grounded in the observation that there have been
many other and frequent collisional events that did
not cause cooling. We suggest that collisions robust enough to cause significant CO_2 drawdown and
cooling are rare for the following reasons:

1. Not all arc terranes are large. The New Guinea
arc terrane, for example, is a composite terrane
made of components assembled previously without
much exhumation in the Pacific Ocean.

2. Not all collisions take place in a tropical climate where temperature, precipitation, and tropical
vegetation enhance weathering.

3. During the life span of an ocean basin, collisions of arcs with large buoyant objects such as
passive continental margins are infrequent.

4. Not all collisions coincide with times of high
atmospheric pCO_2. If pCO_2 is low, then silicate
weathering may not be capable of lowering it
further.

Implications for Carbon Cycle Models

To obtain better estimates of paleo-atmospheric pCO$_2$, we here suggest a way to modify C cycle models. Current C cycle models inadequately account for the C flux due to weathering of nonradiogenic Ca, Mg silicates. Our main recommendation is to split the term for silicate weathering into two terms to account for both the weathering of old K, Rb-rich radiogenic silicates and young Ca, Mg-rich nonradiogenic silicates. An extra term requires an additional constraint. Since Os is coupled to the organic C cycle (and possibly to the inorganic C cycle), we concur with others that marine ^{187}Os/^{186}Os records can provide such a constraint.

Previous modeling of the C cycle has attempted to quantify the C flux due to silicate weathering in different ways. GEOCARB I (Berner, 1990; 1991) improved upon earlier models (see Berner, this volume, chapter 12 for references) by incorporating forcing parameters more realistically. The tectonic parameter, f$_R$, estimates relief as a function of sea level. This parameter probably underrepresents silicate weathering in narrow orogenic belts affected by orographic rain. Raymo et al., (1988) suggested that the marine ^{87}Sr/^{86}Sr record might serve as a proxy for silicate weathering. We agree that seawater ^{87}Sr/^{86}Sr ratios must in part reflect the C flux due to weathering of old K, Rb-rich materials. However, their usual interpretation fails to account for weathering of young Ca, Mg-rich silicates that are potent C sinks. The frequently noted crude correlation between increasing marine ^{87}Sr/^{86}Sr ratios and global cooling (increasing δ^{18}O) during the Cenozoic breaks down when these records are examined in detail. Berner and Rye (1992) showed, by using GEOCARB I to constrain the Sr system, that the ^{87}Sr/^{86}Sr of river water must have varied considerably through time for their model results to match the observed marine ^{87}Sr/^{86}Sr record. Jacobsen (1988) also concluded that the seawater ^{87}Sr/^{86}Sr record is not proportional to erosion rate but reflects variable isotopic composition of river input. To better represent silicate weathering, GEOCARB II (Berner, 1994) incorporates the seawater ^{87}Sr/^{86}Sr record in the parameter for relief (increases in seawater ^{87}Sr/^{86}Sr, because they reflect the exposure of radiogenic silicates, should also reflect the general abundance of silicates available for weathering).

To illustrate how marine ^{87}Sr/^{86}Sr ratios need not increase through intervals of increasing silicate weathering and CO$_2$ consumption, we consider the following two examples. First, if the Sr-isotopic composition of a silicate source is the same as contemporaneous seawater, then variable C flux due to weathering of this source will not be reflected in the marine ^{87}Sr/^{86}Sr record. For example, the exhumation of a large mass of basalt (Ca, Mg silicate) that had been previously equilibrated with seawater would have minimal to no effect on marine ^{87}Sr/^{86}Sr ratios but could focus a large CO$_2$ flux from the atmosphere. Second, if a small, highly radiogenic source and a large, weakly radiogenic source expand in unison (CO$_2$ supply permitting), then marine ^{87}Sr/^{86}Sr ratios will remain constant while the C flux from the atmosphere increases. (As we have suggested, such a scenario may apply to orogens during the early stages of their development. In time, however, old radiogenic materials generally dominate after the young supracrustal marine elements have been removed and sialic crust, thickened during the collision, becomes widely exposed.)

A superior strategy for estimating C fluxes is to solve a system of coupled isotopic mass balance equations. These flux estimates would be useful for assessing the arc exhumation hypothesis, in that discrete pulses of nonradiogenic silicate weathering are predicted to correspond with cooling events. (Unfortunately, even relatively precise flux estimates cannot be used directly to calculate the past size of the atmosphere–ocean C reservoir because of its small size relative to the much larger input and output fluxes.) A system of three coupled isotopic mass balance equations is suggested by Figure 13.7. If values for metamorphic degassing and carbonate weathering are assumed, then this system of equations can be solved for C fluxes due to the weathering of organic C, radiogenic silicates, nonradiogenic silicates, and (if expressed as a dependent variable) burial of organic C. In such a multi-element model, the stoichiometric ratios between corresponding fluxes in the different element systems must be established to enable isotopic measurements to be translated into C fluxes. These coupling ratios constitute a major arena of uncertainty. Carbon and Sr are linked through the weathering of Ca, Mg silicates. The Ca/Sr ratio in silicate rocks shows limited variability (50–200) with the excep-

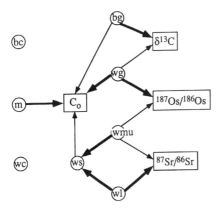

Fig. 13.7. C–Sr–Os carbon cycle model. Boxes represent size of the atmosphere–ocean C reservoir (C_0) and marine isotopic ratios ($^{13}C/^{12}C$, $^{87}Sr/^{86}Sr$, $^{187}Os/^{186}Os$). Circles represent coupled fluxes of C, Sr, and Os that affect the size of the atmosphere–ocean C reservoir and/or marine isotopic ratios. The fluxes include metamorphic degassing (m), burial of carbonate (bc), burial of organic matter (bg), weathering of carbonate (wc), weathering of organic matter (wg), total silicate weathering (ws), weathering of sialic rocks (wl), and weathering of mafic/ultramafic rocks (wmu). Thick arrows cause values in the boxes to increase; thin arrows cause values in the boxes to decrease.

tion of ultramafic rocks (25,000). Thus, ^{86}Sr fluxes are proportional to Ca fluxes, and the burial (and weathering) fluxes of Ca and C are equal in calcium carbonate. Carbon and Os are linked through the weathering (and burial) of organic matter (i.e., the concentration of Os in sediments correlates with the concentration of organic C). Osmium and Sr may be linked during the exhumation of silicate materials, although Os release probably is not limited by CO_2. First, Os and Sr may be weakly linked during the weathering of sialic materials, which release highly radiogenic Sr and Os (but in low concentrations). Second, during the exhumation of arc–ophiolite systems, mafic silicates are an important source of Sr, while ultramafic materials are an important source of Os. Ophiolitic mantle is typically accompanied by a far greater amount of mafic material. In the absence of river geochemical data constraining the proportions of Os, Sr, and C in rivers draining ophiolites, a reasonable assumption may be that the C flux due to weathering of Os-poor mafic rocks (unlikely to influence the marine $^{187}Os/^{186}Os$ record) is an order of magnitude greater than the C

flux due to weathering of Os-rich ultramafic rocks (that might be detected in the marine $^{187}Os/^{186}Os$ record).

The introduction of a term for nonradiogenic Ca, Mg silicate weathering in C cycle models, constrained by multiple isotopic records, may have two additional benefits. First, the model results should reveal independent variations in the weathering of radiogenic and nonradiogenic materials that potentially might be correlated with specific sources (e.g., Himalaya vs. Indonesia). Due to the very low background concentrations of Os in most rocks, marine $^{187}Os/^{186}Os$ records might even be sensitive to the exhumation of particular thrust sheets made of Os-rich lithologies such as ultramafic rocks or black shales. Second, the short residence time of Os in the ocean, while long enough for global homogenization, is much shorter than the residence time of Sr and should therefore yield much higher resolution records of exhumational events (Ravizza, 1993; Reusch et al., 1996).

We are still left with the dilemma of not having an adequate proxy to represent the quantity of Ca, Mg silicates undergoing exhumation. This parameter should correspond to the seafloor spreading rate and pelagic carbonate parameters that control degassing in the GEOCARB models. Perhaps a composite index that incorporates compositional anomalies in olivines, pyroxenes, and plagioclases might reflect the abundance of Ca, Mg silicate in sites of exhumation. Ideally, this proxy would not involve carbonation-type weathering, so that it is independent of atmospheric pCO_2.

Conclusion

The Miocene collision of New Guinea with Australia affords a test of the hypothesis that the transformation of an aggrading volcanic arc system (net CO_2 source) into an exhuming orogen (net CO_2 sink) causes CO_2 drawdown and global cooling. Following the subduction of northern Australia, the composite arc terrane of northern New Guinea was uplifted and deeply exhumed. Carbon dioxide consumption due to silicate weathering is estimated to have increased by 10^{17}–10^{18} moles myr^{-1} (1.4–14% of the modern total silicate weathering flux) due to the increased availability of highly soluble Ca, Mg silicates, high relief, high temperature, high orographic rainfall, and extensive tropical vegetation. A decreased rate of increase of marine $^{87}Sr/^{86}Sr$ ratios overlaps the middle Miocene cooling event and may reflect the weathering of these nonradiogenic silicates.

A stratigraphic analysis of late Precambrian and Phanerozoic records indicates that the timing of changes in seawater composition and tectonic events is consistent with the hypothesis that arc exhumation contributed to CO_2 drawdown and contemporaneous cooling in the late Precambrian, late Ordovician, Permo-Carboniferous, Paleogene, and Neogene.

Materials exhumed early in the development of orogens contain a higher proportion of soluble nonradiogenic Ca, Mg-rich silicates than materials exhumed during later stages of intracontinental thrusting. Current C cycle models do not adequately account for these materials that represent potent CO_2 sinks. We recommend that C cycle models incorporate an additional C flux term for weathering of nonradiogenic Ca, Mg-rich silicates, constrained by both marine $^{87}Sr/^{86}Sr$ and $^{187}Os/^{186}Os$ records, to obtain more accurate estimates of total silicate weathering and atmospheric pCO_2.

Acknowledgments

We appreciate helpful reviews by Bob Berner, Kevin Burke, Ken Caldeira, Tom Crowley, and Lou Derry.

References

Arrhenius, S., 1896, On the influence of carbonic acid in the air upon the temperature of the ground, *Phil. Mag., 41*, 237–276.

Beck, R.A., D.W. Burbank, W.J. Sercombe, T.L. Olson, and A.M. Khan, 1995, Organic carbon exhumation and global warming during the early Himalayan collision, *Geology, 23*, 387–390.

Benjamin, M.T., N.M. Johnson, and C.W. Naeser, 1987, Recent rapid uplift in the Bolivian Andes: Evidence from fission-track dating, *Geology, 15*, 680–683.

Berner, R.A., 1990, Atmospheric carbon dioxide levels over Phanerozoic time, *Science, 249*, 1382–1386.

Berner, R.A., 1991, A model for atmospheric CO_2 over Phanerozoic time, *Am. J. Sci., 291*, 339–376.

Berner, R.A., 1992, Weathering, plants, and the long-term carbon cycle, *Geochim. Cosmochim. Acta, 56*, 3225–3231.

Berner, R.A., 1994, GEOCARB II: A revised model for atmospheric CO_2 over Phanerozoic time, *Am. J. Sci., 294*, 56–91.

Berner, R.A., and G.R. Holdren, Jr., 1977, Mechanism of feldspar weathering: Some observational evidence, *Geology, 5*, 369–372.

Berner, R.A., and K.A.Maasch, 1996, Chemical weathering and controls on atmospheric O_2 and CO_2: Fundamental principles were enunciated by J. J. Ebelmen in 1845, *Geochim. Cosmochim. Acta, 60*, 1633–1637.

Berner, R.A., and D.M. Rye, 1992, Calculation of the Phanerozoic strontium isotope record of the oceans from a carbon cycle model, *Am. J. Sci., 292*, 136–148.

Burke,W.H., R.E. Denison, E.A. Hetherington, R.B. Koepnick, H.F. Nelson, and J.B. Otto, 1982, Variation of $^{87}Sr/^{86}Sr$ throughout Phanerozoic time, *Geology, 10*, 516–519.

Burton, K.W., J.L. Birck, and C.J. Allegre, 1995, Fine scale records of seawater $^{187}Os/^{186}Os$, *Eos Trans. AGU, 76*, S182.

Caldeira, K., 1992, Enhanced Cenozoic chemical weathering and the subduction of pelagic carbonate, *Nature, 357*, 578–581.

Caldeira, K., 1995, Long-term control of atmospheric carbon dioxide: Low-temperature seafloor alteration or terrestrial silicate-rock weathering? *Am. J. Sci., 295*,1077–1114.

Chamberlin, T.C., 1899, An attempt to frame a working hypothesis of the cause of glacial periods on an atmospheric basis, *J. Geol., 7*, 545–584, 667–685, 751–787.

Derry, L.A., A.J. Kaufman, and S.B.Jacobsen, 1992, Sedimentary cycling and environmental change in the Late Proterozoic: Evidence from stable and radiogenic isotopes, *Geochem. Cosmochim. Acta, 56*, 1317–1329.

Derry, L.A., and C. France-Lanord, 1996, Neogene growth of the sedimentary organic carbon reservoir, *Paleoceanography, 11*, 267–275.

Dunne, T., 1978, Rates of chemical denudation of silicate rocks in tropical catchments, *Nature, 274*, 244–246.

Ebelmen, J.J., 1845, Sur les produits de la decomposition des especes minerales de la famille des silicates, *Ann. des Mines, 7*, 3–66.

Edmond, J.M., 1992, Himalayan tectonics, weathering processes, and the strontium isotopic ratio in marine limestones, *Science, 258,* 1594–1597.

Edmond, J. M., M. R. Palmer, C. I. Measures, E. T. Brown, and Y. Huh, 1996, Fluvial geochemistry of the eastern slope of the northeastern Andes and its foredeep in the drainage of the Orinoco in Columbia and Venezuela, *Geochim. Cosmochim. Acta, 60,* 2949–2976.

England, P. and P. Molnar, 1990, Surface uplift, uplift of rocks, and exhumation of rocks, *Geology, 18,* 1173–1177.

Farrell, J.W., S.C.Clemens, and L.P.Gromet, 1995, Improved chronostratigraphic reference curve of late Neogene seawater ^{87}Sr/^{86}Sr, *Geology, 23,* 403–406.

Faure, G., *Principles of Isotope Geology*, John Wiley and Sons, New York, 1986.

Fischer, A.G., The two Phanerozoic supercycles, in Catastrophes and Earth History, W.A. Berggren and J.A. van Couvering (eds.), pp. 129–150, Princeton University Press, Princeton, N.J., 1984.

Frakes, L.A., J.E. Francis, and J.I. Syktus, *Climate Modes of the Phanerozoic,* Cambridge University Press, Cambridge, 1992.

France-Lanord, C., L. Derry and A. Michard, 1993, Evolution of the Himalaya since Miocene time: Isotopic and sedimentologic evidence from the Bengal fan, in Himalayan Tectonics, P.J. Treloar and M.P. Searle (eds.), pp. 605–623, *Geological Society Special Publication London, 74,* 1993.

François, L.M. and J.C.G. Walker, 1992, Modelling the Phanerozoic carbon cycle and climate: constraints from the ^{87}Sr/^{86}Sr isotopic ratio of seawater, *Am. J. Sci., 292,* 81–135.

Gansser, A., The Forgotten TransHimalaya: 10th Himalaya-Karakorum-Tibet Workshop, Abstract Volume, *Mitteilungendus dem geologischen Institut der ETH und der Universitat Zurich, Neue Folge, Nr. 298,* 1995.

Goddéris, Y. and L.M. François, 1995, The Cenozoic evolution of the strontium and carbon cycles: Relative importance of continental and mantle exchanges, *Chem. Geol., 126,* 169–190.

Hamilton, W.B., Tectonics of the Indonesian Region, *U.S. Geological Survey Professional Paper 1078,* 345 pp., 1979.

Hamilton, W.B., 1988, Plate tectonics and island arcs, *Geol. Soc. Am. Bull., 100,* 1503–1527.

Hiscott, R.N., 1978, Provenance of Ordovician deep-water sandstones, Tourelle Formation, Quebec, and implications for initiation of the Taconic orogeny, *Can. J. Earth Sci., 15,* 1579–1597.

Hodell, D.A., and F. Woodruff, 1994, Variations in the strontium isotopic ratio of seawater during the Miocene: Stratigraphic and geochemical implications, Paleoceanography, *9,* 405–426.

Hoffman, P.F., 1991, Did the Breakout of Laurentia Turn Gondwanaland Inside-Out?, *Science, 252,* 1409–1412.

Jacobsen, S.B., 1988, Isotopic constraints on crustal growth and recycling, *Earth Planet. Sci. Lett., 90,* 315–329.

Kaufman, A.J., S.B. Jacobsen, and A.H. Knoll, 1993, The Vendian record of Sr and C isotopic variations in seawater: Implications for tectonics and paleoclimate, *Earth Planet. Sci. Lett., 120,* 409–430.

Keto, L.S., and S.B. Jacobsen, 1987, Nd and Sr isotopic variations of Early Paleozoic oceans, *Earth Planet. Sci. Lett., 84,* 27–41.

Kyte, F.T., M. Leinen, G.R. Heath and L. Zhou, 1993, Cenozoic sedimentation history of the central North Pacific: Inferences from the elemental geochemistry of core LL44-GPC3, *Geochim. Cosmochim. Acta, 57,* 1719–1740.

Leggett, J.K., 1978, Eustacy and pelagic regimes in the Iapetus Ocean during the Ordovician and Silurian, *Earth Planet. Sci. Lett., 41,* 163–169.

Marchant, D.R., G.H. Denton, D.E. Sugden and C.C. Swisher, III, 1993, Miocene glacial stratigraphy and landscape evolution of the western Asgard Range, Antarctica. *Geogr. Ann., 75A,* 303–330.

Mead, G.A. and D.A. Hodell, 1995, Controls on the ^{87}Sr/^{86}Sr composition of seawater from the middle Eocene to Oligocene: Hole 689B, Maud Rise, Antarctica, *Paleoceanography, 10,* 327–346.

Miller, K.G., R.G. Fairbanks, and G.S. Mountain, 1987, Tertiary oxygen isotope synthesis, sea level history, and continental margin erosion, *Paleoceanography, 2,* 1–19.

Mix, A.C., N.G. Pisias, W. Rugh, J. Wilson, A. Morey, and T.K. Hagelberg, Benthic foraminifer stable isotope record from Site 849 (0–5 Ma):

Local and global climate changes, in *Proc. ODP, Sci. Results*, *138*, Pisias, N.G., L.A Mayer, et al. (eds.), pp. 371–412, Ocean Drilling Program, College Station, Tex., 1995.

Nelson, K.D., and J.F. Casey, 1979, Ophiolitic detritus in the Upper Ordovician flysch of Notre Dame Bay and its bearing on the tectonic evolution of western Newfoundland, *Geology*, *7*, 27–31.

Peucker-Ehrenbrink, B., 1996, Accretion of extraterrestrial matter during the last 80 million years and its effect on the marine osmium isotope record, *Geochim. Cosmochim. Acta, 60*, 3187–3196.

Peucker-Ehrenbrink, B., and G. Ravizza, 1996, Continental runoff of osmium into the Baltic Sea, *Geology, 24*, 327-330.

Peucker-Ehrenbrink, B., G. Ravizza, and A.W. Hofmann, 1995, The marine ^{187}Os / ^{186}Os record of the past 80 million years, *Earth Planet. Sci. Lett.*, *130*, 155–167.

Pigram, C.J., and H.L. Davies, 1987, Terranes and the accretion history of the New Guinea orogen, *BMR J. of Aust. Geol. Geophys.*, *10*, 193–211.

Plass, G.N., 1956, The carbon dioxide theory of climatic change, *Tellus*, *8*, 140–154.

Rangin, C., L. Jolivet, and M. Pubellier, 1990, A simple model for the tectonic evolution of southeast Asia and Indonesia region for the past 43 m.y., *Bull. Soc. Geol. France 6*, 889–905.

Ravizza, G., 1993, Variations of the ^{187}Os / ^{186}Os ratio of seawater over the past 28 million years as inferred from metalliferous carbonates, *Earth Planet. Sci. Lett.*, *118*, 335–348.

Ravizza, G., and G. McMurtry, 1993, Os isotopic variations in metalliferous sediments from the East Pacific Rise and the Bauer Basin. *Geochim. Cosmochim. Acta, 57*, 4301–4310.

Raymo, M.E., 1994, The Himalayas, organic carbon burial, and climate in the Miocene, *Paleoceanography*, *9*, 399–404.

Raymo, M.E., W.F. Ruddiman, and P.N. Froelich, 1988, Influence of late Cenozoic mountain building on ocean geochemical cycles, *Geology, 16*, 649–653.

Reusch, D.R., J.D.Wright, K.A.Maasch, and G.Ravizza, 1996, Miocene seawater ^{187}Os/^{186}Os ratios inferred from metalliferous carbonates, *Eos Trans. AGU, 79* F325.

Richter, F.M., D.B.Rowley, and D.J. DePaulo,

1992, Sr isotope evolution of seawater: The role of tectonics, *Earth Planet. Sci. Lett.*, *109*, 11–23.

Rowley, D.B. and W.S.F. Kidd, 1981, Stratigraphic relationships and detrital composition of the medial Ordovician flysch of western New England: Implications for the tectonic evolution of the Taconic orogeny, *J. Geology*, *89*, 199–218.

Searle, M.P., et al., 1987, The closing of Tethys and the tectonics of the Himalaya, *Geol. Soc. Am. Bull.*, *98*, 678–701.

Shackleton, N.J., The carbon isotope record of the Cenozoic: History of organic carbon burial and of oxygen in the ocean and atmosphere, in *Marine Petroleum Source Rocks*, J. Brooks and A.J. Fleet (eds.), pp. 423–434, Geol. Soc., London, 1987.

Selverstone, J., and D.S. Gutzler, 1993, Post-125 Ma carbon storage associated with continent-continent collision, *Geology*, *21*, 885–888.

Stallard, R.F., 1995, Tectonic, environmental, and human aspects of weathering and erosion: A Global Review using a Steady-State Perspective, *Annu. Rev. Earth Planet. Sci., 23*, 11–39.

Stallard, R.F., and J.M. Edmond, 1983, Geochemistry of the Amazon 2: The influence of the geology and weathering environments on the dissolved load, *J. Geophys. Res.*, *88*, 9671–9688.

Turekian, K.T., W.J. Pegram, K.A. Farley, and F.T. Kyte, 1995, Sea water osmium isotopic composition changes around the Eocene-Oligocene boundary: The relation to global tectonics and volcanism, *Eos Trans. AGU, 76*, S188.

Vincent, E., and W.H. Berger, Carbon dioxide and polar cooling in the Miocene: The Monterey hypothesis, in The Carbon Cycle and Atmospheric CO₂: Natural Variations Archean to Present, *Geophys. Monogr. Ser., 32*, E.T. Sundquist and W.S. Broecker (eds.), pp. 455-468, AGU, Washington, D.C., 1985.

Volk, T., 1987, Feedbacks between weathering and atmospheric CO₂ over the last 100 million years, *Am. J. Sci.*, *287*, 763–779.

Walker, J.C.G., 1991, *Numerical Adventures with Geochemical Cycles*, Oxford University Press, N.Y.

Walker, J.C.G., P.B. Hays, and J.F. Kasting, 1981, A negative feedback mechanism for the long-term stabilization of Earth's surface temperature, *J. Geophys. Res., 86,* 9776–9782.

White, A.F., and A.E. Blum, 1995, Effects of climate on chemical weathering in watersheds, *Geochim. Cosmochim. Acta, 59,* 1–19.

Wright, J.D., and K.G. Miller, 1993, Southern ocean influences on late Eocene to Miocene deepwater circulation, *Antarctic Res. Ser., 60,* 1–25.

Zachos, J.C., W.A. Berggren, M-P. Aubry, and A. Mackensen, in *Proc. ODP, Sci. Results, 120,* Wise, S.W., Jr., Schlich, R., et al. (eds.), pp. 839–854, Ocean Drilling Program, College Station, Tex., 1992.

INDEX